proceedings of the 4TH CANADIAN CONFERENCE on GENERAL RELATIVITY and RELATIVISTIC ASTROPHYSICS

proceedings of the 4TH CANADIAN CONFERENCE on GENERAL RELATIVITY and RELATIVISTIC ASTROPHYSICS

G. KUNSTATTER
D. E. VINCENT
J. G. WILLIAMS

UNIVERSITY OF WINNIPEG
16 — 18 MAY, 1991

World Scientific
Singapore • New Jersey • London • Hong Kong

Published by

World Scientific Publishing Co. Pte. Ltd.

P O Box 128, Farrer Road, Singapore 9128

USA office: Suite 1B, 1060 Main Street, River Edge, NJ 07661

UK office: 73 Lynton Mead, Totteridge, London N20 8DH

GENERAL RELATIVITY AND RELATIVISTIC ASTROPHYSICS,
Proceedings of the 4th Canadian Conference

ISBN 981-02-0965-7

Printed in Singapore by JBW Printers & Binders Pte. Ltd.

CONTENTS

Introduction

The Fourth Canadian Conference on General Relativity and Relativistic Astrophysics, held at the University of Winnipeg, had the difficult task of maintaining the high standards set by previous meetings in the series. As in the past, the purpose of the meeting was to bring Canadian physicists up to date on the latest exciting research developments in general relativity, relativistic astrophysics and cosmology. The small size and informal atmosphere of the Conference was successful in initiating and sustaining much fruitful dialogue and collaboration between the participants. In all, approximately 75 physicists attended the meeting, from all over Canada, as well as from Africa, Ireland, Great Britain and the United States.

The format of the meeting was slightly different from that of previous meetings. There were 10 plenary, review lectures on a variety of current topics, given by both Canadian and international researchers. In addition there were 36 shorter, more technical talks and numerous posters. In order to give the posters the attention that they deserved, there was a special poster session, and reception, organized on the second night of the meeting. The technical talks were given in parallel sessions in order to make the most efficient use of the short time available. As much by happy coincidence as by design, the Saturday morning session contained both plenary and technical talks that focussed on a single theme: the nature of time in quantum gravity. All participants who presented papers at the meeting were invited to submit a contribution to the proceedings. One paper that was not presented at the meeting due to illness has also been included in the proceedings.

The Fifth Canadian Conference on General Relativity and Relativistic Astrophysics will be held in the spring of 1993 at the University of Waterloo.

We are grateful to many individuals and organizations for the success of this meetings. First of all, we would like to thank the (bi-)local organizing committee for their monumental efforts. These included Jack Gegenberg (UNB), Dwight Vincent(U. of Winnipeg), Jeff Williams(Brandon U.). We are particularly grateful to the two students Laine Keahey and Jing Wang who did most of the real work both prior to and during the Conference. In addition, thanks go to the National Organizing Committee, F. Cooperstock(Victoria), C.C. Dyer(Toronto), R. McLenaghan(Waterloo), B.O.J. Tupper(New Brunswick) for their constant guidance and support. Of course, such a meeting would have been impossible to host without the substantial assistance and support of the Physics Department at the University of Winnipeg. In this regard, special thanks go to both the Chairman of the Physics Department, Ed Tomchuk, and the departmental secretary, Margaret Hunter. We would also like to thank Bryan Cesmystruk and Shirley Slashinsky from the Mathematics and Computer Sciences Department at Brandon University for their help.

Finally we are grateful to the various organizations and institutions that provided us with financial support. These include:

- The Natural Sciences and Engineering Research Council
- The Winnipeg Institute for Theoretical Physics
- The Canadian Institute for Theoretical Astrophysics
- The University of Winnipeg

G. Kunstatter, D. Vincent, J.G. Williams
University of Winnipeg
May, 1991

Canadian Conference on General Relativity & Relativistic Astrophysics
Thursday May 16 , 1991

8:45		**Opening Remarks - Room 3C00** M. McIntyre - Dean of Arts and Sciences E. Tomchuk - Chairman, Dept. of Physics		
9:00	R.B. Mann	Lowest Dimensional Gravity		
9:45	N. Kaiser	Clustering of Mass on Large Scales		
10:30	1st Floor	Coffee Break Coffee Break		Coffee Break
		Room A = 1L13		**Room B = 1L12**
11:00	B. Dolan	Energy Spectra in Inflationary Models	L. Tarasov	Quantum Aspects of 1+1 Dimensional Gravity
11:20	G.M. Covarrubias	Lie Groups in Cosmology	S. Morsink	Temperature of a 1+1 Dimensional Black Hole
11:40	B. Berger	Remarks on the Dynamics of the Mixmaster Universe	P. Kelly	Expansions of Nonsymmetric Gravitational Theories About a GR Background
12:00	R. Cassidy	Hamiltonian for the Mixmaster Vacuum Universe with Geometry R X S7	J. Palmer	A New Test of Nonsymmetric Gravity: Depolarization of Solar Spectral Lines
12:20		Lunch Break Lunch Break		Lunch Break
2:00	P. Hickson	Room 3C00 Large Scale Structure		
2:45	A. Ashtekar	Loops, Gauge Fields, and Gravity		
3:30	1st Floor	Coffee Break Coffee Break		Coffee Break
		Room A = 1L13		**Room B = 1L12**
4:10	A. Ashtekar	Loops, Gauge Fields, and Gravity	G. Ludwig	Twisting Type N Vacuum Metrics
4:30	J. Gegenberg	Gravitating Topological Matter	H.P. Kunzle	SU(n)-Einstein-Yang-Mills Fields With Spherical Symmetry
4:50	J. Louko	Gravitational Theta-States in Spatially Flat Quantum Cosmology	A. Das	Collapse of a Spherically Symmetric Anisotropic Fluid
5:10	A. Sikkema	Black Hole Mergers and Mass Inflation in a Bouncing Universe	K.L. Duggal	Ricci Inheritance Symmetry in Fluid Space times

Canadian Conference on General Relativity & Relativistic Astrophysics
Friday May 17, 1991

Time	Speaker	Room A = 1L13	Speaker	Room B = 1L12
9:00	R. Webster	Room 3C00 — Determination of Dark Matter from Gravitational Lensing		
9:45	J.B. Hartle	Quantum Mechanics from the Perspective of Quantum Cosmology		
10:30	1st Floor	Coffee Break		Coffee Break
11:00	M.S. Morris	The Wave Equation on a Timehole	S. Richardson	Energy Localization and the Kerr-Newman Metric
11:20	F. Molzahn	Gauge Invariant Derivative Expansion of Schrodinger Propagators on Manifolds	D. Konkowski	Singularities in Colliding Plane Wave Spacetimes
11:40	S.A. Fulling	Group-Theoretical Classification of Polynomial Functions of the Riemann Tensor	D. Salopek	Non-Gaussian Fluctuations for Structure Formation From Nonlinear Gravitational Effects
12:00	G. Williams	Symmetry Operators for Maxwell's Equations in Curved Space-times	M. Morikawa	Luminosity Biasing in Galaxies with Baryonic Dark Matter
12:20		Lunch Break		Lunch Break
2:00	R. McLenaghan	Room 3C00 — Separation of Variables and Constants of Motion for Relativistic Wave Equations on Curved Space-time		
2:45	R. Myers	Two Dimensional Quantum Gravity from Matrix Models		
3:30	1st Floor	Coffee break		Coffee Break
4:10	W.R. Wood	Broken Weyl Invariance and Geometric Particles	G. Fee	Comparison of Algorithms for the Symbolic Computation of the NP Spin Coef. and Curvature Components
4:30	C. Gauthier	Gravitation et Neutrinos	N. Kaloper	Lorentz Chern-Simons Term and the Breakdown of the Cosmic No-Hair Conjecture
4:50	H. King	What is the Angular Momentum of a Closed Universe?	J. Twamley	Wormholes Without a Conserved Charge
5:10	G. Kunstatter	Dirac Versus Reduced Quantization: A Geometrical Approach	A.R. Lyons	Wormholes in String Theory
5:30		Poster / Wine and Cheese Reception — Riddell Cafeteria		Poster / Wine and Cheese Reception

Canadian Conference on General Relativity and Relativistic Astrophysics
Saturday May 18, 1991

		Room A = 1L13
9:00	J. Friedman	Is Physics Consistant with Closed Timelike Curves?
9:20	W. Unruh	External Time and the Interpretation of Quantum Gravity
9:40	G. Hayward	General Laws of Gravitational Thermodynamics
10:00	E. Woolgar	New Demonstration of the Positive Mass Theorem
10:40		Coffee Break Coffee Break Coffee Break
11:00	K. Kuchar	Room 3C00 Time and Interpretation of Quantum Gravity
11:45	M. MacCallum	Computer Algebra for Gravity Theorists - Past, Present, and Future

proceedings of the 4TH CANADIAN CONFERENCE on GENERAL RELATIVITY and RELATIVISTIC ASTROPHYSICS

LOWEST DIMENSIONAL GRAVITY

R.B. Mann
Department of Physics, University of Waterloo
Waterloo, Ontario
N2L 3G1

Gravitation in less than four spacetime dimensions has become a subject of much interest in recent years. Here it is shown how the Einstein equations can be continued into two spacetime dimensions under appropriate limiting circumstances. The resultant theory of gravitation couples the Ricci scalar to the trace of the stress-energy tensor. The basic properties of this theory bear a remarkable resemblance to general relativity, both classically and semiclassically.

1 Introduction

Lower dimensional gravity has proven to be an interesting theoretical laboratory. Since relativistic theories of gravitation in lower dimensions significantly reduce the complexity of $(3+1)$-dimensional general relativity, they would appear to be useful testing grounds for ideas on quantum gravity. In particular, they allow one to explore the various issues associated with short-distance problems, topology change, singularities and the cosmological constant problem.

It is most natural to base a lower-dimensional theory of gravity on the Einstein equations

$$G_{\mu\nu} = 8\pi G_D T_{\mu\nu} \tag{1}$$

where G_D is Newton's constant in D dimensions and the speed of light has been set to unity. In $(3+1)$ dimensions this equation express the conceptual idea that "spacetime curvature"="mass energy". Of course a number of other tensorial equations exist that also express this concept; however from a theoretical standpoint the Einstein equations represent the most attractive choice since the Bianchi identities then imply covariant conservation of the stress-energy tensor in a manner compatible with special relativity in any local freely-falling frame. This theory seems to adequately describe gravitation in our universe, at least in the weak field post-Newtonian approximation.

A $(2+1)$ dimensional theory of gravitation based on (1) is attractive in that it is still founded upon the Einstein equations, thereby embodying both the "spacetime curvature"="mass energy" concept and the covariant conservation of stress energy. Unfortunately the Riemann tensor may be written entirely in terms of the Ricci tensor and the Ricci scalar, yielding the result that spacetime is flat outside of matter. Furthermore, the theory has no Newtonian limit [1], rendering its correspondence with $(3+1)$ dimensional gravity somewhat obscure.

In $(1+1)$ dimensions the Einstein tensor is identically zero for all metrics. The Einstein equations consequently imply that the stress-energy tensor vanishes, a most unattractive prospect for lower dimensional gravity.

However in the last few years it has been shown that lowest dimensional gravity, *i.e.*, gravity in two spacetime dimensions, need not be so trivial [2]. An interesting relativistic theory of gravitation in this context may be formed by setting the Ricci scalar equal to the trace of the stress energy tensor. Despite its simplicity, this theory has a number of remarkable classical and semi-classical features, including a well-defined Newtonian limit [2], black holes [3], a post-Newtonian expansion, gravitational waves, FRW cosmologies, gravitational collapse [4] and black hole radiation [5,6,7]. These features suggest that it is potentially a very useful tool in the study of quantum gravity: since its classical features are so similar to those of $(3+1)$ dimensional general relativity, one might hope that its quantization would bear a similar resemblance to $(3+1)$ dimensional quantum gravity.

Here an overview of this lowest dimensional theory of gravitation is given. Its classical features and quantum properties are reviewed in turn, pointing out their analogs with $(3+1)$ dimensional general relativity. Finally a connection with some recent developments in string theory will be briefly discussed.

2 Foundations

The $(1+1)$ dimensional theory mentioned above may be understood to be the two dimensional Einstein equations in the following way. The traceless part of (1) is

$$R_{\mu\nu} - \frac{1}{D} g_{\mu\nu} R = 8\pi G_D \left(T_{\mu\nu} - \frac{1}{D} g_{\mu\nu} T \right) \tag{2}$$

and the trace is

$$R = 8\pi \frac{G_D}{1 - \frac{D}{2}} T \tag{3}$$

where $R_{\mu\nu}$ is the Ricci tensor, R the Ricci scalar, $T_{\mu\nu}$ the stress energy tensor, T its trace and G_D Newton's constant, all quantities being defined in D dimensions. The system (2,3) still guarantees conservation of the stress-energy tensor

$$\nabla_\nu T^{\mu\nu} = 0 \tag{4}$$

for arbitrary D. Taking $D \to 2$ and defining

$$G_2 \equiv \lim_{D \to 2} \left[\frac{G_D}{1 - D/2} \right] \tag{5}$$

yields $0 = 0$ for (2) and

$$R = 8\pi G_2 T \tag{6}$$

for the trace of the Einstein equations in two dimensions.

These considerations suggest that a relativistic theory of gravitation based on (6) and supplemented with (4) might form a reasonable analog of Einstein's equations in two spacetime dimensions. They embody the "spacetime curvature"="mass energy" concept behind general relativity whilst conserving the stress energy. They have a sensible Newtonian limit (as will be shown) and form a very simple system of one equation in one unknown for the gravitational field.

The system (4,6) may be derived from the action [6]

$$S = \int d^2x \sqrt{-g} g^{\mu\nu} \left(\frac{1}{2} \partial_\mu \Psi \partial_\nu \Psi + \Psi R_{\mu\nu} - 8\pi G T_{\mu\nu} \right) \tag{7}$$

which in turn yields the field equations

$$\frac{1}{\sqrt{-g}} \partial_\mu \left[\sqrt{-g} g^{\mu\nu} \partial_\nu \Psi \right] - R = 0 \tag{8}$$

$$\frac{1}{2} \left(\Psi_{;\mu} \Psi_{;\nu} - \frac{1}{2} g_{\mu\nu} \Psi_{;\alpha} \Psi_{;\alpha} \right) + g_{\mu\nu} g^{\alpha\beta} \Psi_{;\alpha;\beta} - \Psi_{;\mu;\nu} = 8\pi G T_{\mu\nu} \tag{9}$$

plus the equations of motion for the matter fields. Substituting the trace of (9) into (8) yields (6). Conservation of the stress energy tensor (eq. (4)) is guaranteed since the covariant divergence of the left hand side of (9) is identically zero when (8) holds. This also follows from considering the invariance of (7) under infinitesimal co-ordinate transformations, analogous to the $(3+1)$-dimensional case. The classical evolution of the gravity/matter system is independent of the evolution of Ψ. However the Ψ field obeys the equation

$$\frac{1}{2} \left(\Psi_{;\mu} \Psi_{;\nu} - \frac{1}{2} g_{\mu\nu} \Psi_{;\alpha} \Psi_{;\alpha} \right) + \frac{1}{2} g_{\mu\nu} g^{\alpha\beta} \Psi_{;\alpha;\beta} - \Psi_{;\mu;\nu} = 8\pi G (T_{\mu\nu} - \frac{1}{2} g_{\mu\nu} T) \tag{10}$$

and so depends upon the evolution of the gravity/matter system (4,6). Hence Ψ may be treated as an auxiliary field.

The system of equations (8) and (9) is a system of 4 equations with 2 identities, which is equal to the number of unknowns (the metric degree of freedom and Ψ). Reorganized in the form (4,6,10), it is easy to see upon setting $T_{\mu\nu} = g_{\mu\nu}\Lambda$ that they reduce to a proposal considered previously by Jackiw and Teitelboim for two dimensional gravity [8,9]. Such a restriction permits only constant curvature solutions to the field equations. In contrast to this the theory based on (4) and (6) allows curvature and stress-energy to act upon one another in a manner which is very similar to the $(3+1)$ dimensional situation. Moreover, it has a sensible Newtonian limit which is easily seen by choosing co-ordinates so that

$$ds^2 = -\alpha(t,x)dt^2 + \alpha^{-1}(t,x)dx^2 \tag{11}$$

In the static (t-independent) case (6) becomes

$$\frac{d^2}{dx^2}\alpha = -8\pi G T \tag{12}$$

which is *identical* to the Newtonian gravitational field equation in one dimension upon appropriate identification of α with the gravitational potential and T with the mass-density [2]. More generally one can show that a well- defined post-Newtonian expansion exists for this theory [4]. These features are not present in other lower dimensional theories of gravity, which typically involve the incorporation of extended geometrical structures [10,11], non-locality [12], non-linear curvature terms [13], or non-minimal couplings [14]. This last approach is that which results from non-critical string theory; its relationship to the present theory will briefly discussed later.

The field equation (12) illustrates an important distinction between relativistic and non-relativistic gravity. This distinction is not in the relative complexity and non-linearity of the former relative to the latter, since the field equations are identical for this choice of coordinates. Rather it is in the conceptual foundations of the two theories, which gives rise to radically different interpretations of the meaning and role of the coordinates, particularly as regards the role of time. Non- relativistically time is absolute and the gradient of the gravitational potential merely guides particles along trajectories according to their initial conditions. In the relativistic case the gravitational potential is part of the spacetime structure itself and modifies the notion of time via eq. (11). Here is where the distinctions between the two cases are most profound. In two spacetime dimensions, the gravitational potential is the same in each: what differs is the role played by this potential in the context of the curvature and causal structure of space and time. In general the function $\alpha(x)$ will take on both positive and negative values, corresponding respectively to timelike regions (where 'physical' $(1+1)$-dimensional observers live) and spacelike regions (where space and time interchange roles). The points at which $\alpha(x) = 0$ form the boundaries between these regions; these are co-ordinate singularities and locate the event horizons of the spacetime: the black holes.

Of course the non-linearity inherent in the four-dimensional Einstein equations is important and should not be trivialized. It is responsible for a wide variety of the interesting structures and solutions in general relativity and arises as a consequence of the universal coupling of gravitation to all energy, including its own. The possibility of this occuring in a non-trivial way arises only in four or more spacetime dimensions. The system (4,6) has the advantage that the non-linearity is absent for an appropriate choice of coordinates.

3 Classical Features

To solve the field equations (4,6) a concrete model for the matter action is required. This can be either field-theoretic or phenomenological as the situation dictates.

Consider first a phenomenological approach. Taking the stress energy tensor to be that of a perfect fluid

$$T^{\mu\nu} = (\rho + p)u^\mu u^\nu + pg^{\mu\nu} \tag{13}$$

where ρ is the density and p is the pressure, with the 2-velocity u^μ defined so that

$$g_{\mu\nu}u^\mu u^\nu = -1 \quad . \tag{14}$$

then (6) becomes

$$\frac{d^2}{dx^2}\alpha = 8\pi G(\rho - p) \quad . \tag{15}$$

By setting $p = 0$ it is clear upon comparison to Newton's theory

$$\nabla^2\phi = 4\pi G\rho, \tag{16}$$

that the Newtonian potential ϕ is half the relativistic potential α, *i.e.*, $\alpha = 2\phi$.

For the theory given by (4) and (6), the most general solution to the field equations in the absence of matter is given by $\alpha = Kx + C$ which is simply Rindler space. However if a localized distribution of matter is present (*ie* $T_{\mu\nu} \neq 0$ for $r_- < x < r_+$), the most general vacuum metric becomes

$$\alpha = \theta(r_- - x)(K_1 x + C_1) + \theta(x - r_+)(K_2 x + C_2) \tag{17}$$

in which case a global transformation of the vacuum metric to flat space is not possible. It is always possible to choose coordinates so that $r_+ = -r_- \equiv r > 0$ and the constants C_1 and C_2 are of unit magnitude provided they are non-zero. For a symmetric distribution of matter, $K_1 = -K_2$ as a consequence of the matching conditions between the exterior metric (17) and the interior metric given by the solution to (4,6). Furthermore, $K_2 = 4\pi GM$ upon comparison with the Newtonian limit [2], where M is the $(1+1)$ dimensional Newtonian mass parameter. Hence the most general metric exterior to a localized distribution of matter is given by [3]

$$ds^2 = -(2M|x| - \epsilon)dt^2 + \frac{dx^2}{(2M|x| - \epsilon)} \tag{18}$$

where $2\pi G$ has been set to 1 for simplicity and $|\epsilon| = 1$.

Hence the vacuum potential due to any distribution of matter grows with increasing distance from the source. Non-relativistically M is the slope of the potential which corresponds to the strength of the source and ϵ (or $C_1 = C_2$) sets the (irrelevant) zero of energy. Relativistically, $2M$ still sets the strength of the source, but now ϵ determines the signature of the spacetime. For $\epsilon > 0$, the metric flips signature $((-,+) \rightarrow (+,-))$ whenever $|x| < 1/2M$. This is the spacetime corresponding to a black hole. A distribution of $(1+1)$ dimensional matter in hydrostatic equilibrium will typically have $r > \frac{1}{2M}$, just as the radius of a ball of its $(3+1)$-dimensional counterpart will be larger than its Schwarzchild radius. As with the Schwarzchild solution, one can consider the vacuum spacetime given by (18) for all values of x; such a metric is relevant for a description of matter which has undergone complete gravitational collapse, as will be subsequently demonstrated.

It is straightforward to show that it is always possible to solve (4,6) for ρ and p in terms of α [4]. As in general relativity, one can input a variety of densities and pressures that are considered to be reasonable approximations to the matter/energy distribution in the (1+1)-dimensional universe (or parts thereof). The conservation laws (4) for a $(1+1)$ dimensional fluid in hydrostatic equilibrium yield

$$\frac{\alpha'}{2\alpha} = -\frac{p'}{p+\rho} \quad . \tag{19}$$

which has the solution

$$\alpha = \alpha_0 \exp\left(-2\int^{\rho} \frac{\dot{p}(\bar{\rho})}{(p(\bar{\rho})+\bar{\rho})} d\bar{\rho}\right), \tag{20}$$

where $\dot{\alpha} \equiv \frac{d\alpha}{d\rho}$ and where an equation of state relating p and ρ has been given. Inserting this into (6) gives

$$\ddot{\alpha}(\rho')^2 + \dot{\alpha}\rho'' = 8\pi G(\rho - p) \quad . \tag{21}$$

This then yields a solution for ρ in terms of x, which in turn allows one to solve for α. Alternatively, if $\rho(x)$ is specified, then one could solve (4,6) for $p(x)$. Outside of the fluid $p = 0 = \rho$ and the interior solution for α must be matched with (17).

For example, if $\gamma = 1$ in the polytrope equation of state (so that $p = K\rho$) then

$$\alpha = \alpha_0 \rho^{-\frac{2K}{1+K}}, \tag{22}$$

where α is determined by

$$\frac{d^2}{dx^2}\alpha = 8\pi G(1-K)\left(\frac{\alpha}{\alpha_0}\right)^{-\frac{1+K}{2K}}. \tag{23}$$

Note that for $\rho > 0$, $\alpha > 0$ and there are no black holes [2].

The relativistic equation for hydrostatic equilibrium is easily obtained in coordinates where

$$ds^2 = -B^2(y)\,dt^2 + dy^2, \tag{24}$$

and is (with $' \equiv \frac{d}{dy}$)

$$-p'' = 4\pi G(\rho^2 - p^2) - \frac{2p'^2 + p'\rho'}{p+\rho}. \tag{25}$$

which follows upon elimination of $B(y)$ from (4) and (6). In the limit $p \to 0$, (25) becomes

$$-p'' = 4\pi G\rho^2 - \frac{p'\rho'}{\rho} \tag{26}$$

which is the Newtonian equation of stellar equilibrium in two spacetime dimensions [4].

A number of interesting dynamical solutions have been found. Consider the weak field approximation in which

$$g_{\mu\nu} = \eta_{\mu\nu} + h_{\mu\nu}, \tag{27}$$

where $|h_{\mu\nu}| \ll 1$. The field equations then imply [4]

$$h(x,t) = 4\pi G\eta_{\mu\nu} \int dx' \int^{t} dt'\, T(x',\omega) \exp\left[-i\omega(t' - |x - x'|)\right] + \text{c.c.}. \tag{28}$$

for a source oscillating at frequency ω. This solution is the two dimensional analog of a gravitational wave. Such a wave is more properly thought of as a 'pulse' since it is the global nature of the gravitational wave which contains the non-trivial physics. Locally such waves are co- ordinate waves: on either side of the source one may choose a coordinate system which "travels with the wave". However outside of the source the wave is of the form

$$h = H \exp[-iw(t - |x|)], \tag{29}$$

and so such a coordinate transformation cannot be applied globally. An observer crossing the source will see a flip in the wave's direction of propagation.

Lowest dimensional cosmology also bears a resemblance to its (3+1) dimensional cousin [4,15]. The two-dimensional Robertson-Walker metric

$$ds^2 = -dt^2 + \frac{a^2(t)}{1 - kx^2}\, dx^2 \tag{30}$$

where a is the cosmic scale factor, may be rewritten in the form

$$ds^2 = -dt^2 + a^2(t)\, dx^2 \tag{31}$$

where the factor $1-kx^2$ may be absorbed into the definition of the spatial coordinate. Hence although in two dimensions $k = 1$ still describes a closed universe, $k = 0$ a flat universe, and $k = -1$ a hyperbolic universe, this openness or closedness has no effect on the evolution of the cosmic scale factor, unlike the higher-dimensional case. However the basic features of the FRW evolution of the universe are governed by

$$\frac{d}{da}(\rho a) = -p. \tag{32}$$

In a "matter" dominated universe (where p is negligible) $\rho \sim a^{-1}$, whereas in a "radiation" dominated one (where $\rho = p$) $\rho \sim a^{-2}$. This is completely analogous to the $(3+1)$ dimensional case, where $\rho \sim a^{-3}$ for matter and $\rho \sim a^{-4}$, the additional powers of the scale factor being necessary due to the dimensionality of ρ and p. Neither scenario has any particle horizon [15].

As a last example, consider the collapse of $(1+1)$-dimensional dust, which is the analogue of an Oppenheimer-Snyder gravitational collapse [4]. For an initially static distribution of dust with negligible pressure, the dust particles are acted on by purely gravitational forces, so they fall freely and can be used as the physical basis of a comoving coordinate system with Robertson-Walker metric (31). Normalizing the scale factor $a_0 \equiv 1$ at $\tau = 0$, one obtains the metric

$$ds^2 = -d\tau^2 + (1 - b\tau^2)^2\, dx^2 \tag{33}$$

as a solution to (6) in the region $|x| \leq r$, where $x = \pm r$ are the (constant) positions of the edges of the fluid in the comoving coordinate system, and $b \equiv 2\pi G\rho_0)$.

Since the density ρ is inversely proportional to the scale factor a, it is clear that in a finite amount of comoving coordinate time, $\tau = (2\pi G\rho_0)^{-1/2}$, the fluid collapses from a static condition of uniform density to one of infinite density. Matching this to the metric (18) results in collapse to a black hole provided

$$1 - 4br^2 < 0 \Leftrightarrow \rho_0 > \frac{1}{8\pi G r^2}, \tag{34}$$

and so the fluid becomes a black hole at comoving time

$$\tau = \frac{1}{2br} \quad . \tag{35}$$

and at a coordinate time $t \to \infty$. Hence the metric (18) models the endpoint of the gravitational collapse of dust into a black hole. If (34) does not hold, then the dust collapses to a naked point source. Both possibilities may be realized since there is no Birkhoff theorem in $(1+1)$ dimensions [5].

4 Quantum Features

Although full quantization of the "$R = T$" theory has yet to be carried out, a number of its semiclassical properties have been explored.

Consider a static metric of the form (11) with $\alpha = \alpha(x)$. The radiation temperature associated with the event horizon in this spacetime may be most easily deduced by transforming the event horizon to the origin of a polar coordinate system[3], and requiring that the Euclideanized angular 'time' ($\tau \equiv it$) coordinate be periodic. This yields the temperature [3,16]

$$T = \frac{\hbar}{2\pi}\left|\frac{\alpha'(x_H)}{2}\right| \tag{36}$$

where x_H is the location of the horizon ($\alpha(x_H) = 0$). The existence of a meaningful temperature presupposes the existence of both an event horizon and a well-defined value of the slope of α at the horizon. This yields the temperature

$$T = \frac{\hbar M}{2\pi} \tag{37}$$

for the solution (18).

The notion of an entropy associated with a $(1+1)$ dimensional black hole a somewhat more delicate concept to define since the event horizon has no area. From the matching of (33) to (18) one can compute the Newtonian mass parameter from the density of the dust. Hence up to normalization one can *define* the entropy of the black hole via the thermodynamic relation

$$dM = TdS \tag{38}$$

yielding

$$S = \frac{2\pi}{\hbar} \ln\left(\frac{M}{M_0}\right). \tag{39}$$

The temperature of the hole decreases with the mass parameter. As $M \to M_0$ the entropy vanishes, and becomes increasingly negative for $M < M_0$, in a manner similar to what is seen in the isothermal expansion of a perfect gas [3], where:

$$S \sim ln\left(\frac{V}{V_0}\right) \quad . \tag{40}$$

Note that the endpoint of the radiation of the black hole is cold, in contrast to the violent death of a $(3+1)$ dimensional black hole [3].

The logarithmic dependence of the entropy upon the mass can yield other interesting consequences [5]. For system consisting of a box of length L containing a black hole of mass M and thermal radiation at temperature T, the energy and entropy are (setting $\hbar = 1$)

$$S = 2\pi \ln\Big(\frac{M}{M_0}\Big) + \frac{\pi}{3}TL \qquad E = M + \frac{\pi}{6}T^2 L \tag{41}$$

It is easy to shown that for fixed E the entropy is maximized at $T = \frac{M}{2\pi}$ as in (37); in terms of the fixed external parameters E and L, this equilibrium condition becomes

$$ML = -12\pi + 12\pi\sqrt{1 + \frac{EL}{6\pi}} \quad . \tag{42}$$

This condition is non-trivial, since for sufficiently small M the horizons could extend beyond the length L of the box; it is easy to show that such a situation is prevented provided $EL \geq 1.013$.

Consider next the possibility of a phase transition between pure radiation and the combined black-hole/radiation scenarios as in $(3+1)$ dimensions. The relative values of the entropy for the two systems are

$$S_{bh} = 2\pi \ln\Big(\frac{ML}{M_0 L}\Big) + \frac{1}{6}ML \tag{43}$$

for a black hole contained in a box of length L which satisfies the equilibrium condition (42), and

$$S_{rad} = 2\sqrt{\frac{1}{6}\pi EL} \tag{44}$$

for pure radiation contained in a box of the same size. A comparison of (43) and (44) indicates that as in $(3+1)$ dimensions the temperature initially increases with energy and the entropy is maximized by pure radiation. A phase transition is then reached, forming a black hole accompanied by a sharp drop in temperature. A lower limit on the length of the box is provided by the constraint $M_0 L > 0.82$ since for smaller values the lightest black hole (formed at the phase transition) will have $ML < 1$, implying that the horizons cannot be confined to the box.

The constant M_0 appears as a fundamental length in the two dimensional theory, somewhat analogous to the Planck mass in the higher dimensional case. In two dimensions Newton's constant is unitless and so does not imply the existence of a fundamental length. Here however the thermodynamic properties of two dimensional black holes imply the existence of such a length, perhaps signalling a breakdown of the (semi-) classical theory. Some implications of this parameter for black hole radiation have been explored to a certain extent in ref. [5].

More rigorous quantum field-theoretic derivations support the above results. It is possible to obtain exact solutions to the massive Klein- Gordon and Dirac equations in the background spacetime (18) [6]. These equations have a sensible Newtonian/Schroedinger limit. For a naked point source (18) (with $\epsilon = -1$) the energy eigenvalues are quantized; for the black hole (with $\epsilon = 1$) the exact solutions may be used to show that the temperature associated with the radiation of scalars [5] is given by (37) and is the same as that for spinors [7].

Full quantization of the $R = T$ theory has yet to be explored, although some preliminary results have been obtained. Working in a connection dynamics formulation, the action (7) may be rewritten as

$$S = \int d^2x \left[\sqrt{-g}g^{\mu\nu}\frac{1}{2}\partial_\mu\Psi\partial_\nu\Psi + \Psi d\omega + \lambda_a\left(de^a + \omega^a{}_b \wedge e^b\right)\right] + S_M \tag{45}$$

where $e^a = e^a_\mu dx^\mu$ is the dyad, $\omega^a{}_b = \epsilon^a{}_b\omega = \epsilon^a{}_b\omega_\mu dx^\mu$ is the spin-connection and S_M is the matter action. Since the Einstein-Hilbert action is a topological invariant, a Palatini-type variation of it will not yield the correct $(1+1)$ dimensional relationship between the dyad and the spin connection [10]. It is straightforward to show that the field equations which follow from (45) yield the system (4,6) upon elimination of the auxiliary fields Ψ and λ_a.

The Hamiltonian which follows from (45) is a linear combination of constraints

$$H = \int dx \left[eg_{11}^{-1}\hat{D}_1 - g_{01}g_{11}^{-1}D_1 - \omega_0 L_1 + \sum_I X_I K_I\right] \tag{46}$$

where the X_I's are Lagrange multipliers which enforce the vanishing of the primary constraints K_I and where $g_{\mu\nu} = \eta_{ab} e^a_\mu e^b_\nu$. Here

$$
\begin{aligned}
L_1 &= -\left(2\frac{d}{dx}\Psi + \lambda_a \epsilon^a{}_b e^b_1\right) \\
\hat{L}_1 &= -(2\Pi_\Psi + \lambda_a e^a_1) \\
D_1 &= -\Pi_\Psi \frac{d}{dx}\Psi - \left(\frac{d}{dx}\lambda_a + \omega_1 \epsilon^b{}_a \lambda^b\right) e^a_1 \\
\hat{D}_1 &= -\frac{1}{2}\left(\Pi^2_\Psi + (\frac{d}{dx}\Psi)^2\right) - \left(\frac{d}{dx}\lambda_a + \omega_1 \epsilon_a{}^b \lambda^b\right) \epsilon^a{}_c e^c_1 + \text{matter}
\end{aligned}
\tag{47}
$$

where Π_Ψ is the momentum conjugate to Ψ. Upon taking Poisson brackets of the various quantities defined above, it is clear that D_1 and $\hat{D}_1$ form the diffeomorphism algebra of S_1 [17]. The presence of ordinary matter leaves this structure intact, with the eigenvalues of the matter operators modifying those of the gravity operators. Topological matter (*i.e.* matter whose action is of the form $B \wedge DC$, with B a one-form and C a scalar) significantly modifies this, since the topological matter fields enter into the connection/dyad relationship [18].

Finding a representation of this algebra which is suitable for quantization is a project yet to be carried out. Full quantization of this theory (with matter couplings) would be most interesting because it would afford a "toy" version of a self-consistent quantized gravity/matter system whose classical limit closely resembles general relativity.

5 Relationship to String Theory

String-motivated approaches to two-dimensional gravity [14,19,20,21] bear an interesting resemblance to the $R = T$ theory described above. Consider eqs. (8,9). Solving (8) for Ψ gives

$$\Psi = (\nabla^\mu \nabla_\mu)^{-1} R + \Psi_0 \tag{48}$$

where Ψ_0 is the homogeneous solution. Upon setting $\Psi_0 = 0$, insertion of (48) into the action (7) yields, for constant stress-energy $T_{\mu\nu} = \frac{\Lambda}{8\pi G} g_{\mu\nu}$,

$$S = \int d^2x \sqrt{-g} \left[R(\nabla^\mu \nabla_\mu)^{-1} R + \Lambda\right] \quad . \tag{49}$$

This is Polyakov's formulation of two-dimensional gravity, obtained by functionally integrating the bosonic string action over all possible random surfaces [19]. In general it is non-local, but becomes local if one chooses to write the metric in the (local) form $g_{\mu\nu} = e^\phi \eta_{\mu\nu}$, in which case (49) becomes the Liouville action.

Quantization of this theory has been a subject of intense interest in recent years. From the viewpoint of string theory, the action (49) (or its ψ-formulation (7)) is the one-loop effective action of the bosonic string. In general the diffeomorphism algebra develops an anomaly due to the non-locality of this action; this is then cancelled out (if desired) by coupling to (classically) conformally invariant matter which then develops its own anomaly [12,17]. Denoting the anomaly by $\mathcal{A}$ we have in this approach

$$\mathcal{A}_{\rm grav} + \mathcal{A}_{\Psi} = \mathcal{A}_{\rm grav} \neq 0 \tag{50}$$

since the degree of freedom Ψ_0 of the Ψ field has been "frozen out"; introducing matter couplings (with total quantum anomaly $\mathcal{A}_{\rm matter}$) then implies

$$\mathcal{A}_{\rm grav} + \mathcal{A}_{\rm matter} = 0 \tag{51}$$

for a conformally invariant theory at the quantum level. In contrast to this, the $R = T$ theory in the absence of matter has

$$\mathcal{A}_{\rm grav} + \mathcal{A}_{\Psi} = 0 \tag{52}$$

since Ψ is no longer frozen out. Matter is not required *a-priori* to be classically conformally invariant, and will at the quantum level develop its own anomaly. A full study of the gravity–Ψ–matter system has yet to be carried out.

Earlier this year it has been shown that "black holes", *i.e.* spacetimes with the metric

$$ds^2 = -(1 - ae^{-Qx})dt^2 + \frac{dx^2}{1 - ae^{-Qx}} \tag{53}$$

can arise in string theory from two distinct viewpoints. The solution (53) was originally found as a solution to a scale-invariant higher-derivative theory of two-dimensional gravity in a certain limit [13]. Here only the string interpretations will be discussed.

In the first approach one considers a two dimensional (as opposed to 26 dimensional [19]) target space whose geometry is given by an underlying string theory [14,20]. Such a theory may be regarded as a critical string theory in a non-trivial background; the relevant sigma model action is

$$S = \frac{1}{2\pi\alpha_0}\int d^2\xi\sqrt{|\hat{g}|}\left(\frac{1}{2}g_{\mu\nu}\hat{g}^{ab}\partial_a x^\mu \partial_b x^\nu + \Phi(x)\hat{R} + \mathcal{T}(x)\right) \tag{54}$$

where α_0 is the string coupling constant, $\hat{g}$, $\hat{R}$ and (ξ^1, ξ^2) are the worldsheet metric, curvature and coordinates and $g_{\mu\nu}$, Φ, $\mathcal{T}$ and (x^1, x^2) are the target space metric, dilaton, tachyon and coordinates. The effective (Lorentzian) target space action associated with the vanishing of the one-loop beta function computed from (54) is

$$S_{eff} = \int d^2x\sqrt{g}e^{-2\Phi}\left(R - 4(\nabla\Phi)^2 + (\nabla\mathcal{T})^2 + V(\mathcal{T}) + c\right) \tag{55}$$

where $c = 8/\alpha_0$ is the central charge. The field equations for vanishing tachyon field are

$$R_{\mu\nu} - 2\nabla_\mu\nabla_\nu\Phi = 0, \tag{56}$$

$$R + 4(\nabla\Phi)^2 - 4\nabla^2\Phi + c = 0, \tag{57}$$

which yield the metric (53) with dilaton field $\Phi = \frac{Q}{2}x$, $Q^2 = c$. For positive a the solution (53) has one event horizon; it also has a singularity at $x = -\infty$. In this sense the solution has been dubbed a "black hole" [14,20].

From the viewpoint of the $R = T$ theory, the action (55) is the action for a massive scalar non-minimally coupled to curvature,

$$S = \int d^2x\sqrt{|g|}(\varphi^2 R - 4(\nabla\varphi)^2 + \varphi^2(c + (\nabla\mathcal{T})^2 + V(\mathcal{T})) \tag{58}$$

as is easily seen by writing $\varphi = e^{-\Phi}$. The one-sidedness of the "black hole" is somewhat unsatisfactory in this interpretation, since any collapsing matter confined to a spatially finite region will not be able to produce such a structure as its endpoint metric. However it is possible to match the metric (11) with

$$\alpha(x) = 1 - ae^{-Q|x|} \tag{59}$$

to collapsing dust provided appropriate surface stresses are included for the energy-momentum tensor and the dilaton field [16].

The other approach [21] considers solutions to the field equations for the metric of the string worldsheet. The quantum effective action which generalizes the action (49) for the conformally non-invariant case is given by [22]

$$S = \int d^2x\,\sqrt{-\hat{g}}\hat{g}^{\mu\nu}\left[\frac{1}{2}\partial_\mu\psi\partial_\nu\psi + \Psi R_{\mu\nu}(\hat{g}) - \frac{4\lambda}{\sigma}e^{\sigma\psi}\right] \tag{60}$$

where λ and σ are constants and $\hat{g}_{\mu\nu}$ is a fixed background metric; the worldsheet metric is interpreted as $g_{\mu\nu} = e^{\psi}\hat{g}_{\mu\nu}$. For $\hat{g}_{\mu\nu} = \eta_{\mu\nu}$ the field equation which follows from (60) is

$$\partial^2\psi + \lambda e^{\sigma\psi} = 0 \tag{61}$$

which is the Liouville equation. The general solution to this equation is well-known; requiring this solution to be asymptotically flat and static implies $\sigma = 2$; under a coordinate change the metric $g_{\mu\nu} = e^{\psi}\eta_{\mu\nu}$ is given by (53) with Q and a constants of integration [21].

In the $R = T$ interpretation, the ψ degree of freedom of the metric may interpreted as the field Ψ, except that in (60) this field is no longer auxiliary. Instead it contributes to the gravitational field due to the final term in (60).

6 Closing Remarks

The $R = T$ theory has provided yet another example of how rich in content gravitational physics can be in two spacetime dimensions. The classical resemblance between this theory and $(3+1)$ dimensional general relativity and its connections with string-theoretic approaches suggest that its quantization should provide interesting new information concerning quantum gravity.

Acknowledgements

This work was supported by the Natural Sciences and Engineering Research Council of Canada. I would like to thank the organizers of the 4th Canadian Conference on General Relativity and Relativistic Astrophysics for arranging such a stimulating meeting.

References

[1] J. Barrow, A. Burd and D. Lancaster, Classical and Quantum Gravity **3** (1986) 551.

[2] R.B. Mann, Found. Phys. Lett. (to be published).

[3] R.B. Mann, A. Shiekh, and L. Tarasov, Nucl. Phys. **B341** (1990) 134.

[4] A.E. Sikkema and R.B. Mann, Class. Quantum Grav. **8** (1991) 219.

[5] R.B. Mann and T.G. Steele, Class. Quantum Grav. (to be published).

[6] R.B. Mann, S. Morsink, A.E. Sikkema and T.G. Steele, Phys. Rev. **D43** (1991) 3948.

[7] S.M. Morsink and R.B. Mann, Class. Quant. Grav. (to be published).

[8] C. Teitelboim in *Quantum Theory of Gravity*, ed. S. Christensen (Adam Hilger, Bristol 1984), p.327; R. Jackiw, *ibid.*, p. 403; Nucl. Phys. **B252** (1985) 343.

[9] C. Teitelboim, Phys. Lett. **B126** (1983) 41, 46; M. Henneaux, Phys. Rev. Lett. **54** (1985) 959; T. Fukuyama and K. Kimimura, Phys. Lett. **B160** (1985) 259.

[10] J. Gegenberg, P.F. Kelly, R.B. Mann, and D.E. Vincent, Phys. Rev. **D37** (1988) 3463.

[11] P.F. Kelly and R.B. Mann Phys. Rev. **D43** (1991) 1839.

[12] J. Polchinski, Nucl. Phys. **B324** (1989) 123.

[13] H.-J. Schmidt, J. Math. Phys. **32** (1991) 1562

[14] G. Mandal, A.M. Sengupta, and S.R. Wadia, IAS preprint, IASSNS-HEP-91/10.

[15] C.K. Chan, MSc. Report, University of Waterloo (1991).

[16] R.B. Mann, M.S. Morris and S.F. Ross, University of Waterloo preprint, WATPHYS-TH91/04.

[17] C.G. Torre, Phys. Rev. **D40** (1989) 2588; R. Marnelius, Nucl. Phys. **B211** (1983) 14.

[18] S. Carlip and J.D. Gegenberg, Phys. Rev. **D44** (1991) 424.

[19] A.M. Polyakov, Phys. Lett. **B103** (1981) 207.

[20] E. Witten, IAS preprint, IASSNS-HEP-91/12

[21] A.R. Lugo and J.G. Russo, Stanford University preprint, SU-ITP-896

[22] V. Knizhnik, A.M. Polyakov and A.B. Zamolodchikov, Mod. Phys. Lett. **A3** 819 (1988; F. David, Mod. Phys. Lett. **A3** 1651 (1988); J. Distler and H. Kawai, Nucl. Phys. **B321** 509 (1988).

ENERGY SPECTRA IN INFLATIONARY MODELS

Brian P. Dolan
Department of Mathematical Physics, St. Patrick's College
Maynooth, Ireland and Dublin Institute For Advanced Studies
10, Burlington Rd., Dublin, Ireland

ABSTRACT

Curvature effects in inflationary Robertson-Walker models, although negligible at the present day, can drastically alter the energy spectra of non-zero spin quantum fields in $k \neq 0$ universes at early times. This affects the thermal energy densities of quantum fields, and hence the dynamical evolution of the universe. It is argued that, for $k \neq 0$ universes with temperature T and spatial curvature R, if the inflation factor were so large that $RT \lesssim 10^{-6+2k}$ before inflation, then there was never a pre- inflationary era in new inflation when the universe evolved according to the classical Einstein equations, but that the inflationary epoch must have grown directly out of the quantum gravity era.

Inflation offers an intriguing explanation of why the universe appears so flat at the present day. Of the three possibilities for the spatial curvature of a Robertson-Walker universe, $k = 0, \pm 1$, $k = 0$ is a set of measure zero and would require a very special set of initial conditions to be realised. But if $k \neq 0$ the observed value of the spatial curvature is much smaller than any "natural" value. Inflation cures this by reducing the spatial curvature by many orders of magnitude. In chaotic inflation[1] the inflationary era is regarded as emerging directly from the quantum gravity domain. In new[2,3] and extended[4] inflation the universe is usually regarded as undergoing a standard, radiation dominated expansion (or a modified power law for extended inflation) from the time at which it emerges from the mysterious quantum gravity era, at about $10^{-43}s$, until the time when the energy density of the scalar field starts to dominate the dynamics, often taken to be about about $10^{-35}s$, the time at which a GUT phase transition might be expected to start. In this talk it will be argued that if the inflation factor were so large that $RT \lesssim 10^{-6+2k}$ before inflation, then that there was never any such era and inflationary era must have emerged *directly* from the quantum gravity era in new and extended as well as chaotic inflationary models[5].

The basic observation is that the energy spectra of a quantum field with spin s in a $k \neq 0$ static Robertson-Walker universe is not quite of Planckian form, but is distorted by a term of order $(RT)^{-2}$, where R is the scale factor and T is the temperature. In an open Einstein ($k = -1$) universe the energy density is given by[6] (natural units are used in which $\hbar = c = k_B = 1$, where k_B is Boltzmann's constant),

$$\begin{aligned}\rho_- &= \frac{g_s T^4}{2\pi^2}\int_0^\infty \frac{x^3+\left(\frac{s}{RT}\right)^2 x}{e^x-(-1)^{2s}}dx \\ &= \frac{g_s\pi^2 T^4}{96}\left[\left(\frac{15+(-1)^{2s}}{5}\right)+\left(\frac{s}{\pi RT}\right)^2\left(\frac{9+7(-1)^{2s}}{2}\right)\right],\end{aligned} \tag{1}$$

where g_s is the number of degrees of freedom for the field, $g_s = 2$ for photons. $s = 0$ refers only to the conformally coupled wave equation for scalars.

For a closed closed ($k = +1$) Einstein universe it is[7,8]

$$\begin{aligned}\rho_+ &= \frac{g_s}{2\pi^2 R^4}\sum_{n=1}^{\infty}\frac{n^3-s^2 n}{e^{n/(RT)}-(-1)^{2s}}+\frac{a(s)}{\pi^2 R^4} \\ &= \frac{g_s T^4}{2\pi^2}\frac{1}{RT}\sum_{n=1}^{\infty}\frac{x_n^3-\left(\frac{s}{RT}\right)^2 x_n}{e^{x_n}-(-1)^{2s}}+\frac{a(s)}{\pi^2 R^4},\end{aligned} \tag{2}$$

where $x_n = n/(RT)$. The second term here represents the Casimir energy on the three sphere[8]. Again, the $s = 0$ case is only for the conformally coupled scalar wave equation. In particular the spin dependent co-efficients are $a(0) = 1/480$, $a(1/2) = 17/1920$, $a(1) = 11/240$.

For a static spatially flat ($k = 0$) universe, there is no deviation from the usual Planck form and the considerations in this talk are irrelevant, but this will be regarded as being a very special situation requiring special initial conditions.

Of course, in the real world the universe is not static and the Robertson-Walker scale factor, $R(t)$, is not a constant so the thermal spectra are not known exactly in general, but as long as the spectrum is only analysed in the frequency range $\nu >> H$, where $H = \dot{R}/R$ is Hubble's "constant", then one would expect the spectrum to be given reasonably accurately by its static form.

For the present day cosmic microwave backround, with $T = 2.7^oK \sim (1mm)^{-1}$ and $R \gtrsim 10^{28}$cm, $RT \gtrsim 10^{29}$ and so the distortion from the Planckian spectrum is less than one part in 10^{58} and is completely negligible. Without inflation, the distortions are negligible at all earlier times. This follows simply from the adiabatic expansion of the standard Big Bang model. Since the total entropy, S, is constant we have $S = sR^3 \sim T^3 R^3 =$ constant, where $s \sim T^3$ is the entropy density. Thus as long as the total entropy remains constant, TR remains constant and the distortions are always negligible.

However if the total entropy of the universe at an earlier epoch was ever smaller than it is today, $1/(RT)$ would have been larger and the distortions from the Planck spectrum correspondingly larger. This is exactly what inflation does. It is designed to hold T fixed while R increases by at *least* twenty nine orders of magnitude, thus since $RT \gtrsim 10^{29}$ after inflation, we might expect to have $RT \lesssim 1$

before inflation, and the distortion from a Planckian spectrum for fields with non-zero spin is significant, and even dominant if $RT << 1$.

The shape of the spectra for photons is very different from the Planckian form if $RT << 1$. For $k = -1$ the energy density per unit frequency is non-zero at zero frequency (specifically, it is $T^4/(\pi RT)^2$) and there is much more energy density at low frequencies than in the Planck spectrum, the distortion being larger for smaller values of RT. For values of RT less than 0.619 there is no longer a peak and the spectrum is monotonically decreasing.

For $k = +1$ the thermal spectrum is discrete and there is less energy at low frequencies than in the Planck spectrum. For $RT << 1$, the thermal spectrum is strongly supressed and almost all the energy is in the Casimir term.

For values of $RT << 1$ the thermal history of the universe before inflation is also modified. For the Planck spectrum $\rho_0 = \frac{\pi^2}{30} N_{eff} T^4$ where $N_{eff} = N_b + \frac{7}{8} N_f$ is the effective number of degrees of freedom (1 per bosonic degree of freedom and 7/8 per fermionic degree of freedom). But for $k \neq 0$, $\rho_\pm >> \rho_0$ when $RT << 1$ (by a factor of order $(RT)^{-2}$ for $k = -1$ and $(RT)^{-4}$ for $k = +1$ - in the latter case, the Casimir energy is the dominant contribution).

For example in new inflationary models during the pre-inflation, radiation dominated era the age of the universe goes like $t \propto 1/\sqrt{\rho}$. Thus, for a given temperature, a $k \neq 0$ universe is much younger than a $k = 0$ one would be (by a factor of order $(RT)^{-1}$ for $k = -1$ and $(RT)^{-2}$ for $k = +1$). If the inflation factor in a $k = +1$ universe is so large that $RT \lesssim 10^{-4}$ before inflation then the universe is younger by a factor $\gtrsim 10^8$ than is usually assumed. So instead of setting in at $10^{-35}s$ inflation would have started at $10^{-43}s$, i.e. directly from the quantum gravity era! For a $k = -1$ universe a pre-inflation value of $RT \lesssim 10^{-8}$ gives the same conclusion.

We have seen that, if the inflation factor is large enough to give a pre-inflation value of $RT \lesssim 10^{-6+2k}$, then there was never a pre-inflationary era in new inflation during which the universe evolved classically. Such inflationary factors are not at all uncommon in these models. Thus, in contrast to the usual assumption, the thermal history of the very early, pre-inflation, universe is very sensitive to the value of k.

For models of extended inflation, there are also power law solutions of the Einstein equations in the pre-inflationary era, with $t \sim \rho^{-\alpha}$ where α is model dependent. For these models similar conclusions can be drawn.

Of course, as stated earlier, the thermal spectra quoted in Eqs. 1 and 2 are only for *static* Robertson-Walker space times and are not to be trusted for frequencies smaller than or of the same order as the Hubble constant. Nevertheless, one can still expect drastic modifications of the Planck spectrum for large inflation factors.

To summarise it has been argued that, while distortions of thermal spectra for fields of non-zero spin due to curvature effects in $k \neq 0$ universes are negligible at the present day, they would have been very significant in any theory with a large production of entropy $\gtrsim 10^{87}$ at any time later than a few Planck times. In particular for new or extended inflationary models these effects would be expected to be important before the era of inflation. It would appear that the assumption of an essentially classical post-quantum gravity but pre-inflation era is not tenable.

REFERENCES

1. A.D. Linde Phys. Lett. **123B** 1983 177
2. A.D. Linde Phys. Lett. **108B** 1982 389
   A.D. Linde Phys. Lett. **114B** 1982 431
3. A. Albrecht and P.J. Steinhardt Phys. Rev. Lett. **48** 1982 1220
4. D. La and P.J. Steinhardt Phys. Lett. **62B** 1989 376
5. *Thermal Spectra In The Early Universe*
   B.P. Dolan Dublin Institute for Advanced Studies pre-print 91-05 1991
6. P. Candelas and J.S Dowker Phys. Rev. D **19** 1979 2902
7. B. Mashhoon Phys. Rev. D **8** 1973 4279
8. L.H. Ford Phys. Rev. D **14** 1976 3304

LIE GROUPS IN COSMOLOGY

G.M. COVARRUBIAS

Departamento de Física
Universidad Autónoma Metropolitana-Iztapalapa
P.O. Box. 55-534, México D.F. 09430, México

ABSTRACT

The Lie group technique for differential equations is applied to the cosmological equations of the Jordan and Brans-Dicke gravitational theory for the Robertson-Walker universes containing dust and for the vacuum. The problem of solving two coupled first order and second degree differential equations is reduced to that of solving a single Abel equation and a quadrature in each case. All previously known exact solutions are obtained. It is proved that the known solutions for the 3-space flat slices are the general solution.

1. INTRODUCTION

The Jordan and Brans-Dicke gravitational theory has been the object of new interest recently. This theory is the low energy limit of some superstring theories[1]. Also some of them are related with inflation[2] without additional assumptions on the lagrangian of the matter as in the case of general relativity, and Hawking singularity theorems are not valid in this theory so we can have "eternal" and black hole free universes. On the other hand, the use of the Lie Group technique for differential equations has been shown to be a powerful method for finding and extend exact solutions in gravitational theories[3,4]. We do so in this work for the homogeneous and isotropic Robertson-Walker Universes for the vacuum and dust equations of state. It is known that two first order and second degree coupled equations follow in each case (the case $\phi\alpha R^2$ has to be excluded [5]). We reduce the problem of solving these equations to that of solving a single Abel equation and performing a quadrature for each case. All previously known exact solution follow and we prove that all 3-space flat slice universes are known.

2. THE COSMOLOGICAL EQUATIONS

With the Robertson-Walker line element, the cosmological equations for the dust universes ($p = 0,\ \rho R^3$ =const.) are

$$\left(\frac{\dot{R}}{R} - \frac{1}{2}\frac{\dot{\kappa}}{\kappa}\right)^2 + \frac{\varepsilon}{R^2} = \beta\frac{\dot{\kappa}}{\kappa}\left(\frac{1}{4}\frac{\dot{\kappa}}{\kappa} - \frac{\alpha}{\alpha t + \delta}\right), \tag{1}$$

$$\frac{\dot{\kappa}}{\kappa^2}R^3 = -(\alpha t + \delta), \qquad \frac{\kappa}{R^2} \neq const., \tag{2}$$

and for the vacuum universes ($p = 0 = \rho$).

$$\left(\frac{\dot{R}}{R} - \frac{1}{2}\frac{\dot{\kappa}}{\kappa}\right)^2 + \frac{\varepsilon}{R^2} = \frac{1}{4}\beta\left(\frac{\dot{\kappa}}{\kappa}\right)^2, \tag{3}$$

$$\frac{\dot{\kappa}}{\kappa}R^3 = -\gamma, \qquad \frac{\kappa}{R^2} \neq const., \tag{4}$$

where $\epsilon = 1, 0, -1$ for the 3-space slices of positve, cero and negative curvature respectively; κ is related to the scalar field of the theory by

$$\kappa = 8\pi\frac{(2\omega + 4)}{(2\omega + 3)}\phi^{-1}; \tag{5}$$

ω is the free available parameter of the theory, and

$$\alpha := \frac{R_0^3\rho_0}{2\omega + 4}, \quad \beta := 1 + \frac{2}{3}\omega, \quad \delta = const. \quad ; \tag{6}$$

where ρ_0 and R_0 are the present values of the mass density and radius of curvature of the universe.

3. THE REDUCTION

If we make

$$H := \frac{R^3}{\kappa} + \frac{1}{4\alpha}(\alpha t + \delta)^2 \tag{7}$$

for the dust universes, and

$$H := \frac{R^3}{\kappa} + \frac{1}{2}\gamma t \tag{8}$$

for the vacuuum universes, we can prove that the equations

$$(9\alpha\beta H - H'^2) = \frac{3}{4}(H - \frac{1}{4}\alpha\tau^2)(9\alpha\beta - 2H'') \quad , \tag{9}$$

$$\frac{\dot{\kappa}}{\kappa} = \frac{\alpha t + \delta}{\frac{1}{4\alpha}(\alpha t + \delta)^2 - H}. \tag{10}$$

and equation 7, with the constraints $H'^2 = 9\alpha\beta H$ for 3-space flat slice spaces, and $H'^2 \neq 9\alpha\beta H$ for 3-space curved slice spaces, along with the subsidiary equation

$$R^2 = 9\varepsilon \frac{(H - \frac{1}{4}\alpha\tau^2)^2}{(9\alpha\beta H - H'^2)}. \tag{11}$$

are completely equivalent to the original system (1) and (2) for the dust universes, where $\tau = t + \frac{\delta}{\alpha}$.

The same logical relations hold for the equations

$$\frac{3}{2}H''(H - \frac{1}{2}\gamma\tau) = (H'^2 - \frac{9}{4}\beta\gamma^2), \tag{12}$$

$$\frac{\dot{\kappa}}{\kappa} = \frac{\gamma}{\frac{1}{2}\gamma t - H}. \tag{13}$$

Equation 8, and

$$R^2 = 9\varepsilon \frac{(H - \frac{1}{2}\gamma\tau)^2}{\frac{9}{4}\beta\gamma^2 - H'^2} \quad , \tag{14}$$

for the vacuum universes where $\tau = t$.

Now, equation 9 is invariant under the continuous Lie Group of transformations

$$\tilde{H} = a^2 H, \quad \tilde{\tau} = a\tau \tag{15}$$

and equation 12 is invariant under the continuous Lie Group of transformations

$$\tilde{H} = aH, \quad \tilde{\tau} = a\tau, \tag{16}$$

so, according to Lie's Theory[6] equation 9 is equivalent to the equations

$$\frac{dv}{du} = \frac{4v^2 - 6uv + \frac{3}{2}\alpha v - 9\alpha\beta u - \frac{27}{4}\alpha^2\beta}{6(u - \frac{1}{4}\alpha)(v - 2u)}. \tag{17}$$

$$\frac{d\tau}{\tau} = \frac{du}{v - 2u}. \tag{18}$$

where

$$u = \frac{H}{\tau^2} = const. \quad , \tag{19}$$

$$v = \frac{H'}{\tau} = const. \tag{20}$$

and equation 12 is equivalent to equations

$$\frac{dv}{du} = \frac{2}{3}\frac{(v^2 - \frac{9}{4}\beta\gamma^2)}{(v-u)(u-\frac{1}{2}\gamma)} \quad , \tag{21}$$

$$\frac{d\tau}{\tau} - \frac{du}{v-u} \quad , \tag{??}$$

where

$$u = \frac{H}{\tau}. \tag{23}$$

$$v = H'. \tag{24}$$

It is apparent that the problem of solving equations 9 and 12 is reduced to that of solving equations (17) and (21) which are Abel–type equations.

4. SOLUTIONS

It follows trivially from equations (9) and (12) that the cases $H'^2 = 9\alpha\beta H$ and $H'^2 = 9\frac{9}{4}\beta\gamma^2$ (where $H' \neq 0$) solve the equations and give all universes with 3-space flat slices of the dust and vacuum universes respectively. All these solutions are known[7,8,9]. So, the general solution for the $\epsilon = 0$ case is completely solved.

The known solution for the closed space[10] and the dust universe corresponds to a path curve of the Lie Group which is also a solution of Eq. 9.

There are more detailed treatments of this work[11,12],

5. REFERENCES

1. K. Maeda, *Phys. Letters* **166 B** (1986). 59.

2. N. Morales and L.O. Pimentel, *Internal Report* (Universidad Autónoma Metropolitana).

3. I. Häuser and F.J. Ernst, *J. Math. Phys.* **21** (1980) 1126.

4. C. Hoenselaers and W. Dietz (Eds) *Lecture Notes in Physics* (Springer-Verlag, N.Y.) p. 205.

5. G. Lessner, *Astrophys. Space Sci.* **98** (1984) 101.

6. G.W. Bluman and J.D. Cole, *Similarity Methods for differential equations* (Springer-Verlag, N.Y., 1974).

7. L.E. Gurevich, A.M. Finkelstein and V.A. Ruban, *Astrophys. Space Sci.* **22** (1973) 231.

8. C. Brans and R.H. Dicke, *Phys. Rev.* **124** (1961) 925.

9. J. O'Hanlon and B.O.J. Tupper, *Nuovo Cimento* **7b** (1972) 305.

10. H. Dehnen and O. Obregón,*Astrophys. Space Sci.* **14** (1971) 454.

11. G.M. Covarrubias and O. Obregón, *Astrophys. Space Sci.*. In print.

12. G.M. Covarrubias, O. Obregón and J. Cervantes-Cota, *Internal Report* (Universidad Autónoma Metropolitana).

SPINOR PARTICLES AND BLACK HOLES IN 1+1 DIMENSIONS

Sharon M. Morsink
Department of Physics
University of Waterloo
Waterloo, Ontario
N2L 3G1

June 11, 1991

Abstract

The Hawking radiation of massive spinor particles from a $1+1$ dimensional black hole is examined so that the temperature of the black hole can be determined. Using the solution of Dirac's equation in the $1+1$ black hole metric the vacuum expectation value of the spinor stress-energy tensor is calculated. The energy density is compared with the energy density of black body radiation. This allows an identification to the black hole of a temperature of $\frac{M}{2\pi}$. The temperature found here is the same found when a similar analysis is done on massive scalar particles.

1 Introduction

The study of quantum field theory in curved space time led to the theoretical discovery that black holes radiate energy [1]. The radiation of scalar particles has been studied in great detail in the past. The study of massive spinor particles has been hampered by an absence of exact solutions to Dirac's equation in a black hole metric. In this paper exact solutions to Dirac's equation for massive spinors will be found in the metric of a $1+1$ dimensional black hole. These solutions will be used to calculate the vacuum stress-energy outside of the black hole. Past work in this $1+1$ theory of gravity has shown that it has many of the same qualitative features of $3+1$ general relativity. Some of its features include gravitational collapse, gravitational waves, a Newtonian limit and black holes [2,3,4].

2 The Classical $1+1$ Black Hole

The simplest non-trivial theory of gravity in $1+1$ dimensions has the field equation

$$R = 8\pi G T\,, \tag{1}$$

where R is the Ricci scalar, G is Newton's constant and T is the trace of the stress-energy tensor of the gravitational source. The black hole solution to (1) is [4]

$$ds^2 = \alpha(x)dt^2 - \frac{1}{\alpha(x)}dx^2\,, \tag{2}$$

$$\alpha(x) = 2M|x| - 1\,, \tag{3}$$

where M is the mass of the black hole which has event horizons at $|x| = \frac{1}{2M}$.

3 Solution of the Dirac Equation

An exact solution to the Dirac equation for massive spinor particle can be found in the space-time metric (2,3) of the $1+1$ dimensional black hole. The covariant form of Dirac's equation for spinor particles of mass m is [1,5]:

$$i\gamma^\mu \nabla_\mu \Psi - m\Psi = 0, \tag{4}$$

where ∇_μ is a spinor covariant derivative defined by $\nabla_\mu = \partial_\mu - \frac{1}{4}e^\beta_d e_{\beta c;\mu}\epsilon^{cd}\gamma^0\gamma^1$.

The dyad formalism is used to relate the curved space to a locally flat space. The dyad is defined by $g^{\mu\nu} = e^\mu_a e^\nu_b \eta^{ab}$ so that

$$e^\mu_a = \begin{pmatrix} \frac{1}{\sqrt{\alpha}} & 0 \\ 0 & \sqrt{\alpha} \end{pmatrix}. \tag{5}$$

The curved space matrix γ^μ is related to the flat space matrix γ^a by the relation $\gamma^\mu = \gamma^a e^\mu_a$. The chiral γ matrix representation used to solve the Dirac equation is

$$\gamma^0 = \begin{pmatrix} 0 & 1 \\ 1 & 0 \end{pmatrix} \quad \text{and} \quad \gamma^1 = \begin{pmatrix} 0 & -1 \\ 1 & 0 \end{pmatrix} \tag{6}$$

which obeys the anti-commutation relation $\{\gamma^a, \gamma^b\} = 2\eta^{ab}$.

In this $1+1$ system Dirac's equation becomes (see [5] for details)

$$i\gamma^a e^\mu_a \partial_\mu \Psi + \frac{i}{2}\gamma^a e^\mu_{a;\mu}\Psi - m\Psi = 0. \tag{7}$$

The solution to (7) when regularity at infinity is imposed is [5]

$$\Psi = \int_0^\infty d\omega N_\omega \left\{a_\omega e^{-i\omega t}U_\omega(y) + b^\dagger_\omega e^{+i\omega t}V_\omega(y)\right\} \tag{8}$$

where $y = \frac{m}{M}\sqrt{\alpha}$ and

$$U_\omega(y) = \begin{pmatrix} K_{\frac{i\omega}{M}-\frac{1}{2}}(y) \\ -iK_{\frac{i\omega}{M}+\frac{1}{2}}(y) \end{pmatrix} \quad V_\omega(y) = \begin{pmatrix} K_{\frac{i\omega}{M}+\frac{1}{2}}(y) \\ -iK_{\frac{i\omega}{M}-\frac{1}{2}}(y) \end{pmatrix}. \tag{9}$$

The function $K_{i\mu+\nu}$ is a modified bessel function of the second kind of complex order. The mode solutions (9) have been shown [5] to be orthonormal and complete when the normalization function $N_\omega = \frac{1}{\pi}\sqrt{\frac{m}{M}\cosh\frac{\pi\omega}{M}}$ is used.

4 Temperature of the Black Hole

The "Schwarzschild" vacuum outside the black hole is defined by $a_\omega|0\rangle = 0$ and $b_\omega|0\rangle = 0$. The expectation value of the stress-energy tensor in this vacuum can now be calculated using a point splitting regularization technique. The point-split stress-energy tensor is [6,7]

$$\langle 0|T_{\mu\nu}|0\rangle = \frac{i}{4}\lim_{x'\to x} \text{Trace}\left\{(\nabla' - \nabla)_{(\nu}\gamma_{\mu)}(S(x',x) - S(x,x'))\right\} \tag{10}$$

where the Feynman propagator is defined as

$$S(x,x') = \langle 0|\overline{\Psi}(x')\Psi(x)|0\rangle. \tag{11}$$

Substituting the mode decomposition (8) into (11) and using the canonical anti-commutation relations and vacuum definition the propagator becomes [7]

$$S = \frac{m}{\pi^2 M}\int_0^\infty d\omega \cosh\frac{\pi\omega}{M}\,\overline{V}_\omega(y')\,V_\omega(y)\,e^{-i\omega\Delta t}, \tag{12}$$

$$\overline{V}_\omega(y')V_\omega(y) = \begin{bmatrix} iK_{i\frac{\omega}{M}+\frac{1}{2}}(y)K_{i\frac{\omega}{M}+\frac{1}{2}}(y') & K_{i\frac{\omega}{M}+\frac{1}{2}}(y)K_{i\frac{\omega}{M}-\frac{1}{2}}(y') \\ K_{i\frac{\omega}{M}-\frac{1}{2}}(y)K_{i\frac{\omega}{M}+\frac{1}{2}}(y') & -iK_{i\frac{\omega}{M}-\frac{1}{2}}(y)K_{i\frac{\omega}{M}-\frac{1}{2}}(y') \end{bmatrix} \tag{13}$$

The integral combination formula [8] for these Bessel functions is

$$K_\mu(y)K_\nu(y') = -\frac{1}{2}\int_{-\infty}^{\infty} dse^{-(\mu+\nu)s}\xi^{\frac{1}{2}(\mu-\nu)}K_{\mu-\nu}(\beta), \tag{14}$$

$$\xi = \frac{ye^{s/2} + y'e^{-s/2}}{ye^{-s/2} + y'e^{s/2}} \qquad \beta^2 = y^2 + (y')^2 + 2yy'\cosh s. \tag{15}$$

The ω integral diverges unless $\Delta t = \Delta t - i\tau$ where $M\tau > \pi$. In the limit $\lim_{\tau\to 0}$ the ds integral is deformed to a contour C consisting of two sections, the entire real s axis and a loop encircling the pole at $s = M\Delta t + i\pi$. It is not difficult to verify that the pole term is infinite when the limit $(y',t') \to (y,t)$ is taken. It has been shown elsewhere [7] that the pole term approaches infinity in a covariant manner. The propagator can be regularized by subtracting out the infinite pole term.

$$\begin{aligned} S^{reg} &= S - S^{pole} \\ &= -\frac{m}{2\pi^2}\int_{-\infty}^{\infty} ds \frac{i(M\Delta t - s)}{\pi^2 + (M\Delta t - s)^2}\begin{bmatrix} ie^{s/2}K_0(\beta) & \xi^{1/2}K_1(\beta) \\ \xi^{-1/2}K_1(\beta) & -ie^{-s/2}K_0(\beta)\end{bmatrix} \end{aligned} \tag{16}$$

The regularized Feynman propagator (16) can now be used to calculate the energy density of the vacuum (10). The covariant derivative in (10) is

$$\gamma_t \nabla_t = \frac{My}{m}\gamma^0 \partial_t \,. \tag{17}$$

Since $S(t,t')$ depends only on $(t'-t)$, (10) becomes

$$\begin{aligned} T_{tt} &= -i\frac{M}{m}\lim_{\substack{y'\to y \\ t'\to t}} \mathrm{Trace}\{y\partial_t\gamma^0 S(y,t;y',t')\} \\ &= -\frac{M^2}{\pi^2}y\int_{-\infty}^{\infty} ds \frac{\pi^2 - s^2}{(\pi^2+s^2)^2}K_1(2y\cosh s/2) \end{aligned} \tag{18}$$

This integral can be simplified in the region near the event horizon where $y \to 0$. The integral representation of K_1 is [8]

$$K_1(az) = K_{-1}(az) = \frac{1}{2a}\int_0^{\infty} du \exp(-\frac{z}{2}(u + \frac{a^2}{u})) \sim \frac{1}{az} \,. \tag{19}$$

This representation of the Bessel function can be used in all regions outside of the black hole as long as the backreaction of the particle's mass on the curvature can be ignored, so that the energy density becomes

$$T_{tt} = -\frac{M^2}{2\pi^2}\int_{-\infty}^{\infty} ds \frac{\pi^2 - s^2}{(\pi^2+s^2)^2\cosh\frac{s}{2}} \tag{20}$$

$$= -\frac{M^2}{24\pi} \tag{21}$$

where the contour integral in (20) is evaluated using residues [7].

In 1+1 the energy density of black body radiation for fermions is $\frac{\pi T^2}{6}$ (including contributions from particles and anti-particles). A comparison of the energy density in (21) with the black body energy density yields a temperature of $\frac{M}{2\pi}$ for the $1+1$ black hole.

5 Conclusions

It is of interest to compare the vacuum expectation value of the stress-energy tensor for the spinor field with the stress-energy tensor calculated previously [9] for the scalar field. The energy density for a scalar field in the same black hole metric is $T_{tt}^{scalar} = -\frac{M^2}{24\pi}$. Since the energy density for scalar black body radiation in $1+1$ dimensions is $\frac{\pi T^2}{6}$, it can be seen that the temperature calculated for spin-$\frac{1}{2}$ radiation is the same found earlier for scalar radiation.

The temperature identified to the $1+1$ dimensional black hole for both types of measurements is $\frac{M}{2\pi}$, which also agrees with earlier work using a Wick rotation[9]. Further analysis has shown that all components of the stress-energy tensor are the same for both scalar and spinor particles. It should be noted that while the solutions to the Dirac equation are infinite at the event horizon, the stress-energy tensor is finite. This work should make it possible to examine other quantum aspects of spinors near black holes. In particular, it should be possible to calculate the effect of the particle's mass on the curvature of space-time.

6 Acknowledgments

I would like to thank Robb Mann for his help and suggestions in this project.

References

[1] N.D. Birrell, P.C.W. Davies, *Quantum Fields in Curved Space*, (Cambridge University Press, 1982).

[2] R.B. Mann, University of Waterloo preprint WATPHYS-TH89/06.

[3] A.E. Sikkema and R.B. Mann, Class. Quant. Grav. **8** 219.

[4] R.B. Mann, A. Shiekh, and L. Tarasov, Nucl. Phys. **B341** (1990) 134.

[5] R.B Mann, S.M. Morsink, A.E. Sikkema and T.G. Steele, Phys. Rev. **D** (in press).

[6] S.M. Christensen, Phys. Rev. **D17** (1978) 946.

[7] S.M. Morsink and R.B. Mann, University of Waterloo preprint WATPHYS-TH91/02.

[8] A. Erdelyi et al, *Higher Transcendental Functions Vol. 2*, (McGraw-Hill, 1953).

[9] R.B. Mann and T.G. Steele, University of Waterloo preprint WATPHYS-TH90/02.

REMARKS ON THE DYNAMICS OF THE MIXMASTER UNIVERSE

BEVERLY K. BERGER
Physics Department, Oakland University
Rochester, Michigan 48309, USA

ABSTRACT

A single generic Mixmaster model is followed numerically toward the singularity to illustrate the following features of the dynamics: (1) the validity of the related 1-dimensional map as an approximation to the solution and its explicit relation to the minisuperspace picture; (2) the nature of consecutive single epoch eras; (3) deviations from the idealized behavior assumed in the approximation; (4) identification of minisuperspace proper time as the time coordinate to use in computation of Liapunov exponents since it most closely relates the dynamics to the map.

Recent computations of vanishing Liapunov exponents (LE's) for the supposedly chaotic Mixmaster cosmologies[1] have motivated a reexamination of the behavior of the models. The chaotic nature of the models may be inferred from the known properties of a 1-dimensional map which has been used to parametrize the solution to Einstein's equations.[2,3] The (vacuum, diagonal) Mixmaster universe approaches the singularity through an infinite sequence of Kasner epochs. Each Kasner epoch can be described either in terms of the anisotropic expansion indices p_i parametrized by the BKL variable u or as straight lines in the Ω, $\beta_\pm$ minisuperspace (MSS) parametrized by θ, the angle of the trajectory with respect to the β_+-axis. For $1 \le u \le \infty$ and $\frac{2}{3}\pi \le \theta \le \pi$, the parametrizations are related by

$$u = -\frac{2 + \cos\theta + \sqrt{3}\sin\theta}{1 + 2\cos\theta} \quad . \tag{1}$$

The Kasner epoch changes when the system bounces off a wall of the triangular MSS potential. The BKL (or Ryan) bounce laws relating u (or θ) from epoch to epoch are easily derived by approximating Einstein's equations by a scattering off the dominant potential wall with the Kasner epochs as the asymptotic regimes. BKL obtained the rule $u' = u - 1$ for the next parameter value. If the Kasner indices $p_1 < p_2 < p_3$ are associated respectively with the metric scale factors $a > b > c$ as the singularity is approached (since $p_1 < 0$ so that a increases toward the singularity), then $u \to u - 1$ corresponds to $p_1 < p_2 < p_3 \to p_2 < p_1 < p_3$. The sequence of these epochs describes oscillations of a and b with monotonic decrease of c. If $1 < u < 2$, the next epoch starts a new era with $u' = (u-1)^{-1}$. The transition to a new era has

$$p_1 < p_2 < p_3 \rightarrow p_2 < p_1 < p_3 \rightarrow p_2 < p_3 < p_1 \quad ,$$

$$a > b > c \rightarrow b > a > c \rightarrow b > c > a \ . \tag{2}$$

Thus in the next era, b and c oscillate while a decreases monotonically.

In MSS, the Ryan bounce law plus the rotations which transform the next target potential wall into the original wall show that the epochs of a given era describe bounces outward along a corner of the MSS potential. A new era begins when the trajectory moves to a new corner. Fig. 1 illustrates this behavior for a presumably generic single trajectory followed numerically toward the singularity. Einstein's equations were integrated numerically from the maximum of expansion at $\Omega = 1.94$ to $\Omega = -7.5 \times 10^7$ where $\Omega = -\infty$ at

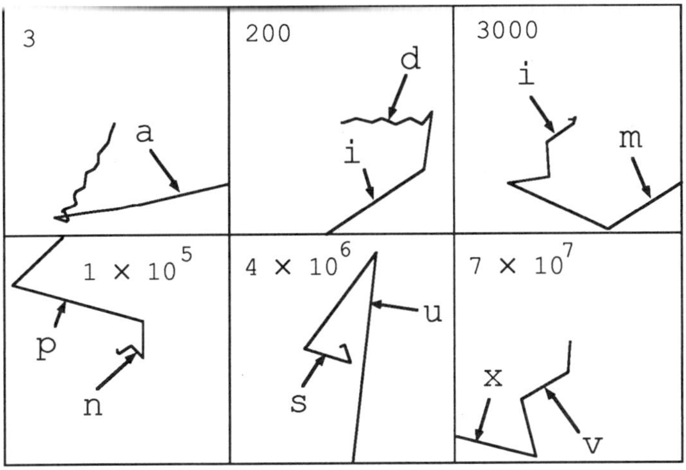

Fig. 1. Consecutive segments of a single trajectory are displayed from left to right and top to bottom row. In all cases the horizontal axis is β_+ and the vertical β_- with the origin at the center of each square. The letters label selected epochs. The numbers refer to the ranges of the $\beta_\pm$ -axes in each square.

the singularity. The first epochs of eras (and their u-values) are a (7.2327976), h (4.29557600), l (3.38322467), o (2.60943537), q (1.64086308), r (1.56039571), s (1.7844557), t (1.2747693), u (3.6394167), and x (1.56392533). The u-values agree with the values predicted from the Gauss map[3]

$$u_{n+1} = (u_n - [u_n])^{-1} \tag{3}$$

where [ ] denotes the integer part and n refers to the era number. To compute the u- parameter from the trajectory, first the angle θ of the trajectory in the anisotropy plane is found from its tangent. The Kasner indices are then constructed from θ with the u-parametrization of the minimum index inverted analytically. The value of the Hamiltonian constraint is monitored as a check on the accuracy of the numerically obtained trajectory. The accuracy is further verified by the fact that the same trajectory is generated either toward the singularity or (by reversing the initial velocities) toward the maximum of expansion starting anywhere on the trajectory.

A u-value cannot be identified prior to the epoch a because the influence of the MSS potential is too great for the required asymptotic Kasner regimes to exist. In principle, this anomalous behavior could recur closer to the singularity if the trajectory is within an angle $\varepsilon = \ln 2 / (4\sqrt{3}\, |\beta_+|)$ of the center of the potential corners so that the oscillator-like corner potential dominates the triangular wall. This requires $u \geq \sqrt{3} / \varepsilon$. However, the invariant probability[3] $\mu(x) = [\ln 2\,(1 + x)]^{-1}$ of the map given by Eq. 3 where $x = 1/u$ can be used to show that, for large u, the probability of a given value of $u \propto u^{-2}$. Since $|\beta_+| \approx |\Omega|$ and $|\Omega| \to \infty$ as the singularity is approached, $\varepsilon \to 0$ so the required larger and larger u-values become less and less likely to occur.

We note that epochs q - t are single epoch eras (SEE's). There will be at least s successive SEE's if $u \in [\, z_1^s \,,\, z_2^s \,]$ where

$$z_1^s = \left(\frac{f_{s+j}}{f_s}\right)^j \quad , \quad z_2^s = \left(\frac{f_s}{f_{s-j}}\right)^j \tag{4}$$

with $f_0 = 1, f_1 = 1, f_2 = 2, \ldots$ the Fibonacci numbers and $j = (-1)^s$. The probability for at least s SEE's is

$$P(s) = \frac{j}{\ln 2} \ln \frac{f_{s+1}^2 + j}{f_{s+1}^2} \quad . \tag{5}$$

Fig. 2 is a schematic diagram illustrating the behavior of the logarithms of the metric components during epoch and era changes. Deviations from the idealized behavior of Eq. 2 are seen in the relative magnitudes. For example, in the era beginning with o, the monotonically decreasing component is not the smallest until p. This also occurs in the SEE q, but not in r or s. The bounce law derivation requires only that the increasing component dominate which occurs in all cases so that the Kasner indices and thus u behave as predicted.

The LE for the map (Eq. 3) is $\pi^2 / 6(\ln 2)^2$ thus signalling chaotic behavior. LE's for the numerical trajectories depend on the time variable. Hobill *et al* and Burd *et al* found zero LE's using τ (defined by $dt = abc\, d\tau$ for comoving proper time t) and Ω respectively. Pullin[4] using essentially $\ln \Omega$ found non-zero LE's. LE's for any time variable can be found with the method of Wolf *et al.*[5] Any convenient coordinate time is used to compute the trajectories (and their rate of divergence) with the time variable of interest evaluated separately in terms of the coordinate time. From Fig. 1 (since $|\beta_{\pm}| \approx |\Omega|$), we see that the duration of epochs increase roughly exponentially in Ω as the singularity is approached.

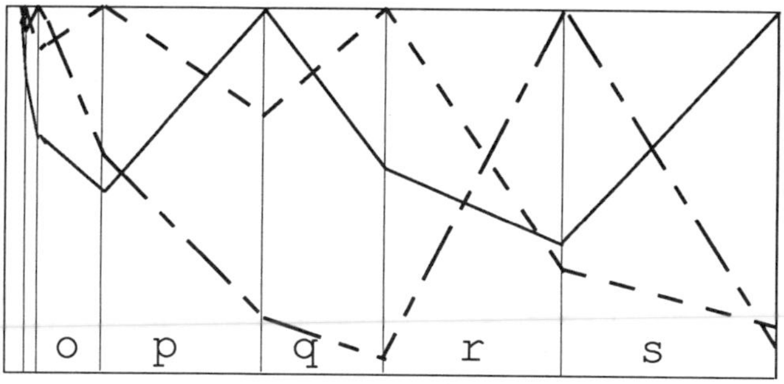

Fig. 2. The three segmented curves are the logarithms of the metric components during the indicated epochs. The horizontal axis is an arbitrarily rescaled time. The relative magnitudes and slopes are accurately represented. Actual values and epoch durations are not.

The map LE is based on unit time intervals between iterations. We therefore argue that MSS proper time λ defined by

$$d\lambda = \left| - p_\Omega{}^2 + p_+{}^2 + p_-{}^2 \right|^{1/2} d\tau \tag{6}$$

be used to compute the LE's. The same derivation which yields the bounce laws gives $\Delta\lambda = \pi / 4\sqrt{3}$ for every bounce while $\Delta\lambda = 0$ during a Kasner regime. Since an infinite number of bounces are expected, $\lambda = \infty$ at the singularity.

Fig. 3 shows the LE's computed for (part of) the trajectory of Fig. 1 using λ as the time both from all the degrees of freedom and from just the θ degree of freedom (since θ also changes only during a bounce). The LE's do not vanish. Of course, only 12 - 14 epochs are represented in Fig. 3 reaching $\lambda \approx 6.8$ rather than infinity. Strictly speaking, conclusions about the LE's cannot be made. However, the corresponding epochs yield vanishing LE's when computed in terms of τ which reaches $\approx 10^6$.

As a final remark, we note that the map[6]

$$u_{n+1} = \begin{cases} u_n - 1 & u_n > 2 \\ \dfrac{1}{u_n - 1} & 1 < u_n < 2 \end{cases} \tag{7}$$

is a closer analog of the trajectories since both epochs and eras appear. The invariant probability for this map is not normalizable but may be expressed as

$$\mu(x) = \frac{1}{x \ \ln u_{max}} \tag{8}$$

with u_{max} the largest u on a given trajectory. The LE is then $\pi^2 \, [\, 6 \ln 2 \ln u_{max} \,]^{-1}$ which is trajectory dependent. The LE's shown in Fig. 3 would also change their magnitude for a different trajectory, but would never vanish.

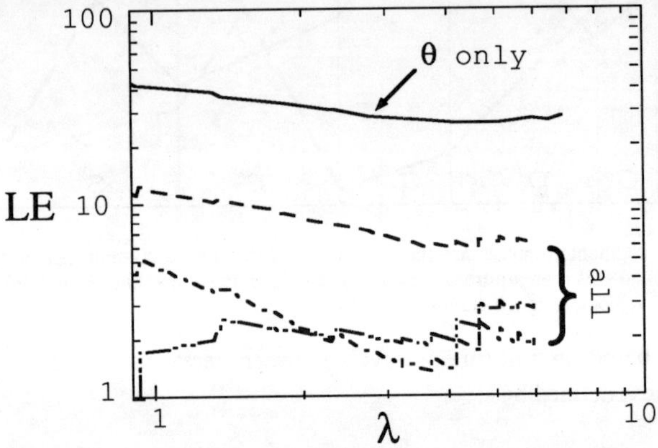

Fig. 3. The (positive) LE's for MSS proper time. The LE's for all the variables include epochs a - l. The computation for the θ degree of freedom requires fewer equations and can be carried to epoch n.

This work has been supported in part by National Science Foundation Grant PHY-8904847 to Oakland University. I am extremely grateful for the hospitality provided by the Institute for Geophysics and Planetary Physics of the Lawrence Livermore National Laboratory. I wish to thank Adrian Burd and David Hobill for useful discussions. The computations reported here were performed primarily with the facilities of the National Center for Supercomputing Applications of the University of Illinois.

1. A .B. Burd, N. Buric, and G.F.R. Ellis, *Gen. Rel. Grav.* **22** (1990) 349–363; D. Hobill, D. Bernstein, M.Welge, and D. Simkins, preprint; S. E. Rugh and B.J.T. Jones, *Phys. Lett.* **A147** (1990) 353–355.
2. V. Belinskii and I. Khalatnikov, *Zh. Eksp. Teor. Fiz.* **56** (1969) 1700; V.A. Belinskii , I.M. Khalatnikov, and E.M. Lifshitz,*Adv. Phys.* **31** (1982) 639–667. (These authors will be referred to as BKL.) C.W. Misner, *Phys. Rev.* **186** (1969) 1319; M. Ryan, *Hamiltonian Cosmology* (Springer-Verlag, Berlin, 1972).
3. J. Barrow, *Phys. Rep.* **85** (1982) 1.
4. J. Pullin, preprint.
5. A. Wolf, J. Swift, H. Swinney, and J. Vastano, *Physica* **16D** (1985) 285.
6. P.K.-H. Ma and J. Wainwright , in *Proceedings of the Third Hungarian Relativity Workshop*, in press.

THE HAMILTONIAN FOR THE MIXMASTER VACUUM UNIVERSE WITH GEOMETRY $R \times S^7$

R. M. Cassidy
Departments of Physics and Mathematics, Elizabethtown College
Elizabethtown, Pennsylvania 17022, USA

and

W. M. Stuckey
Department of Physics, Elizabethtown College
Elizabethtown, Pennsylvania 17022, USA

ABSTRACT

The Hamiltonian for the Mixmaster vacuum universe with geometry $R\times S^7$ is constructed using the formalism first proposed by Arnowitt, Deser, and Misner. A preliminary verification of the integrity of the basis vectors tangent to the S^7 in an E^8 embedding space is accomplished. The intrinsic curvature is calculated by means of the embedding space definition followed by a calculation of the extrinsic curvature. The constraint equation for the Hamiltonian then requires the lapse function be a function of time and position. Therefore a global Mixmaster analysis of $\Re\times S^7$ can not be done as in the original $\Re\times S^3$ case.

1. Introduction

The possiblility of performing a Mixmaster analysis[1] on $\Re\times S^7$ is pursued. The S^7 was chosen because spaces of the form S^n with n linearly independent vector fields exist only for n=1,3,7.[2] The $\Re\times S^3$ model has received extensive study and the $\Re\times S^1$ case is relatively uninteresting.

A set of orthonormal vector fields existing for all $p\in S^7$ is given by DeWitt[3]. It is shown in this paper that these vector fields are in fact orthonormal and tangent to the S^7. Furthermore, it is shown that these fields are tangent to great circles on the sphere. These properties correspond to those pertaining to the vector fields of the S^3 in Misner's original model.

However, the elements of the vector basis on the S^3 are in one-to-one correspondence with the elements of the Lie algebra of SO(3). No similar correspondence exists for the basis elements on the seven sphere[4]. Therefore, the intrinsic curvature of the anisotropized S^7 is spatially dependent, and the lapse function is forced to be spatially dependent.

2. Basis tangent to S^7

The Einstein summation convention is used throughout this paper with greek indices denoting space and time components of tensors (0,1,2,...,7) and latin indices denoting space components of tensors (1,2,...,7) on S^7 and (1,2,...,8) in E^8.

The basis used on the S^7 in an E^8 embedding space is given by[3]

$$\mathbf{l}_i = -x^i\,\mathbf{e}_8 + x^8\,\mathbf{e}_i + a_{ij}{}^k\,x^j\,\mathbf{e}_k \qquad (1)$$

where

$$a_{ij}{}^k = \begin{cases} +1 \text{ for even permutations of ijk} \\ -1 \text{ for odd permutations of ijk} \\ 0 \text{ for combinations of ijk not shown} \end{cases} \qquad (2)$$

$$\text{ijk} \rightarrow 123,\ 516,\ 624,\ 435,\ 471,\ 673,\ 572$$

This basis is in fact an orthogonal set with respect to the standard metric of the E^8 embedding space, i.e.

$$\mathbf{l}_i \cdot \mathbf{l}_j = \delta_{ij} \qquad (3).$$

It must then be verified that this basis is tangent to the S^7. This is done by showing the basis to be normal to the radial vector, $\mathbf{r}$, on the S^7 embedded in E^8. Again through the use of the standard dot product with

$$\mathbf{r} \equiv x^i\,\mathbf{e}_i \qquad (4)$$

we have

$$\mathbf{l}_i \cdot \mathbf{r} = 0 \qquad (5).$$

There is one more property of interest. Using the covariant derivative in E^8, we have

$$\nabla_{\mathbf{l}_i} \mathbf{l}_i = -\,\mathbf{r} \qquad (6).$$

Therefore the $\mathbf{l}_i$ are tangent to geodesics on S^7.

3. Intrinsic scalar curvature

The metric used is

$$ds^2 = -N^2\,dt \otimes dt + g_{ij}\,\omega^i \otimes \omega^j \qquad (7)$$

where g_{ij} is diagonal, chosen to be a function of time only, and

$$\omega^i = -x^i\,dx^8 + x^8\,dx^i + a_{ijk}\,x^j\,dx^k \qquad (8)$$

are the one-forms dual to the basis vectors $\mathbf{l}_i$. [Note: $a_{ijk} = a_{ij}{}^k$] The commutation coefficients,

$$c_{ij}{}^k \equiv \omega^k\left[\,\hat{\mathbf{l}}_i, \hat{\mathbf{l}}_j\,\right] \qquad (9)$$

are

$$
\begin{aligned}
c_{12}{}^{3} &= 2\left[-(x^1)^2 - (x^2)^2 - (x^3)^2 + (x^4)^2 + (x^5)^2 + (x^6)^2 + (x^7)^2 - (x^8)^2 \right] \\
c_{12}{}^{4} &= 4\left[x^8x^5 - x^1x^6 + x^2x^7 - x^3x^4 \right] \\
c_{12}{}^{5} &= 4\left[-x^8x^4 - x^1x^7 - x^2x^6 - x^3x^5 \right] \\
c_{12}{}^{6} &= 4\left[-x^8x^7 + x^1x^4 + x^2x^5 - x^3x^6 \right] \\
c_{12}{}^{7} &= 4\left[x^8x^6 + x^1x^5 - x^2x^4 - x^3x^7 \right] \\
c_{13}{}^{4} &= 4\left[x^8x^6 + x^1x^5 + x^2x^4 + x^3x^7 \right] \\
c_{13}{}^{5} &= 4\left[x^8x^7 - x^1x^4 + x^2x^5 - x^3x^6 \right] \\
c_{13}{}^{6} &= 4\left[-x^8x^4 - x^1x^7 + x^2x^6 + x^3x^5 \right] \\
c_{13}{}^{7} &= 4\left[-x^8x^5 + x^1x^6 + x^2x^7 - x^3x^4 \right] \\
c_{14}{}^{5} &= 4\left[x^8x^2 + x^1x^3 - x^4x^6 - x^5x^7 \right] \\
c_{14}{}^{6} &= 4\left[x^8x^3 - x^1x^2 + x^4x^5 - x^6x^7 \right] \\
c_{14}{}^{7} &= 2\left[-(x^1)^2 + (x^2)^2 + (x^3)^2 - (x^4)^2 + (x^5)^2 + (x^6)^2 - (x^7)^2 - (x^8)^2 \right] \\
c_{15}{}^{6} &= 2\left[(x^1)^2 - (x^2)^2 - (x^3)^2 - (x^4)^2 + (x^5)^2 + (x^6)^2 - (x^7)^2 + (x^8)^2 \right] \\
c_{15}{}^{7} &= 4\left[x^8x^3 - x^1x^2 - x^4x^5 + x^6x^7 \right] \\
c_{16}{}^{7} &= 4\left[-x^8x^2 - x^1x^3 - x^4x^6 - x^5x^7 \right] \\
c_{23}{}^{4} &= 4\left[-x^8x^7 - x^1x^4 + x^2x^5 + x^3x^6 \right] \\
c_{23}{}^{5} &= 4\left[x^8x^6 - x^1x^5 - x^2x^4 + x^3x^7 \right] \\
c_{23}{}^{6} &= 4\left[-x^8x^5 - x^1x^6 - x^2x^7 - x^3x^4 \right] \\
c_{23}{}^{7} &= 4\left[x^8x^4 - x^1x^7 + x^2x^6 - x^3x^5 \right] \\
c_{24}{}^{5} &= 4\left[-x^8x^1 + x^2x^3 + x^4x^7 - x^5x^6 \right] \\
c_{24}{}^{6} &= 2\left[(x^1)^2 - (x^2)^2 + (x^3)^2 - (x^4)^2 + (x^5)^2 - (x^6)^2 + (x^7)^2 - (x^8)^2 \right] \\
c_{24}{}^{7} &= 4\left[-x^8x^3 - x^1x^2 - x^4x^5 - x^6x^7 \right] \\
c_{25}{}^{6} &= 4\left[x^8x^3 + x^1x^2 - x^4x^5 - x^6x^7 \right] \\
c_{25}{}^{7} &= 2\left[(x^1)^2 - (x^2)^2 + (x^3)^2 + (x^4)^2 - (x^5)^2 + (x^6)^2 - (x^7)^2 - (x^8)^2 \right] \\
c_{26}{}^{7} &= 4\left[x^8x^1 - x^2x^3 + x^4x^7 - x^5x^6 \right] \\
c_{34}{}^{5} &= 2\left[-(x^1)^2 - (x^2)^2 + (x^3)^2 + (x^4)^2 + (x^5)^2 - (x^6)^2 - (x^7)^2 + (x^8)^2 \right] \\
c_{34}{}^{6} &= 4\left[-x^8x^1 - x^2x^3 + x^4x^7 + x^5x^6 \right] \\
c_{34}{}^{7} &= 4\left[x^8x^2 - x^1x^3 - x^4x^6 + x^5x^7 \right] \\
c_{35}{}^{6} &= 4\left[-x^8x^2 + x^1x^3 - x^4x^6 + x^5x^7 \right] \\
c_{35}{}^{7} &= 4\left[-x^8x^1 - x^2x^3 - x^4x^7 - x^5x^6 \right] \\
c_{36}{}^{7} &= 2\left[(x^1)^2 + (x^2)^2 - (x^3)^2 + (x^4)^2 + (x^5)^2 - (x^6)^2 - (x^7)^2 - (x^8)^2 \right] \\
c_{45}{}^{6} &= 4\left[x^8x^7 + x^1x^4 + x^2x^5 + x^3x^6 \right] \\
c_{46}{}^{7} &= 4\left[x^8x^5 + x^1x^6 - x^2x^7 - x^3x^4 \right] \\
c_{45}{}^{7} &= 4\left[-x^8x^6 + x^1x^5 - x^2x^4 + x^3x^7 \right] \\
c_{56}{}^{7} &= 4\left[-x^8x^4 + x^1x^7 + x^2x^6 - x^3x^5 \right]
\end{aligned}
\qquad (10)
$$

The above list is complete because the $c_{ij}{}^{k}$ are totally antisymmetric in i,j,k. These commutation coefficients then give

$$
{}^{7}R = \frac{-1}{4} \sum_{ijk} \left[g^{ii} \left(g_{kk} g^{jj} - 2 \right) \left(c_{ij}{}^{k} \right)^2 \right] \qquad (11).
$$

Examining the isotropic case where $g_{ii} = 1$, and using

$$\left(\sum_j (x^j)^2\right)^2 = \sum_j (x^j)^2 = 1 \tag{12}$$

Eq. 15 yields the desired result of 42. Eq. 15 can also be applied to the case of the S^3 (using the analogous basis set) giving a scalar curvature of 6. The S^7 and S^3 scenarios differ, however, as $c_{ij}{}^k$ for the S^3 are SO(3) structure constants but on S^7 they are functions of spatial position.

4. Extrinsic curvature

The extrinsic curvature **K** is given by

$$\mathbf{K} = \frac{-1}{2} L_n \mathbf{g} \tag{13}$$

where L is the lie derivative, **g** is the metric on the S^7, and

$$\mathbf{n} = \frac{1}{N}\frac{\partial}{\partial t} \tag{14}.$$

This gives

$$\mathbf{K} = \frac{-1}{2}\left[n^\alpha g_{ij,\alpha} - g_{mj}\left(n^\beta c_{\beta i}{}^m\right) - g_{im}\left(n^\beta c_{\beta j}{}^m\right)\right] \omega^i \otimes \omega^j \tag{15}.$$

From

$$[\,\mathbf{e}_0, \mathbf{l}_i\,] = \frac{-N_{,i}}{N}\frac{\partial}{\partial t} \tag{16}$$

we have

$$c_{0i}{}^m = 0 \tag{17}.$$

And thus

$$K_{ij} = \frac{-1}{2}\frac{1}{N}\frac{\partial g_{ij}}{\partial t} \tag{18}.$$

5. Hamiltionian

In vacuum we have for the action[5]

$$S \equiv \frac{1}{16\pi} \int \left[\pi^{ij} \frac{\partial g_{ij}}{\partial t} - N\, H(\pi^{ij},\, g_{ij}) \right] d^8x \tag{19}$$

where the Hamiltonian is

$$H(\pi^{ij}, g_{ij}) = g^{-1/2} \left(\mathrm{Tr}\, \pi^2 - \frac{1}{2} (\mathrm{Tr}\, \pi)^2 \right) - g^{1/2}\, {}^7R \tag{20}$$

and

$$\pi^{ij} = g^{1/2} \left(g^{ij}\, \mathrm{Tr}\, \mathbf{K} - K^{ij} \right) \tag{21}.$$

The constraint equation obtained from $\delta S/\delta N = 0$ is

$$H = 0 \tag{22}.$$

Therefore, with ${}^7R$ a funtion of spatial position, we have that N must be a function of spatial position.

6. Conclusion

There exists an orthonormal basis for all $p \in S^7$, each element of which is tangent to a geodesic of S^7. These characteristics are analogous to those of the basis on S^3, used for the original Mixmaster universe. However, because the commutation coefficients of the S^7 are not structure constants for any Lie group, we cannot do a global Mixmaster analysis as in the original $\mathfrak{R} \times S^3$ case.

7. Acknowledgements

We wish to thank L. Shepley, C. DeWitt, and J. Milnor for their valuable time and effort expended in helping us to complete this project.

8. References

1. C. Misner, *Phys. Rev. Lett.* **22** (1969) 1071.
2. J. F. Adams, *Annals of Mathematics* **75** (1962) 603.
3. Y. Choquet-Bruhat and C. DeWitt-Morette, *Analysis, Manifolds and Physics* (North-Holland, New York, 1989) p. 265.
4. ibid, p. 267.
5. C. Misner, K. Thorne, and A. Wheeler, *Gravitation* (W. H. Freeman, San Francisco, 1973) p. 521.

POLARIZATION SHIFTS IN NONSYMMETRIC THEORIES OF GRAVITY

J. H. Palmer† and R. B. Mann
Department of Physics, University of Waterloo
Waterloo, Ontario, N2L 3G1, CANADA

and

M. D. Gabriel and M. P. Haugan
Department of Physics, Purdue University
West Lafayette, IN 47907, USA

ABSTRACT

Theories of gravity that include a totally antisymmetric component of the metric coupled to electromagnetism predict that light will propagate through a gravitational field with a polarization dependent velocity. The observed polarization depends upon the initial polarization and the path traveled by the light. For extended sources of polarized light within magnetically active regions on the Sun, this leads to an observed depolarization. Observations of the polarization of solar spectral lines near the limb of the Sun are used to place limits on the nonsymmetric charge of the Sun in Moffat's Nonsymmetric Gravitation Theory (NGT).

1. Introduction

Nonsymmetric theories of gravity introduce a totally antisymmetric tensor identified as the antisymmetric part of a general metric.[1,3] The nonsymmetric metric then determines the connection, including macroscopic torsion, through the application of metric compatibility. (Metric compatibility arising as a field equation from the variation of an action with respect to the connection, or through the requirement that the inner product of two vectors is invariant under parallel transport). These theories are strongly related to metric-tensor theories through interpreting the antisymmetric part of the metric as an additional tensor field. However, the way in which nonsymmetric theories introduce the antisymmetric tensor leads naturally to its interpretation as the antisymmetric part of a general metric. The interpretation of the antisymmetric field as a gravitational field is bolstered by coupling the full metric to matter. In this way, the antisymmetric part of the metric is coupled to mass/energy as expected in a theory of

† **Presenter. This paper is based upon work by Gabriel, Haugan, Mann and Palmer to be published in Phys. Rev. Lett. Modifications made here to reflect the presentation are the responsibility of J.H. Palmer.**

gravitation. All nonsymmetric theories which couple the full metric to electromagnetism predict polarization dependent bending of light and radar echo time delay.[2] Furthermore, the observed polarization of light depends upon the path of propagation, as will be shown here.

The existence of path dependent polarization is general for all nonsymmetric theories which couple the full metric to electromagnetism. In order to make a physical prediction, however, it is necessary to work within one theory. The obvious nonsymmetric theory to choose is Moffat's Nonsymmetric Gravitation Theory (NGT).[3] This theory has been studied through its evolution over the past ten years. It is the unique nonsymmetric theory with hermitian symmetry which survives a multitude of theoretical and physical criteria. Other nonsymmetric theories can be created by relaxing the hermitian symmetry condition, but these theories have yet to be fully investigated.[1]

2. General Formalism

We begin with the general action coupling electromagnetism with gravitation,

$$I_{em} = \frac{-1}{16\pi}\int d^4x\,\sqrt{-g}\mathcal{F}g^{\mu\alpha}g^{\nu\beta}\{ZF_{\mu\nu}F_{\alpha\beta} + (1-Z)F_{\alpha\nu}F_{\mu\beta} + YF_{\mu\alpha}F_{\nu\beta}\} \quad (1)$$

where $F_{\mu\nu} \equiv A_{\nu,\mu} - A_{\mu,\nu}$ and the matrix $g^{\mu\nu}$ is the inverse of the nonsymmetric gravitational field $g_{\mu\nu}$ defined by $g^{\mu\alpha}g_{\nu\alpha} = g^{\alpha\mu}g_{\alpha\nu} = \delta^{\mu}_{\nu}$. The new contractions allowed in nonsymmetric theories are parametrized by the constants Y and Z. In addition, a scalar coupling function, $\mathcal{F} = \mathcal{F}(\sqrt{-g}/\sqrt{-\gamma})$, is introduced where $g \equiv \det g_{\mu\nu}$ and $\gamma \equiv \det g_{(\mu\nu)}$.

The Sun's essentially static, spherically symmetric gravitational field admits, in theories like NGT, a representation in spherical coordinates $g_{00} = -T(r)$, $g_{(0i)} = 0$, $g_{(ij)} = H(r)\delta_{ij}$, $g_{[0i]} = L(r)n_i$, and $g_{[ij]} = 0$, where $n_i \equiv x_i/r$ and $r \equiv |\vec{x}|$. Note that we have chosen to diagonalize the symmetric part of the covariant metric rather than the symmetric part of the contravariant metric in order to retain a sensible limiting velocity for the motion of particles. Generally in nonsymmetric theories $g^{(\mu\nu)}$ and $g_{(\mu\nu)}$ cannot be diagonalized simultaneously when $g^{\mu\alpha}g_{\mu\beta} = \delta^{\alpha}_{\beta}$. Using this metric in Eq. (1) and identifying $E_i = F_{0i}$ and $F_{ij} = \epsilon_{ijk}B_k$ gives

$$I_{em} = (8\pi)^{-1}\int d^4x\,\left\{\mathbf{E}\cdot\mathbf{E} + X\alpha\epsilon(\hat{\mathbf{n}}\cdot\mathbf{E})^2 - \frac{1}{\mu}(\mathbf{B}\cdot\mathbf{B} - \Omega(\hat{\mathbf{n}}\cdot\mathbf{E})^2)\right\} \quad (2)$$

where $\epsilon = \mathcal{F}(H/T)^{1/2}(1-\Omega)^{-1/2}$, $\mu = \mathcal{F}^{-1}(H/T)^{1/2}(1-\Omega)^{1/2}$, $\alpha = 2\Omega(1-\Omega)^{-1}$, $\Omega = L^2/TH$, and $X = 1 - Z - Y$.

We now choose $\mathcal{F} = \sqrt{-g}/\sqrt{-\gamma}$ to remove the electrostatic energy dependence of the rate of free-fall analyzed by Will.[4] We also choose $X = 0$ to keep electrostatics unmodified. We expect that experiments using atomic transitions to test the isotropy of local Lorentz frames will tightly constrain the $X\alpha$ term of Eq. (2). With these choices, Eq. (2) becomes

$$I_{em} = (8\pi)^{-1} \int d^4x \left\{ \epsilon E^2 - \frac{1}{\mu}(B^2 - \Omega(\hat{n}\cdot\vec{B})^2) \right\}, \tag{3}$$

where $\epsilon = \mu = (H/T)^{1/2}$, and $\Omega = L^2/TH$.

Varying Eq. (3) with respect to A_μ yields the gravitationally modified Maxwell equations. Substituting locally plane electromagnetic waves, $\vec{E} = \vec{A}_E e^{i\Phi}$ and $\vec{B} = \vec{A}_B e^{i\Phi}$ where $\partial_\mu \Phi = (\partial\Phi/\partial t, \vec{\nabla}\Phi) = (-\omega, \vec{k})$, into the Maxwell equations and applying the geometric optics approximation gives the Eikonal equation,[2]

$$k^2\vec{A}_B - \epsilon\mu\omega^2\vec{A}_B - \Omega(\hat{n}\cdot\vec{A}_B)(k^2\hat{n} - \vec{k}(\hat{n}\cdot\vec{k})) = 0. \tag{4}$$

The Eikonal equation describes the propagation of light in the spherical coordinate system being used. When $\Omega \neq 0$, the propagation is dependent upon the polarization through the $\hat{n}\cdot\vec{A}_B$ term and upon the direction of propagation through the $\hat{n}\cdot\vec{k}$ term. Linearly polarized light with $\hat{n}\cdot\vec{A}_B = 0$ has a coordinate phase velocity of $(1/\epsilon\mu)^{1/2}$. When the light is linearly polarized with $\vec{A}_B$ in the plane spanned by $\vec{k}$ and $\hat{n}$, the coordinate phase velocity is $((1-\Omega\sin^2\psi)/\epsilon\mu)^{1/2}$ where ψ is the angle between the propagation vector and the unit radial, $\hat{k}\cdot\hat{n} = \cos\psi$. By perturbing the Eikonal equation about the path taken by light in the vacuum, it is possible to derive the total phase shift between the two polarizations. To first order in Ω and $(1-\epsilon\mu)$ the total phase difference integrated over the unperturbed path is

$$\Delta\phi = \frac{\omega}{2}\int \Omega \sin^2\psi \, dt \tag{5}$$

where t parametrizes the path.

Eq. (5) is general for all nonsymmetric theories. Now substituting the NGT solution for $\Omega = L^2/(TH)$ into Eq. (5) will allow the integration to be performed. To post-Newtonian order, NGT gives $\Omega = l_\odot^4/r^4$ where $l_\odot$ is the nonsymmetric charge of the Sun. Choosing the path taken by light traveling from the surface of the Sun of radius R to the Earth at infinity we perform the integral and obtain

$$\Delta\phi(\mu) = \frac{\pi l_\odot^4}{\lambda R^3}\left\{ \frac{3\pi}{16(1-\mu^2)^{3/2}} - \frac{\mu}{4} - \frac{3\mu}{8(1-\mu^2)} - \frac{3}{8(1-\mu^2)^{3/2}}\arcsin(\mu) \right\} \tag{6}$$

where μ (not to be confused with the magnetic permeability of Eq. (3)) is the standard solar physics notation for the cosine of the heliocentric angle and indicates the location of the origin of the polarized light on the surface of the Sun.

When the surface of the Sun is observed with a small aperture (5 arcsec diameter), the observed area is filled with magnetic and nonmagnetic areas. The magnetic field exists in small fluxtubes sprinkled throughout the aperture. Wherever light is emitted or absorbed within the magnetic field, Zeeman splitting can be observed. The partially circularly polarized σ components of the Zeeman split line, as recorded in measurements of Stokes V, provides the source of polarized light for our study. The value of Stoke's V observed depends strongly upon complicated processes occurring at the point of emission or absorption as well as upon the structure of the magnetic fields. Imagine making a number of identical copies of one such magnetic structure and sprinkling those copies across the solar disk from the centre of the disk to the limb. Eq. (6) shows that the observed polarization would vary from structure to structure as μ changes. One structure might be observed to be linearly polarized, the next circularly polarized, even though the physics of emission is identical. If the variation in the type of polarization occurs quickly enough with change in μ to allow a number of polarizations to exist within our small aperture, then there will be a wash-out of polarizations and a resultant depolarization. This can be modeled mathematically by assuming an initial polarization, introducing the gravitationally induced phase shift, Eq. (6), and calculating the observed Stoke's parameters at any point within the aperture. Since we cannot resolve structures smaller than the aperture, we then make use of the additive property of Stokes parameters for incoherent addition of light and integrate the Stoke's parameters over the magnetic area within the aperture. This will give the observed Stokes parameters.

3. Results and Discussion

We have written code which performs trapezoidal integration over the magnetic areas within the aperture. Assumptions made on the initial polarization of the light were chosen to be as conservative as possible. Since nonsymmetric theories will cause a decrease in circularly polarized light, we assume in all cases that the initial light is 100% circularly polarized. This assumption is extremely conservative since it results in the largest possible upper bound on the theoretically observable values of Stokes V. Furthermore, in the most conservative case we assume that all of the circularly polarized light is coming from one fluxtube with 1 arcsec diameter and that the light from this fluxtube is so intense as to totally dominate the light output within the aperture. This has the effect of giving a small range of μ values over which the polarization is allowed to vary and ensuring that the initial light is strongly polarized. Thus, significant gravitationally induced depolarization will only be achieved with relatively large values of $l_{\odot}$. In the second case we make a less conservative assumption on the total size of the magnetically active area by using the estimates of Solanki *et al.*[5] They have studied the polarization measurements made by Stenflo *et al*[6] and have found that, for the observation in question, 12.9% of the area observed within the aperture emits polarized light. Once again we assume that the polarized light is so intense as to totally dominate the light output.

We have run the code to find the theoretical upper limit of observed Stokes V as a function of the cosine of the heliocentric angle μ. In Figure I we present the results of this calculation by plotting $P_e = \sqrt{U^2+V^2}/I$ against μ. Case 1 uses the extremely conservative assumption of a single small fluxtube and is calculated for light of wavelength 5250Å. Case 2 assumes 12.9% of the observed area is magnetically active and is calculated for light of wavelength 5247Å. The two Stokes V data points are at the two different wavelengths and come from the polarization survey done by Stenflo *et al.*[6] Since the theoretical curves of Figure I exclude the observational data, we can assign an upper bound to the solar NGT charge: in case 1 $l_\odot < 1200$km, and in case 2 $l_\odot < 600$km. These are factors of 2.7 and 5.4, respectively, below the presently accepted limit of 3230km. Since the static, spherically symmetric metric in NGT depends upon $l_\odot^4$, the limits on the metric are tighter by factors of 50 and 850 respectively. With an increase in the number of observations of Stokes V near the limb of the Sun, these limits may be further substantiated and improved.

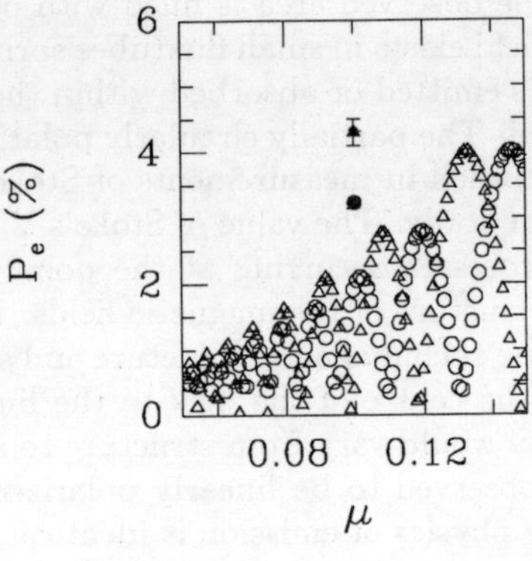

Fig. I. Triangles represent Case 1. Circles represent Case 2. Solid points represent observational data.

4. References

1. R.B. Mann, Class. Quantum Grav. **1**, 561 (1984); **6**, 41 (1989).
2. M.D. Gabriel, M.P. Haugan, R.B. Mann and J.H. Palmer, Phys. Rev D. **43**, 308 (1991).
3. J.W. Moffat in *Gravitation 1990: A Banff Summer Institute*, eds. R.B. Mann and P.S. Wesson (World Scientific: to be published).
4. C.M. Will, Phys. Rev. Lett. **62**, 369 (1989); R.B. Mann, J.H. Palmer and J.W. Moffat, Phys. Rev. Lett. **62**, 2765 (1989).
5. J.O. Stenflo, S.K. Solanki and J.W. Harvey, Astron. Astrophys. **171**, 305 (1987).
6. S.K. Solanki, C. Keller, and J.O. Stenflo, Astron. Astrophys. **188**, 183 (1987).

GALAXY COUNTS, MERGERS AND COSMOLOGY

PAUL HICKSON
Department of Geophysics and Astronomy,
University of British Columbia,
2219 Main Mall, Vancouver, B.C. V6T 1Z4, Canada

ABSTRACT

This paper reviews some recent observational results on galaxy counts and considers their interpretation in the light of current cosmological models. It is found that galaxy mergers introduce a considerable uncertainty in the interpretation of the counts. Detailed studies of the effects of mergers and other evolutionary processes are needed to before the observational data can strongly constrain cosmological models.

1. Introduction

In recent years, important developments have occurred in both theory and observations relating to the structure of the universe on very large scales. On one hand, the inflationary scenario[1,2,3] provides a natural resolution of many questions unanswered by the standard hot big bang model, notably the horizon and flatness problems[4]. On the other hand, new observations of the distributions of galaxies, combined with increasingly accurate measurements of the spectrum and isotropy of the microwave background radiation[5,6,7], have come into conflict with the simplest models for the formation of structure in the universe[8,9].

One of the most powerful observational tools for the study of structure of the universe on the largest scales is deep imaging surveys from which one obtains counts of distant galaxies. Advances in instrumentation and techniques have now made it possible to obtain data of sufficient quality to constrain possible cosmological models.

This review examines some recent observational results involving galaxy counts, and considers their implications for cosmology. Several other recent reviews[10,11,12] provide information on other aspects of large scale structure and the theoretical implications.

2. Theoretical Perspective

The universe is generally believed to be homogeneous and isotropic when averaged over large enough scales. Its geometry and dynamics are then determined by the dimensionless parameters $\Omega \equiv 8\pi G\rho/3H^2$ and $\lambda \equiv \Lambda/3H^2$ where ρ is the mean mass density, Λ is the cosmological constant and H is the Hubble parameter, related to the scale factor a by $H = \dot{a}/a$. Here we assume units in which the speed of light $c = 1$. The redshift z is an observable quantity, related to the ratio of the scale factors at the time of emission and absorption of the light $1+z=a/a_0$, where the subscript $_0$ denotes the present value of a parameter. In terms of these definitions, Einstein's equations give the Friedmann relation

$$\Omega+\lambda=1+k/H^2a^2 \tag{1}$$

where the curvature constant k takes values -1, 0, +1. An important consequence of all inflationary models is that space is now essentially flat ($k=0$) which gives the inflation condition $\Omega+\lambda=1$.

Standard big-bang nucleosynthesis[13] places a strong upper limit on the density of baryons $\Omega_b<0.035h^{-2}$, where h is the Hubble parameter in units of 100 km s^{-1} Mpc^{-1}, thought to be in the range $0.5<h<1$. In the simplest models where $\lambda=0$, the additional matter needed to make $\Omega=1$ is therefore presumed to be non-baryonic. In fact, it is not necessary to appeal to inflation to justify a significant non-baryonic component. Velocity dispersions in rich clusters are sufficiently high as to require a present mean mass density $0.2<\Omega_0<0.35$, which already exceeds the baryonic limit[14].

The last term in Eq. 1 is roughly the squared ratio of the horizon distance to the radius of curvature of the universe. If this ratio is of order unity or larger one would expect to observe curvature related optical phenomena (such as quasar pairs with 180° separation, and focussing of light from the antipole). The absence of such effects gives a rough upper bound $\lambda_0\leq 2-\Omega_0$.

Theories describing the development of structure in the universe may be classified according to the temperature of the dark matter when it decouples. In hot dark matter (HDM) theories[15], the dark matter consists of particles, neutrinos for example, which are relativistic at decoupling. Free streaming of these particles erases adiabatic density fluctuations on small scales. As a result, large structures with masses comparable to superclusters are the first to collapse. Gas hydrodynamics results in the formation of sheet-like "pancake" structures which later fragment into galaxies.

In cold dark matter (CDM) theories[16] no such damping occurs and the distribution of density fluctuations $\delta\equiv\delta\rho/\rho$ is determined largely by the primordial spectrum. One assumes a spectrum of the form $P_k\propto k^n$ where P_k is the power spectrum of density fluctuations

$$P_k=<|\delta_k|^2> \tag{2}$$

where

$$\delta_k=\int\delta(\mathbf{x})e^{i\mathbf{k}\cdot\mathbf{x}}d^3x \tag{3}$$

and angular brackets denote a spatial average. Harrison[17] and Zeldovich[18] have advocated the choice $n=1$ as this leads to a constant "scale invariant" spectrum of mass fluctuations entering the horizon $|x|=ct$, and is therefore free of divergences on both large and small scales. This spectrum is also found to arise naturally in inflationary models from quantum fluctuations of the scalar field[19,20,4]. The CDM spectrum is modified on small scales as fluctuations entering the horizon cannot grow while the universe is radiation dominated. As a result the CDM spectrum is modified on small scales, gradually steepening to the primeval spectrum at a characteristic comoving scale corresponding to the horizon size at the time of equality of the matter and radiation energy densities. Because the spectrum contains more power on small scales than large, small structures collapse first and a pattern of hierarchical clustering develops[21]. Galaxies form by the merging of smaller objects.

3. Galaxy Counts

Galaxy counts provide one of the oldest methods of investigating large scale structure and cosmology, and is the only method which is able to sample large numbers of galaxies to very faint magnitude limits. The observational technique is relatively simple, requiring only the measurement of galaxy fluxes. Remarkable progress has been made with the advent of electronic detectors, and modern surveys now detect several million galaxies.

The number of galaxies detected as a function of magnitude is given by

$$N(m) \propto \int_0^\infty \frac{dV}{dz} dz \int_{L(m,z)}^\infty \phi(L) dL \tag{4}$$

where V is the volume of space with redshift $\leq z$ and ϕ, the luminosity function, is the comoving density of galaxies per luminosity interval. At least at low redshifts, ϕ may be represented by the functional form

$$\phi(L) = \phi^*(L/L^*)^\alpha exp(-L/L^*) \tag{5}$$

where ϕ^*, L^* and α may be functions of z. The galaxy counts thus depend sensitively on both the geometry of space, through $V(z)$ and $L(m,z)$, and on evolutionary effects which modify ϕ. It has not yet been possible to clearly disentangle these two dependences, and most work in the past decade has been directed towards understanding the evolutionary effects. Two primary effects are possible: the luminosities of individual galaxies may change with time due to stellar evolution and star formation (evolution in L^*), and the number of galaxies per unit comoving volume may change due to galaxy mergers (evolution in ϕ^*). If either of these effects depend on galaxy mass, α may also change. Of course the actual evolution may be a combination of the above.

Figure 1. shows combined galaxy counts from several deep surveys[22,23,24,25] in blue (B) and near infrared (I) bands. The smooth curve corresponds to a model in which the parameters ϕ^*, L^* and α are assumed to be independent of z for $\Omega_0 = 1$ and $\lambda_0 = 0$.

The counts, particularly in the blue band, are clearly rising faster than the no evolution model. In fact Koo[26] and others have argued that **no** plausible model involving pure luminosity evolution can produce such a steep rise in the counts, in a universe with $\Omega_0 = 1$, although models with $\Omega_0 < 0.2$ can be reconciled with the data.

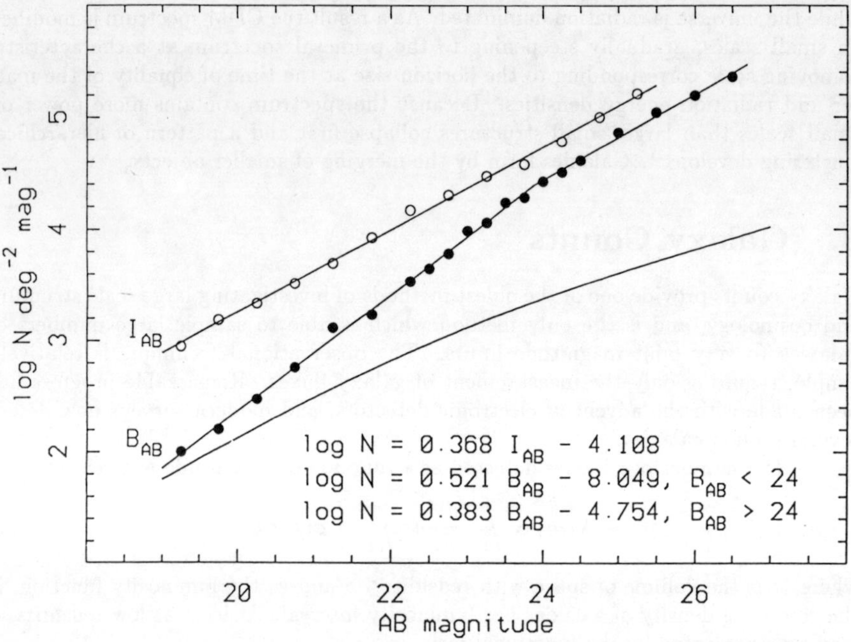

Figure 1. Combined galaxy counts. The number of galaxies per square degree per magnitude interval is plotted for blue (B) and infrared (I) bands. The smooth curve is the behavior expected (to within an arbitrary multiplicative constant) for a $\Omega_0 = 1, \lambda_0 = 0$ universe with no galaxy evolution.

A fundamental difficulty with the interpretation of these data is that the redshifts of these faint galaxies are not known. The excess counts might arise from a population of distant, highly luminous galaxies, or a population of relatively nearby, low luminosity galaxies. Recent spectroscopic studies[27] of galaxies brighter than $B \simeq 22$ indicates that the redshift distribution of these galaxies is consistent with the no-evolution model, even though the galaxy counts are not. One should note, however, that this redshift distribution is also consistent with plausible models which include evolution[28,29]. It is possible that this problem may resolved in the next few years as the observations are extended to fainter magnitudes, better sampling the excess population.

If the steep slope of the galaxy counts cannot be explained by the effects of normal stellar evolution within galaxies, what are the alternatives? One possibility is by the adoption of a nonstandard cosmological model. A positive cosmological constant has the effect of increasing the volume available at high redshifts, leading to a steepening of the galaxy counts. Fukugita *et al.* [30] have recently argued that a model with $\Omega_0 = 0.1$ and $\lambda_0 = 0.9$ (thus satisfying the inflation constraint) is consistent with the observations. It is important to note, however, that this is a purely empirical result. There is no a priori reason, theoretical or otherwise, that one would expect $\lambda_0 \sim 1$.

4. Galaxy Mergers

It now seems almost certain that the number density of galaxies varies significantly with redshift. Observations of galaxies in small groups[31] provide ample evidence of gravitational interactions. Simple dynamical considerations[32], confirmed by detailed numerical simulations[33] indicate that galaxy mergers may be commonplace even at the present epoch. With physical galaxy densities increasing as $(1+z)^3$, it is possible that galaxy merging was widespread in the past.

What would be the effect of galaxy mergers on the number counts? Clearly, mergers result in a decrease in the comoving number density. However, it is not immediately obvious that this would explain the steepness of the galaxy counts. If, as expected, the number of stars is conserved by the merger process, the luminosities of the merger products will be larger than those of the individual progenitor galaxies. These more luminous galaxies would be visible to greater distances so the galaxy counts might actually increase rather than decrease.

The effect of mergers on the number counts is further complicated by the fact that the observations are not limited by galaxy magnitude alone. Because of the difficulty of detecting faint galaxies against the relatively bright background of the night sky, galaxy surface brightness also plays an important role. It is easy to show that mergers lead to a decrease in the mean galaxy surface brightness, making them more difficult to detect. This effect, which does not seem to have been considered in the analysis of galaxy counts, may offset the increase in luminosity and lead to a steepening of the galaxy counts.

The argument is simple: for a galaxy in virial equilibrium the total internal energy is $E = -GM^2/2R$ where M is the galaxy mass and R a characteristic radius. Conservation of energy and mass during mergers then requires that R grow at least as fast as $R \propto M$ (faster if the galaxies collide with appreciable velocity). If the mass to light ratio remains constant, the mean surface brightness should decrease as $I \propto M/R^2 \propto L^{-1}$.

A rough estimate of the merger rate may be obtained in the hierarchical clustering scenario. Assume that galaxies form from the peaks of the density fluctuations $\delta(\mathrm{x})$. Fluctuations in the mass within a sphere of radius R are then given by

$$\delta_M^2 = < (\delta M/M)^2 > \propto \int P_k W(kR) k^2 dk \propto \int W(kR) k^{n+2} dk \qquad (6)$$

where $M = 4\pi\rho R^3/3$ and $W(kR)$ is a weight function. This leads to

$$\delta_M \propto R^{-(n+3)} \propto M^{-(n+3)/3} \quad (7)$$

It is common to assume that the frequency distribution of density fluctuations is Gaussian:

$$f(\delta) = \frac{1}{\sqrt{2\pi}\sigma} exp(-\delta^2/2\sigma^2) \quad (8)$$

Small fluctuations grow linearly, in proportion to the scale factor

$$\sigma \propto a \propto (1+z)^{-1}. \quad (9)$$

High density regions collapse first, forming galaxies. If these galaxies are contained within larger regions with mean density greater than some critical value then will merge within a given time. Let M be the mass contained within such a region (*i.e.* $\delta_M = \delta_{crit}$). The distribution of M will then be determined by the distribution of δ_M (which is also Gaussian):

$$f(M)dM = f(\delta_M)d\delta_M = f(\delta_M)\frac{d\delta_M}{dM}dM. \quad (10)$$

Recalling Eq. 7 we write

$$\delta_M = (M/M^*)^{-(n+3)/3}\sigma \quad (11)$$

where M^* is a characteristic mass, and obtain the mass distribution

$$f(M)dM = \frac{n+3}{6\sqrt{2\pi}}(M/M^*)^{(n-3)/6} exp(-\frac{1}{2}(M/M^*)^{(n+3)/3})dM. \quad (12)$$

Eq. 12 was first obtained by Press and Schechter[34] and motivated the form of Eq. 5.

From Eq. 9 we obtain the redshift dependence of the characteristic mass

$$M* \propto (1+z)^{-6/(n+3)} \quad (13)$$

which, by mass conservation, gives the dependence of the comoving number density

$$n \propto (1+z)^{\eta}, \qquad \eta = 6/(n+3). \quad (14)$$

In the CDM model, $n \simeq -1.5$ on galaxy scales so we would expect rather strong density evolution due to mergers, $\eta \simeq 4$.

It must be emphasized that this analysis applies to all the matter, which is mostly nonluminous. The merger rate of actual luminous galaxies might be significantly smaller due to nonlinear effects[35]. For comparison, empirical estimates of η from galaxy counts require $\eta \simeq 2$, ignoring selection effects[36].

5. Conclusions

We have seen that galaxy counts have the potential to provide strong constraints on cosmological models. However, may uncertainties remain. The luminosity evolution of galaxies is still not well understood. More redshifts are needed for faint galaxies. In addition to this, a better understanding of the rate of galaxy mergers, and their effects is needed. The standard picture of galaxy formation predicts rapid evolution of the galaxy number density with redshift. Given this large potential evolution, and our lack of a detailed understanding of the effects of mergers on the galaxy counts, it is certainly premature to suggest that the present data favor a nonzero cosmological constant. With a better understanding of these effects, galaxy counts will become a more powerful tool with which to probe the universe.

I thank B. Gibson for helpful comments. Financial supported from the Natural Sciences and Engineering Research Council of Canada is gratefully acknowledged.

References

1. A. H. Guth, *Phys. Rev.*, **D23** (1981), 347.
2. A. D. Linde, *Phys. Lett.*, **108B** (1982), 389.
3. A. Albrecht and P. J. Steinhardt, *Phys. Rev. Lett.*, **48** (1982), 1220.
4. A. D. Linde, *Particle Physics and Inflationary Cosmology*, (Harwood, London, 1990).
5. J. C. Mather *et al.* , *Astrophys. J. Lett.*, **354** (1990), L37.
6. H. P. Gush, M. Halpern and E. Wishnow, *Phys. Rev. Lett.*, **65** (1990), 537.
7. S. S. Meyer, E. S. Cheng and L. A. Page, *Astrophys. J. Lett*,**371** (1991), L7.
8. W. Saunders *et al.* ., *Nature*, **349** (1991), 32.
9. S. J. Maddox, G. Efstathiou, W. J. Southerland and J. Loveday, *Mon. Not. R. A. S.*, **242** (1990), 43p.
10. A. Kashlinsky and B. J. T. Jones, *Nature*, **349** (1991), 753.
11. F. D. A. Hartwick and D. Schade, *Ann. Rev. Astr. Ap.*, **28** (1990), 437.
12. J. P. Huchra in *Observational Tests of Inflation*, ed. T. Shanks, (NATO ARW), in press.
13. J. Yang, M. S. Turner, G. Steigman, D. N. Schramm and K. A. Olive, *Astrophys. J. Let.* **301** (1984), L1.
14. P. J. E. Peebles, *Nature*, **321** (1986), 27.
15. Ya. B. Zel'dovich, *Astr. Ap.*, **5** (1970), 84.
16. M. Davis, G. Efstathiou, C. S. Frenk and S. D. M. White, *Astrophys. J.*, **292** (1985), 241.
17. E.R. Harrison, *Phys. Rev.*, **D1** (1970), 2726.
18. Ya.B. Zeldovich, *Month. Not. R.A.S* **160** (1972), 1P.
19. G. W. Gibbons and S. W. Hawking, *Phys. Rev.*, **D15** (1977), 2738.

20. R. H. Brandenberger, *Rev. Mod. Phys.*, **57** (1985), 1.
21. P. J. E. Peebles, *Astrophys. J.*, **142** (1965), 1317.
22. J. A. Tyson, *Astron. J.*, **96** (1988), 1.
23. L. R. Jones, R. Fong, T. Shanks, R. S. Ellis and B. A. Peterson, *Month. Not. R. A. S.*, **249** (1991), 481.
24. N. Metcalfe, T. Shanks, R. Fong and L. R. Jones, *Month. Not. R. A. S.* **249** (1991), 498.
25. S. J. Lilly, L. L. Cowie and J. P. Gardener, *Astrophys. J.*,**369** (1991), 79.
26. D. C. Koo, *Astr. Soc. Pac. Conf. Ser.*, **10** (1990), 268.
27. M. Colless, R. S. Ellis, K. Taylor and R. N. Hooke, *Mon. Not. R. A. S.*, **244** (1990), 408.
28. B. Guiderdoni and B. Rocca-Volmerange, *Astron. Ap.*, **227** (1990), 362.
29. Y. Yoshii and B. A. Peterson, *Astrophys. J.*, **372** (1991), 8.
30. M. Fukugita, F. Takahara, K. Yamashita and Y. Yoshii, *Astrophys. J. Lett.*, **361** (1990), L1.
31. P. Hickson and H. J. Rood, *Astrophys. J. Lett.*, **331** (1988), L69.
32. P. Hickson, D. O. Richstone and E. L. Turner, *Astrophys. J.*, **213** (1977), 323.
33. J. Barnes, *Nature*, **338** (1989), 123.
34. W. H. Press and P. Schechter, *Astrophys. J.*, **187** (1974), 425.
35. R. G. Carlberg *Astrophys. J. Lett.*, **359** (1990), L1.
36. B. Rocca-Volmerange and B. Guiderdoni, *Mon. Not. R. A. S.*, **247** (1990), 166.

LOOPS, GAUGE FIELDS AND GRAVITY

ABHAY ASHTEKAR
Physics Department, Syracuse University, Syracuse, NY 13244-1130

ABSTRACT

The current status of a program for non-perturbative canonical quantization of general relativity is briefly summarized.

1. Introduction

In my talk I presented some recent results on canonical quantization of the gravitational field. By now there are several detailed reviews of the program I summarized (see, e.g., Refs. 1-6). In particular, there is a recent book[1] which treats the subject in detail paying special attention to mathematical subtleties and also a more compact review article[2] addressed to standard physics audiences. Therefore, rather than repeating the material, here I will only summarize the main ideas and provide (especially the more recent) references where the details can be found.

The program is motivated by the fact that, inspite of the perturbative non-renormalizability of quantum general relativity, we do not yet know if the theory exists non-perturbatively in 3+1 dimensions. Since there is now a growing conviction in almost all quarters that the problem of quantum gravity must be faced non-perturbatively, it seems natural to examine the situation in the context of general relativity which is the simplest of the viable classical theories of gravitation. Thus, the immediate goal of the program is to find out if non-perturbative quantum general relativity exists as a mathematically consistent theory. Of course, even if it turns out that it *is* consistent, there is no a priori guarantee that it would be "the" correct physical theory. Nonetheless, since many of the conceptual problems of quantum gravity stem from the fact that one does not have a classical background space-time at one's disposal[7] and since this is precisely the key obstacle faced by quantum general relativity, the insights offered by non-perturbative quantum general relativity would be valuable irrespective of what the "ultimate" theory turns out to be. That is, one can argue that the qualitative picture of quantum space-times provided by non-perturbative quantum general relativity would continue to be useful even when this theory is eventually replaced by a more complete, deeper theory. (For details on motivation, see chapter 1 in Ref. 1 and Introduction in Ref. 7.)

Our non-perturbative approach falls in the framework of canonical quantization which is perhaps the oldest of quantization methods used in field theories including general relativity. However, the specific program that is being pursued is based on three ingredients which were absent in the older treatments: i) the observation that any theory of connections can be recast in the language of loops; ii) the fact that general relativity can be regarded as a theory of "connection dynamics"; and, iii) the realization that the Dirac method of quantization of constrained systems can be extended to incorporate the peculiarities of Hamiltonian general relativity based on connections and loops. In section 2, I will discuss these three points in some detail. Section 3 will summarize the current status of the program.

2. The main ideas

2.1 From connections to loops

For simplicity, let us first consider the Maxwell theory in Minkowski space. In the Hamiltonian formulation, one normally uses as one's canonical variables the pairs $(A_a(\vec{x}), E^a(\vec{x}))$ consisting of vector potentials and electric fields on a space-like 3-plane. However, all the gauge invariant information in the potential $A_a(\vec{x})$ can be coded in the holonomies $Q[\alpha](A) := i \exp \oint_\alpha dl^a A_a$ while that in the divergence-free electric fields can be captured by the functionals $P[f] := \int d^3\vec{x}\, E^a(\vec{x}) f_a(\vec{x})$, where f_a is *any* 3-form modulo an exact 3-form. (Note that if f_a is exact, $E(f)$ vanishes identically.) Thus, the set of functionals $(Q[\alpha], P[f])$ forms a complete (or, more precisely, an overcomplete) set of configuration and momentum variables on the phase space. It is easy to check that the set is in fact *closed under the Poisson bracket*: $\{Q[\alpha], P[f]\} = i(\oint dl^a f_a) Q[\alpha]$. This Poisson algebra of closed loops and 1-forms can be used as the point of departure for quantization of the Maxwell field a *la* Dirac.

In this scheme, the basic quantum operators are labelled just by closed loops and 1-forms and one seeks representations of the associative algebra they generate subject to the (generalized) CCRs: $[\hat{Q}[\alpha], \hat{P}[f]] = -\hbar \oint dl^a f_a \hat{Q}[\alpha]$. It turns out[8] that the states in the representation space can also be taken to be functionals of closed loops. (See, however, Ref. 9 for subtleties.) Furthermore, one can write down a functional transform which maps the states in the conventional connection representation to those in this *loop representation.* The transform is faithful so that no information is lost if one works just with loop states. Thus, while one generally works with connections both for the purpose of constructing quantum operators and quantum states, the entire theory can be reformulated in terms of loops (and 1-forms.)

These general ideas go over also to the non-Abelian gauge theories both in the continuum and lattice.[10,11] Furthermore, the resulting loop representation is particularly convenient for certain calculations in lattice QCD.[12] We will see below that the loop representation is well suited also for quantum general relativity; it is almost tailored for implementing the diffeomorphism invariance of the theory.

2.2 General relativity as connection dynamics

Traditionally, the configuration space of general relativity was taken to be the space of 3-metrics, or, of 3-geometries. A dynamical trajectory in the phase space, when projected down to the configuration space, then tells us how the 3-geometries "evolve" in time. Hence, in this Hamiltonian formulation, general relativity can be regarded as a theory of "geometrodynamics".[13,14] All of the early work in canonical gravity falls in this framework. A key idea in the new framework is to shift the emphasis from the 3-geometry to 3-connections and regard general relativity as a theory of "connection dynamics".

The idea of formulating gravitational theory in terms of a connection is of course not new: Since all other basic interactions are mediated by gauge bosons, described classically by a connection, it is natural to try to describe gravity also in terms of a non-Abelian connection. In the previous attempts, however, one was led to *new* theories of gravity, generally based on an action of the Yang-Mills type. In the present approach, on the other hand, the underlying theory is still general relativity; it is *Einstein's field equation that is interpreted as governing the dynamics of a connection.*[*]

The non-triviality of the construction therefore lies in the choice of the specific connection that is to play the role of the configuration variable. One's first impulse is to rewrite general relativity in the Palatini form and use the (pull-back to the 3-surface of the) spin connection as the configuration variable. Unfortunately, this strategy does not work: it leads to a Hamiltonian framework with second class constraints which, when solved, lead one directly back to the Arnowitt-Deser-Misner phase space of geometrodynamics[15]. (See chapters 3,4 of Ref. 1.) A better choice, it turns out, is the self dual (or anti-self dual) part of the spin connection.[16] With this choice, second class constraints are avoided. The configuration variables are complex-valued, $SO(3)$ connection 1-forms $A_a^i(\vec{x})$ on the spatial 3-manifold Σ and their conjugate momenta $E_i^a(\vec{x})$ are vector densities

---

⋆ The approach does, however, lead one to a slight extension of Einstein's theory in which the metric is allowed to be degenerate. Although this extension seems trivial at the classical level, it does play an important role in the quantum theory.

of weight 1. Thus, the phase space of general relativity is now embedded in the phase space of Yang-Mills theory. The momenta E_i^a, however, now have the interpretation of (density weighted) *spatial triads* and therefore, unlike Yang-Mills electric fields, are *not* components of the curvature of the 4-connection. Similarly, the physical dimensions of A_a^i and E_i^a are now different from those in the Yang-Mills theory. In particular, it is GA_a^i, where G is Newton's constant, that has the traditional dimensions (length^{-1}) of a connection. Inspite of these differences, the use of these new canonical variables does bring out a hidden *mathematical* similarity between Yang-Mills and Einstein theories thereby making it possible to import into quantum general relativity concepts and techniques from QCD. This is a key achievement of the new canonical formulation.

The system has a set of 7 first class constraints. The first 3 are the familar Gauss constraints, $\mathcal{G}_i := \mathcal{D}_a E_i^a = 0$, which tell us that the $SO(3)$ rotations on the "internal" indices should be thought of as gauge transformations. (Here, $\mathcal{D}$ is the gauge covariant derivative operator defined by A_a^i.) The next three are the vector constraints –i.e. the"$0i$"component of Einstein's equation– $C_a := E_i^a F_{ab}{}^k = 0$, where $F_{ab}{}^k$ is the field strength of the connection A_a^i. The canonical transformations they generate correspond to diffeomorphisms on the spatial 3-slice Σ. The last is a scalar constraint –the "00-part" of Einstein's equation– $C := E_i^a E_j^b F_{abk}\epsilon^{ijk} = 0$. It generates ("bubble") time evolutions. All the constraints are simple, low order polynomials in the basic canonical variables. It is this technical simplification that has made it possible to find explicit solutions to the exact constraint equations in both classical and quantum theory.

This simplicity is, however, achieved at a cost: the Hamiltonian framework, in its simplest form, refers to *complex* general relativity and one must impose certain reality conditions to recover the real theory. In the early work, there was some confusion on this issue. It is now clear, however, that the procedure of first analysing the equations of the complex theory and then imposing the reality conditions *is consistent* both at the classical and quantum levels. In the quantum theory, these conditions become restrictions on the choice of the inner product and, at least in simplified models, suffice to pick out the inner product uniquely.

The details of the new Hamiltonian framework appeared in a spinorial language in Ref. 17 and a BRST formulation appeared in Ref. 18. Several authors independently recast this framework in the language of triads. (See, especially, Ref. 19.) The Lagrangian basis for this formulation was obtained in Ref. 20 and some subtleties regarding the role of the "reality conditions" are discussed in chapters 8 and 10 of Ref. 1. Extension to supergravity is achieved in Ref. 21. Coupling to matter fields is discussed in Ref. 22. It turns out that, contrary to one's first expectations, the key features of the source-free theory –the close relation to Yang-Mills theory and simplicity of the constraint equations– are maintained even when matter couplings are introduced. In a later series of papers[23] the formulation of general relativity as a theory of connection-dynamics is taken to its logical extreme: Using a number of highly imaginative techniques, the entire theory is built from connections, metric being a genuinely derived object. (This framework has also provides the "generic" solution to the vector and the scalar constraints. Furthermore, unlike in the York procedure, the prescription to construct solutions from the free data is algebraic; no differential equations have to be solved. To find the generic initial data for general relativity, however, it remains to impose the Gauss constraint as well as the reality conditions which are non-trivial in the -+++ case.) The geometrodynamical and connection dynamical frameworks are compared and contrasted in detail in Ref. 24.

2.3 Extension of the Dirac program

The discussion of the previous two sub-sections indicates that one should be able to construct quantum operators and states based on loops also for general relativity. To implement this idea, however, one needs an extension of the Dirac program of quantization of constrained systems. Some aspects of the extension are "universal" in the sense that they are needed in, and will be useful in the implementation of, *any* canonical quantization scheme for general relativity, including, e.g., the one based on geometrodynamics. Other features of the extension are needed specifically because we wish to use complex connections and phase space variables based on loops.

The incompleteness of the Dirac program stems from the fact that it provides no guidelines for the selection of the Hermitian inner product on the space of *physical* states, i.e. states that are annihilated by the quantum constraints. For systems with a finite number of degrees of freedom or for fields in Minkowski space, additional physical inputs are generally available to select the inner product. For example, in Maxwell theory in Minkowski space, the imposition of quantum constraints tells us that the states $\Psi(A)$ must be gauge invariant. To obtain the correct inner product on the resulting space of physical states, one can first select the vacuum state by requiring it to be Poincaré invariant, and then compute the vacuum expectation values of suitable operators. In non-perturbative general relativity, however, we cannot invoke the Poincaré group. The only group available is the diffeomorphism group. However, the imposition of constraints requires that *all* physical states are diffeomorphism invariant. Therefore we cannot hope to select the vacuum state –or, alternatively, the inner product– by invoking diffeomorphism invariance. A new guideline is needed. It turns out that, for a large class of systems such a principle is in fact available. Roughly, it states that the inner product be selected by requiring that the complex-conjugation relations between a complete set of physical observables become Hermitian-adjointness relations in the quantum theory. In the Maxwell theory as well as linearized gravity, for example, this principle provides us the correct inner product without having to invoke the Poincaré invariance (at least explicitly).[8,25] In non-perturbative canonical gravity then, the "reality conditions" which, in the classical theory restrict us to the real slice of the complex phase space, are to determine the inner product in the quantum theory. This strategy for selecting the inner product does indeed work in 2+1-gravity (see Ref. 26 and 27 and chapter 17 of Ref. 1) as well as in some mini-superspace models in 3+1 dimensions. One therefore wishes to use it also in the full, 3+1 dimensional theory.

We will see in section 3 that the formulation of general relativity as a theory of connection dynamics naturally leads one to two representations: the connection representation in which the states are holomorphic functionals of the complex connections A_a^i and the loop representation where they are functionals of closed loops on the spatial 3-manifold. In the connection representation, a small extension of the standard canonical quantization procedure is needed because while the connection is complex (even in real, Lorentzian general relativity), its conjugate momentum E_i^a, the triad, is real. In the loop representation, a strategy is needed to incorporate the fact that the holonomies constitute an overcomplete set of ($SO(3)$-gauge invariant) configuration variables, i.e., to accommodate the algebraic (Mandlstam) identities that exist between holonomy functionals.* A systematic strategy is needed to incorporate these relations in the quantum theory. An extension of the Dirac program is called for to address these issues as well.

A possible extension which resolves all these problems is presented in chapter 10 of Ref. 1. The program has been successfully carried out in a number of model systems which capture various peculiarities of general relativity [28].

3. Current status

The three main ideas discussed in section 2 have been combined to to carry out canonical quantization of general relativity. I will now summarize the current status of this effort. A more complete report on the construction of the loop

---

* This overcompleteness occurs in *any* system in which the configuration space is a genuine manifold. For example, it is present in the Yang-Mills theory because the reduced configuration space –the quotient of the space of connections by the group of local gauge transformations– is a manifold. Thus, the problem is not peculiar to the use of loop variables.

representation and the results that have ben obtained in that framework can be found in Ref. 2 and 5.

The complex connections provide us with phase space variables which are analogous to the complex coordinate $z = q - ip$ often used on the phase space of a particle. Therefore, one may mimic the procedure used in the construction of the Bargmann representation in non-relativistic quantum mechanics where quantum states arise as holomorphic functions of z. This leads us to the connection representation where states are holomorphic functionals of A_a^i, the operator $\hat{A}_a^i$ acts by multiplication and the conjugate operator $\hat{E}_i^a$ acts by functional differentiation ($\hat{E}_i^a = \hbar\delta/\delta A_a^i$). This representation is rather similar to the one generally used in the canonical treatment of Yang-Mills theories and has therefore proved to be useful to explore the structure of θ-vacua and the related possibility of CP-violation in quantum gravity.[29] It has also been used successfully in the context of certain simplified models, including Bianchi superspaces[30] and the weak field truncation of quantum general relativity. Perhaps the most interesting application[31] is the demonstration that, in this limit, the scalar constraint equation of quantum general relativity reduces precisely to the Schrödinger equation, once the appropriate component of the connection is identified as the "time variable" (i.e., as the internal clock) with respect to which the other, dynamical components "evolve".

As far as the full, non-truncated theory is concerned, most of the progress has occurred in the framework of the loop representation. As outlined in section 2.1, a theory of connections can be recast in the language of loops. This is true, in particular, of the connection dynamics version of general relativity. The key idea[32] is to first introduce on the phase space of general relativity a set of variables based on loops which is complete in the appropriate sense and also closed under the Poisson bracket, use it to construct an algebra of quantum operators and then represent this algebra on the space of appropriate functionals of loops. As we shall see, the advantage of this formulation is that the "loop basis" is particularly well-suited for solving the quantum constraint equations.

One begins with the observation that the gauge invariant information in connections is coded in the traces of holonomies. (For precise statements and subtleties that arise because the connection 1-forms are complex, see Ref. 33.) Therefore one can use as "configuration variables" the functionals $Q[\alpha](A) := tr\ \mathcal{P}\ \exp G \oint_\alpha dl^a A_a \equiv trU_\alpha$, parametrized by closed loops α. (Here $\mathcal{P}$ stands for "path-ordered" and U_α is the group element defined by the holonomy around the loop α.) To introduce the appropriate momentum variables –i.e. phase space functionals which are linear in momenta– we need a new notion, that of (closed) strips. A strip is a mapping from $S^1 \times (-\epsilon, \epsilon)$ to the (spatial) 3-manifold Σ.

Thus, a strip can be coordinatized by two parameters, (u, v) where $u \in [0, 2\pi]$ and $v \in [-\epsilon, \epsilon]$. (The coordinates can be changed by a reparametrization. The invariant structure of a strip is simply that it has a topology $S^1 \times R$ and is foliated by a family of circles.) With each strip S, we associate a momentum variable $P[s](A, E) := \int_S dS^{ab} \eta_{abc}\, trU_v E^c$ where U_v is the group element representing the holonomy of the connection A_a^i around the loop labelled by the specific value of v. These loop-strip variables turn out to be closed under the Poisson bracket: the bracket of any two loop variables vanishes; that of a loop and a strip variable is a linear combination of two loop variable and that between two strip variables is a linear combination of two strip variable. The bracket vanishes if the loops and/or strips in question do not intersect. If they do intersect, the loops and strips that appear on the right side of the bracket are obtained by simple geometric constructions involving the loops and strips that feature in the bracket. (For details, see Refs. 34, 32 and 35.)

In quantum theory, then the basic operators are $\hat{Q}[\alpha]$ and $\hat{P}[S]$, associated with loops and strips. We impose on them the commutation relations that reflect the Poisson brackets. This is the quantum algebra that we seek to represent. It turns out that if we choose, as our states, functionals $\Psi[\gamma]$ of closed loops, it is easy to write down concrete operators that realize the algebraic relations between $\hat{Q}[\alpha]$ and $\hat{P}[S]$. Again, the definitions involve simple operations of breaking and joining loops, rather similar to the operations encountered in Maxwell[8] and Yang-Mills[10,11] theories. There is thus a "universality" in the basic structures that arise in the loop description of all these interactions. In all these cases, one can write a heuristic transform to pass between the connection and the loop representations.[32,34] Work is in progress to make this heuristic relation precise using the Gel'fand spectral theory.[35]

The next step is to impose the quantum constraints. Since everything that is written down in the loop description is (internal) gauge invariant, the Gauss constraint is automatically satisfied by all loop states. The vector constraint operator can be expressed as a limit of the strip operator. A state $\Psi[\gamma]$ is annihilated by this operator if and only if $\Psi[\gamma] = \Psi[\gamma']$ whenever γ and γ' are related by a diffeomorphism. Since the diffeomorphism class of closed loops defines a knot, we conclude that the physical states depend on individual loops *only* through the knot class they define. Thus, the mathematics of non-perturbative quantum gravity is closely related to knot theory! Furthermore, this argument has given us the *general solution* to the Gauss and the vector constraints. In the geometrodynamical variables, on the other hand, while one can write out an infinite dimensional family of solutions to these constraints, it is hard to obtain the *general* solution because one does not have a simple characterization of the equivalence classes of 3-metrics under diffeomorphisms. Note also that since knot classes form a discrete set, a discrete basis is now naturally available on the space of all solutions to the Gauss and the diffeomorphism constraints. Finally, we wish to emphasize that this result does not depend on technical details of Einstein's equations. Rather, it stems from quite general considerations: Einstein's theory can be formulated in terms of connections; a theory of connections can be recast in the language of loops; and, a knot characterizes a diffeomorphism equivalence class of loops. Therefore, one expects that the close relation to knot theory is one of the qualitative features of the present description that would survive if and when quantum general relativity is replaced by a more complete theory.

The status of the scalar constraint is not so satisfactory. First of all, there is some ambiguity in writing out the operator in the loop representation[32,36]. Recent unpublished work of Brügmann and Pullin suggests that one may soon be able to fully understand –and hopefully resolve– these ambiguities. However, as the matter stands, there do exist two inequivalent transcriptions of the scalar constraint operator. Within each scheme, an infinite dimensional family of solutions is available[32,36,37]. Furthermore, the overall situation is rather similar in the two schemes: the scalar constraint can be solved because its functional form is rather simple in connection dynamics and because, in the loop representation, it requires simply that the support of the wave functionals be restricted in certain ways. Thus, the loop representation is well suited to solve *all* quantum constraints. These general features are quite attractive. However, there are strong indications that the the solutions to the scalar constraint that have been discovered so far constitute only a *partial* set. The issue of completeness is completely open. In fact, in the present non-perturbative context, the *precise* meaning of "completeness" is itself not entirely clear!

In addition to the question of completeness, there also exist several other unresolved issues both on the technical and conceptual sides. In particular, although the strategy of using the "reality conditions" to select the correct inner product on physical states is available, its implementation is far from being straightforward in the full 3+1 theory. Similarly, relatively little is known as to how one might interpret the physical states and the operators that act on them.

To gain further understanding of the mathematical structures that have emerged, therefore, the framework was applied to two simplified models: full, non-linear gravity in 2+1 dimensions[26,27] (see also chapter 17 of Ref. 1) and linearized gravity in 3+1 dimensions[25]. The general program could be completed successfully in both cases. In the first application, the knot classes of the 3+1 theory are replaced by the homotopy classes of closed loops on the spatial 2-manifold. There are thus close parallels between our incomplete description of the 3+1 theory and the analogous parts of the complete description we now have of the 2+1 theory. The second application has yielded, in particular, a loop description of (linear) gravitons for which we have good physical intuition. The experience gained from these two applications is now being applied to the incomplete 3+1 theory. Results on 2+1 gravity are providing insight into the problems whose origin lies in non-linearity and diffeomorphism invariance of the classical theory while those on linearized gravity are offering hints on some problems associated with the presence of an infinite number of degrees of freedom as well as on issues of interpretation.

Perhaps the most significant result to emerge from these efforts is an insight into the micro-structure of spatial geometry in the full theory. One might ask if the full theory admits "non-linear coherent states" which can be interpreted as representing a classical geometry "on a large scale". The simplest case is that of flat geometry. Let us fix a flat metric q_0^{ab} and ask for a loop state which approximates this metric on large scales where "scales" refer to q_0^{ab}. (Of course, one does not expect the quantum state to be unique; if such a state does exist, one would expect that there would also be other quantum states which approximate the same geometry on a macroscopic scale but which differ from one another microscopically.) It turns out that such states do exist but *they display a discrete structure precisely at the Planck length.* [1,2,5] These states are called *weaves.* The calculation[38] leading to the discreteness is rather involved since one must regularize the metric operator $\hat{q}^{ab}$ which, in this framework, is constructed from products of the triads, $\hat{q}^{ab} := \hat{E}_i^a \hat{E}^{bi}$. The final conclusion may not seem surprising at first. After all, there is a widespread expectation that "something strange should happen at the Planck length" and a number of proposals exist in which one *begins* by postulating the existence of a discrete structure at this scale. Note, however that the status of the result just reported is quite different: here, the existence of such a structure is *derived* from the basic principles of quantum general relativity.* Finally, one can do perturbation theory around one of these quantum states. This provides an alternate description of (linear) gravitons. Roughly, the graviton states can be pictured as "embroideries" around the background weave. One might hope that the discreteness at the Planck scale would cure at least some of the divergences of perturbative quantum general relativity. Work is now in progress to see if this is indeed the case.

---

* The states considered here are not annihilated by the diffeomorphism constraint simply because we are approximating a given metric q_0^{ab} rather than a given 3-geometry. If we wish to approximate only the 3-geometry –i.e. the diffeomorphism equivalence class– to which q_0^{ab} belongs, we can construct states which are in the kernel of the vector constraint by replacing loops by knots.

In this review, I have restricted myself to issues directly related to quantization. There are a number of other directions in which work is in progress. These include the study of degenerate metrics allowed by connection dynamics, relation between Yang-Mills and Einstein instantons, canonical descriptions based on null hypersurfaces, relation between connection dynamics and geometrodynamics in 2+1 dimensional gravity, etc. Eventually, Some of this work will also have concrete applications to quantum gravity.

Acknowledgements

It is a pleasure to thank Carlo Rovelli and Lee Smolin for collaboration and for innumerable stimulating discussions. This work was supported in part by the NSF grant PHY90-16733 and by research funds provided by Syracuse University.

References

1 A. Ashtekar, *Non-perturbative Canonical Gravity* (World Scientific, Singapore, 1991).

2 C. Rovelli, *Class. & Quantum Grav.* **8** (1991) 1613-75.

3 A. Ashtekar, *New Perspectives in Canonical Gravity* Bibliopoulos, Naples, 1988).

4 C. Rovelli, In: *Recent Advances in General Relativity* ed A I Janis and J R Porter (Birkhäuser, Boston 1991).

5 L. Smolin, In: *The Proceedings of the 22nd GIFT International Seminar on Theoretical Physics* ed j Solà (World scientific, Singapore 1992).

6 G.T. Horowitz, In:*Strings and Symmetries* (World Scientific, Singapore, 1992).

7 *Conceptual Problems of Quantum Gravity* ed A Ashtekar and J Stachel (Birkhäuser, Boston 1991).

8 A. Ashtekar and C. Rovelli, A loop representation for the quantum Maxwell field (pre-print, 1991).

9 A. Ashtekar and C.J. Isham Inequivalent observable algebras: Another ambiguity in field quantisation (pre-print, 1991).

10 R. Gambini and A. Trias *Phys. Rev.* D**23** (1981) 553, **39** (1989) 3137, *Phys. Rev. Lett.* **53** (1984) 2359; *Nucl. Phys.* B**278** (1984) 436.

11 C. Rovelli and L. Smolin, Loop representation for lattice gauge theory (preprint, 1990), Rovelli C The continuum limit of the Wilson loop representation of Yang-Mills theories (unpublished notes, 1989).

12 B. Brügmann, *Phys. Rev.* D**43** (1991) 566.

13 J.A. Wheeler *Geometrodynamics* (Academic Press, New-York, 1962).

14 K. Kuchař, In: *Quantum Gravity 2* ed C J Isham, R Penrose and D W Sciama (Oxford University Press, Oxford, 1981).

15 R. Arnowitt, S. Deser and C.W. Misner, In: *Gravitation: An Introduction to Current Research*, ed L. Witten (John Wiley, New York, 1962).

16 A. Ashtekar, *Phys. Rev. Lett.* **57** (1986) 2244.

17 A. Ashtekar, *Phys. Rev.* D**36** (1987) 1587.

18 A. Ashtekar, P. Mazur and C. Torre, *Phys. Rev.* D**36** (1987) 2955.

19 J.N. Goldberg, *Phys. Rev.* D**37** (1988) 2116, M. Henneaux, J.E. Nelson and C. Schombold, *Phys. Rev.* D**39** (1989) 434.

20 J. Samuel, Pramana J Phys. **28** (1987) L429, T. Jacobson and L. Smolin *Phys. Lett.* **B196** (1987) 39; *Class. & Quantum Grav.* **5** (1988) 583.

21 T. Jacobson, *Class. & Quantum Grav.* **5** (1988) 923; N.N. Gorobey and A.S. Lukyanenko, *Class. & Quantum Grav.* **7**(1990) 67.

22 A. Ashtekar, J.D. Romano and R.S. Tate *Phys. Rev.* D**40** (1989) 2572.

23 R. Capovilla, J. Dell and T. Jacobson, *Phys. Rev. Lett.* **63** (1989) 2325, *Class. & Quantum Grav.* **8** (1991) 59; R. Capovilla, J. Dell, T. Jacobson and L. Mason, *Class. & Quantum Grav.* **8** (1991) 41.

24 J.D. Romano, Physics Rep. (in press, 1991).

25 A. Ashtekar, C. Rovelli and L. Smolin, *Phys. Rev.* D**44** (1991) 1740.

26 A. Ashtekar, V. Husain, C. Rovelli, J. Samuel and L. Smolin *Class. & Quantum Grav.* **6** (1989) L185.

27 A. Ashtekar and J.D. Romano *Phys. Lett.* **B229** (1989). 56; L. Smolin, In: *The Proceedings of the 12th Johns Hopkins Workshop* (In press, 1991).

28 A. Ashtekar and R.S. Tate, An extension of Dirac quantization of constrained systems: Examples (pre-print, 1991).

29 A. Ashtekar, A.P. Balachandran and S. Jo, *Int. J. Theo. Phys.* **A4** (1989) 1493.

30 H. Kodama, *Prog. Theo. Phys.* **80** (1988) 1024, *Phys. Rev.* D**42** (1990) 2548.

31 A. Ashtekar, In: *Conceptual Problems of Quantum Gravity* op.cit.

32 C. Rovelli and L. Smolin, *Nucl. Phys.* **B331** (1990) 80.

33 J.N. Goldberg, J. Lewandowski and C. Stornaiolo, Degeneracy in loop variables (pre-print, 1991).

34 R. Gambini and A. Trias, *Nucl. Phys.* B**278** (1986) 436.

35 A. Ashtekar and C.J. Isham, Representations of holonomy algebras of gravity and non-Abelian gauge theories (pre-print, 1991), C.J. Isham, Loop algebras and canonical quantum gravity (to appear in the Proceedings of the 1991 Seattle AMS conference).

36 M. Blencowe, *Nucl. Phys.* B**341** (1990) 213.

37 B. Brügmann and J. Pullin, *Nucl. Phys.* B**363** (1991) 221, On the constraints of quantum gravity in the loop representation (pre-print 1991); B. Brügmann, R. Gambini and J. Pullin, Knot invariants as non-degenerate states of four dimensional quantum gravity (pre-print, 1991).

38 A. Ashtekar, C. Rovelli and L. Smolin, Low energy physics from loop space quantum gravity (pre-print, 1991).

TWISTING TYPE N VACUUM METRICS

GARRY LUDWIG
Department of Mathematics, University of Alberta
Edmonton, Alberta, Canada T6G 2G1

ABSTRACT

Hauser's metric is the only explicitly known twisting type N solution of Einstein's vacuum gravitational field equations. It is shown how to construct other metrics of this kind in terms of a real function satisfying a nonlinear third-order ordinary differential equation.

1. The Problem

Gravitational radiation from a bounded source, far from the source, is described by an asymptotically flat type N vacuum solution of Einstein's gravitational field equations, with twisting principal null rays. Hauser's solution[1] is the only explicitly known solution that possesses all of these properties except for asymptotic flatness. It is therefore of great interest to find more twisting type N vacuum metrics.

A partial "solution" of Einstein's field equations, namely the radial dependence of the metric and of various Newman-Penrose[2] quantities, was obtained long ago[3,4] for all algebraically special spacetimes and for most Ricci tensors. The formalism was recently rewritten[5] in a manifestly covariant form with respect to the remaining coordinate-tetrad freedom. All quantities depend only on certain functions of retarded time u, of a complex combination ζ of angular variables, and of $\bar{\zeta}$. For type N vacuum spacetimes there are just two such functions, $P(u,\zeta,\bar{\zeta})$ and $L(u,\zeta,\bar{\zeta})$, where P is real and L is complex. They can be interpreted as "initial data" on a local $\mathcal{J}^+$, i.e. as data at the "far end" of each outgoing radial null line. The initial data are subject to certain "reduced" equations obtained from the general formalism referred to by putting the tetrad components of the Ricci tensor equal to an appropriate source (zero in the case of vacuum) and by demanding that the metric be of a certain Petrov type. For type N vacuum solutions there is one real and one complex reduced equation. Furthermore, two inequalities assure that the spacetime is of type N rather than Minkowski space, and that the repeated principal null rays are twisting.

In general, it is even difficult to write down the reduced equations *explicitly* in terms of P and L. To solve them without some further simplifying assumptions appears to be impossible. Therefore, to make the problem more tractable we assume that in *some* frame, i.e. in some tetrad and coordinate system permitted by our general formalism, P and L take on the form

$$P = \frac{1}{2}(\zeta + \bar{\zeta})^{m+1} f(u), \qquad L = (\zeta + \bar{\zeta})^m f(u)\,\ell(u) \tag{1}$$

for some functions $f(u)$ and $\ell(u)$ and some constant m. These include the initial data for the Hauser metric for which $m = 5/2$, $\ell(u) = -2(u + i)$, and $f(u)$ is a solution of

$$(u^2 + 1)\ddot{f} + \frac{3}{16} f = 0. \tag{2}$$

The advantage of the Ansatz given by Eq. (1) is that the field equations reduce further to ordinary differential equations (o.d.e.).

The freedom in the choice of frame is narrowed considerably by the demand that under a permissible change the Ansatz of Eq. (1) be left invariant as well. We may still make a coordinate change

$$u' = g(u), \qquad \zeta' = \zeta, \qquad r' = r/\dot{g}(u), \tag{3}$$

with the inverse of $g(u)$ being denoted by $h(u')$, provided we change the tetrad appropriately, and provided we choose new initial data related to the old by

$$f'(u') = f\big(h(u')\big)\,\dot{h}(u'), \tag{4}$$

$$\ell'(u') = \ell\big(h(u')\big)/\dot{h}(u'). \tag{5}$$

In order to keep the reduced equations and the inequalities as simple as possible we exploit this freedom in the choice of frame to arrange for the function $f(u)$ to take on the constant value 2. Since $\ell(u)$ is complex we make a polar decomposition

$$\ell(u) = H(u)\,e^{i\varphi}. \tag{6}$$

Of the three real reduced equations obtained, one turns out to be (essentially) redundant in the sense that it is automatically satisfied once the other two are. Therefore, the problem to be solved consists of two o.d.e. for the two unknowns $H(u)$ and $\varphi(u)$, and of two inequalities that must be satisfied as well.

2. The Solution

The first step is to check that $\varphi =$ constant will not satisfy the conditions of our problem. Once this is done we may assume that $\dot{\varphi}(u) \neq 0$ and use φ as an *auxiliary* coordinate. However, we do not change the frame (for which u is the coordinate), so as not to disturb the equations. If we let

$$B(\varphi)\sin\varphi = H(u(\varphi))\,\dot{\varphi}(u(\varphi)), \tag{7}$$

the two reduced equations eventually take on the form

$$a_1 \frac{\dddot{\varphi}}{\dot{\varphi}^3} + a_2 \frac{\ddot{\varphi}^2}{\dot{\varphi}^4} + a_3 \frac{\ddot{\varphi}}{\dot{\varphi}^2} + a_4 = 0, \tag{8}$$

$$b_1 \frac{\dddot{\varphi}}{\dot{\varphi}^3} + b_2 \frac{\ddot{\varphi}^2}{\dot{\varphi}^4} + b_3 \frac{\ddot{\varphi}}{\dot{\varphi}^2} + b_4 = 0, \tag{9}$$

where the coefficients are certain functions of φ, $B(\varphi)$, $B'(\varphi)$ and $B''(\varphi)$. Long, weary arguments show that there is no solution $B(\varphi)$, satisfying both equations and both inequalities, for which a_1 or b_1 is zero. We may therefore replace Eq. (8) by

$$(a_1 b_2 - b_1 a_2)\frac{\ddot{\varphi}^2}{\dot{\varphi}^4} + (a_1 b_3 - b_1 a_3)\frac{\ddot{\varphi}}{\dot{\varphi}^2} + (a_1 b_4 - b_1 a_4) = 0. \tag{10}$$

There are now three possibilities. The case where $B(\varphi)$ is such that the first of the coefficients in Eq. (10) vanishes but the second does not, again leads to no solution. The second case, for which $B(\varphi)$ is such that all three coefficients vanish, leads to the Hauser solution. By putting these coefficients equal to zero it is found that $m = 5/2$ and $B(\varphi) = 2$. Eq. (9) then becomes

$$\frac{\dddot{\varphi}}{\dot{\varphi}^3} - 3\left(\frac{\ddot{\varphi}}{\dot{\varphi}^2}\right)^2 + 6\cot\varphi\,\frac{\ddot{\varphi}}{\dot{\varphi}^2} - \frac{99}{16\sin^2\varphi} + 8 = 0. \tag{11}$$

The initial data are therefore

$$P = (\zeta + \bar{\zeta})^{7/2}, \qquad L = 4(\zeta + \bar{\zeta})^{5/2} \sin\varphi(u)\,\dot{\varphi}^{-1}(u) e^{i\varphi(u)}, \tag{12}$$

where $\varphi(u)$ is subject to Eq. (11). The inequalities hold for the solution found. Letting the above be the primed frame and changing the frame by means of Eqs. (3)–(5), with $h(u') = F\big(\varphi(u')\big)$, transforms Eq. (11) to

$$\frac{\ddot{f}}{f}(F')^2 + \frac{\dot{f}}{f}[3F'' + 6\cot\varphi\,F'] + \left[\frac{F'''}{F'} + 6\cot\varphi\frac{F''}{F'} + \frac{99}{16\sin^2\varphi} - 8\right] = 0. \tag{13}$$

The choice $F(\varphi) = \cot\varphi$ (and hence $u = \cot\varphi(u')$) eliminates the middle term and reduces Eq. (13) to the Hauser equation, Eq. (2). The initial data of Eq. (12) become the Hauser initial data[6]

$$P = \frac{1}{2} f(u)(\zeta + \bar{\zeta})^{7/2}, \qquad L = -2 f(u)(u + i)(\zeta + \bar{\zeta})^{5/2}. \tag{14}$$

The third possibility is that $B(\varphi)$ is such that the first coefficient in Eq. (10) is nonzero. If we solve this equation algebraically for $\ddot{\varphi}/\dot{\varphi}^2$ and substitute the result into Eq. (9), we get a third-order nonlinear o.d.e. for $B(\varphi)$. In particular, for $m = 1$ we find

$$\frac{3}{2}B(B-1)\sin\varphi\, S' + S[(B-1)(B-2)\cos\varphi - 2BB'\sin\varphi] - S^{3/2}$$
$$- 3S^{1/2}(B-1)[BB'\sin\varphi\cos\varphi - 3B(B-1)\sin^2\varphi + (B-1)(B-5)\cos^2\varphi] = 0 \tag{15}$$

where

$$S = -3B^2(B-1)B''\sin^2\varphi + (B')^2B(B+3)\sin^2\varphi - 2B'B(B-1)(2B-1)\sin\varphi\cos\varphi$$
$$- 3(B-1)^2(B-2)(5B-3)\sin^2\varphi + (B-1)^2(B^2-B+7)\cos^2\varphi. \tag{16}$$

It is interesting to note that recently Chinea[7] has also arrived at a complicated nonlinear third-order o.d.e. while looking for twisting type N vacuum spacetimes possessing two noncommuting Killing vectors.

The second-order o.d.e.

$$\pm S^{1/2} = 3(B-1)\cos\varphi \tag{17}$$

is a first integral of Eq. (15). Unfortunately, any of its solutions will violate the inequality assuring that the space is not Minkowski space. The more obvious solution $B(\varphi) = 1$ violates the other inequality and gives zero twist.

But any other solution of Eq. (15) obeys both inequalities. Once such a solution has been obtained, Eq. (10) may easily be integrated for $\varphi(u)$. The initial data may then be written down, expressed perhaps in a more convenient frame, and the metric may be constructed from the general formalism [5]. Further details will appear elsewhere.

3. Acknowledgements

I would like to thank my graduate student, Y. Brian Yu, for helping me work out some of the details of this paper. I am also grateful for useful comments made by Drs. W. Allegretto, W. Israel and H. Künzle, and for the financial support of the Natural Science and Engineering Council of Canada.

4. References

1. I. Hauser, *J. Math. Phys.* **19** (1978) 661.
2. E. Newman and R. Penrose, *J. Math. Phys.* **3** (1962) 566.
3. D. W. Trim and J. Wainwright, *J. Math. Phys.* **15** (1974) 535.
4. G. Ludwig, *Gen. Rel. Grav.* **9** (1978) 1009.
5. G. Ludwig, *Int. J. Theor. Phys.* **29** (1990) 45.

6. D. Kramer, H. Stephani, M. MacCallum, and E. Herlt, *Exact Solutions of Einstein's Field Equations,* ed. E. Schmutzer (VEB Deutscher Verlag der Wissenschaften, Berlin, 1980) p. 274.
7. F. J. Chinea, *Phys. Rev.* **D37** (1988) 3080.

Gravitating Topological Matter

S. Carlip
Department of Physics, University of California
Davis, CA 95616
U.S.A.
and
J. Gegenberg
Department of Mathematics and Statistics,
University of New Brunswick
Fredericton, New Brunswick
Canada E3B 5A3

Abstract

We study topological matter minimally coupled to gravity in 2+1 dimensions. We show that the resulting system has a finite dimensional physical phase space that can be exactly quantized. The model exhibits the mixing of "gravity" and "matter" degrees of freedom, and the impossibility of treating them independently.

1. Introduction

As Witten [3] has shown, (2+1)-dimensional gravity [1]–[7] can be exactly quantized. On a manifold with the topology $\mathbb{R}\times\Sigma$, where Σ is a compact genus g surface, the gravitational Hilbert space is the space of square-integrable functions on the moduli space of Σ. The Hamiltonian is zero, and observables can be described explicitly in terms of holonomies of the *dreibein* and the spin connection. The resulting model has been used to explore several conceptual issues of quantum gravity, including the role of topology change [8], the meaning of diffeomorphism-invariant observables [9], and the nature of time in quantum gravity [10].

An important step would be to couple this model to matter, allowing the exploration of such issues as the effect of quantum fluctuations of spacetime on ultraviolet divergences. Unfortunately, when realistic matter is coupled to (2+1)-dimensional gravity, many of the problems of 3+1 dimensions reappear [11]. The combined phase space of matter and gravity is no longer finite dimensional, the space of exact solutions becomes much more complex, and questions of nonrenormalizability again arise (although it is possible that some models are renormalizable in the $1/N$ expansion [12]). Following the philosophy that led us to look at (2+1)-dimensional gravity in the first place, we might therefore begin by examining gravity coupled to models of matter with finitely many degrees of freedom — "topological matter."

A first step in this direction was taken by Gegenberg et al. [13], who looked at one simple form of topological matter nonminimally coupled to gravity. Here I will describe another such system, in which the coupling of matter and gravity is more conventional.

2. Action, Symmetries, and Field Equations

We take as our action functional[1]

$$S = \int d^3x\, \epsilon^{\mu\nu\pi}\left[\frac{1}{2}E_{i\pi}R^i_{\mu\nu}(A) + B_{i\pi}D_{[\mu}C^i_{\nu]}\right], \tag{1}$$

where $R^i_{\mu\nu}(A)$ is the curvature two-form corresponding to an SO(2,1) connection A^i_μ, and the "matter" fields $B_{i\mu}$ and C^i_μ are SO(2,1)-valued one-forms. The first term in (1) is the standard Einstein gravitational action in first order form [3, 7]; in particular, $E_{i\pi}$ is an SO(2,1)-valued one-form that, if nonsingular, can be interpreted as the *dreibein* of the spacetime metric. In the absence of matter, the connection A^i_μ is compatible on shell with E^i_μ, and so can be identified with the ordinary Lorentzian spin-connection. As we shall see, this is no longer the case when the matter fields B^i_μ and C^i_μ are non-trivial. The covariant derivative D_μ in the second term of (1) is with respect to the connection A^i_μ. On shell, the latter is flat, and this term is precisely the action for a topological field theory of the type considered in references [14] and [15]. Our action therefore represents topological matter minimally coupled to gravity.

The equations of motion are obtained by varying the four fields A^i_μ, E^i_μ, B^i_μ, and C^i_μ independently, and, in differential form notation, are

$$R^i := dA^i + \tfrac{1}{4}\epsilon^{ijk}A_j\wedge A_k = 0, \tag{2a}$$
$$DE^i + \tfrac{1}{2}\epsilon^{ijk}B_j\wedge C_k = 0, \tag{2b}$$
$$DC^i = 0, \tag{2c}$$
$$DB^i = 0. \tag{2d}$$

Here D denotes the exterior covariant derivative on spacetime,

$$DB^i := dB^i + \frac{1}{2}\epsilon^{ijk}A_j\wedge B_k, \tag{3}$$

where d is the ordinary exterior derivative on spacetime. Hatted $\hat{d}$'s and $\hat{D}$'s will appear later, where they will denote exterior derivatives on two-dimensional space.

We see from (2*b*) that A^i and E^i are not in general compatible. By (2*c*–*d*), on the other hand, A^i *is* compatible with both B^i and C^i, and it is tempting to reinterpret one of these fields as the spacetime *dreibein.* This was the interpretation given by Horowitz [14] in a 3+1 dimensional analogue of the model (1). However, as we shall show below, it is the spatial projection of E^i, not B^i or C^i, that is canonically conjugate to the spatial projection of A^i, so this interpretation is not possible in the quantum theory.

We remark here that this model is an example of a "teleparallel theory" [16, 17], as is the (3+1)-dimensional version discussed by Horowitz [14].

---

[1] Greek indices take the values 0, 1, 2 and denote the components of tensors, while lower case Latin indices are SO(2,1) algebra-valued, and are raised and lowered by the SO(2,1) metric η_{ij}. Upper case Latin indices, which appear below, are spatial indices with range 1, 2. The Levi-Civita density $\epsilon^{\mu\nu\pi}$ satisfies $\epsilon^{012} = +1$.

The action (1) is invariant under the following (infinitesimal) gauge transformation, with twelve gauge parameters ρ^i, λ^i, ξ^i, and τ^i:

$$\delta B^i = -D\rho^i - \tfrac{1}{2}\epsilon^{ijk}B_j\tau_k, \tag{4a}$$

$$\delta C^i = -D\lambda^i - \tfrac{1}{2}\epsilon^{ijk}C_j\tau_k, \tag{4b}$$

$$\delta E^i = -D\xi^i - \tfrac{1}{2}\epsilon^{ijk}\left(E_j\tau_k + B_j\lambda_k + C_j\rho_k\right), \tag{4c}$$

$$\delta A^i = -D\tau^i. \tag{4d}$$

We now assume that spacetime is topologically $\mathbb{R}\times\Sigma$ with an induced 2+1 splitting of the fields given by $\epsilon^{0AB} = \epsilon^{AB}$ = the Levi-Civita density on Σ. The 0-components of the fields are Lagrange multipliers enforcing the constraints

$$F^i := *\left(\hat{d}\hat{A}^i + \tfrac{1}{4}\epsilon^{ijk}\hat{A}_j\hat{A}_k\right), \tag{5a}$$

$$G^i := *\left(\hat{D}\hat{E}^i + \tfrac{1}{2}\epsilon^{ijk}\hat{B}_j\hat{C}_k\right), \tag{5b}$$

$$J^i := *\left(\hat{D}\hat{C}^i\right), \tag{5c}$$

$$K^i := *\left(\hat{D}\hat{B}^i\right), \tag{5d}$$

where $*$ is the spatial Hodge dual and hatted quantities are projections onto Σ. The momenta canonically conjugate to the 0-components of the fields are all constrained to be zero; the remaining momenta are $\tilde{\Pi}_i^B := \frac{\partial L}{\partial \dot{A}_B^i} = \tilde{E}_i^B, \tilde{P}_i^B := \frac{\partial L}{\partial \dot{C}_B^i} = \tilde{B}_i^B$. The Hamiltonian is weakly zero, as expected in a generally covariant theory.

The algebra of the constraints (5) may be recognized as the Lie algebra of the inhomogenized Poincare group I(ISO(2,1)), that is, the semidirect product of the Poincare group ISO(2,1) with its Lie algebra $\mathbb{R}^6$. I(ISO(2,1)) can be viewed as the cotangent bundle of ISO(2,1), just as ISO(2,1) is itself the cotangent bundle of the Lorentz group SO(2,1). We see that gravitational and matter fields are mixed in a manner reminiscent of supergravity.

3. Quantization

Let us now canonically quantize the system on a spacetime with the topology $\mathbb{R}\times\Sigma$, where Σ is a genus g surface. As in any constrained system, we have two options: we may first quantize and then impose the constraints as operator conditions on the states, or we may solve the constraints and quantize the physical phase space.[2] Following Witten [3], we adopt the latter procedure. Our first task is therefore to study the space of solutions of the constraints ($5a$–d).

We now recall that the phase space of pure (2+1)-dimensional gravity has the structure of a cotangent bundle $T^*\mathcal{M}(\Sigma)$, where $\mathcal{M}(\Sigma)$ is the moduli space of flat SO(2,1) connections *modulo* SO(2,1) gauge equivalence.

In our model, the pair $(\hat{A}^i, \hat{C}^i)$ is a point in the configuration space and it is easy to see that the latter has the structure of a cotangent bundle $T^*\mathcal{A}$. If $((\widehat{\delta A})^i, (\widehat{\delta C})^i)$ is a cotangent

[2] "Solving the constraints" has a double meaning here: we must restrict ourselves to fields that obey the constraint equations, but we must also identify fields that differ by gauge transformations generated by the constraints.

vector at $(\hat{A}^i, \hat{C}^i)$, the constraints (5a) and (5c) then imply that

$$\begin{aligned} \hat{d}(\widehat{\delta A})^i + \tfrac{1}{2}\epsilon^{ijk}\hat{A}_j \wedge (\widehat{\delta A})_k &= 0 \\ \hat{d}(\widehat{\delta C})^i + \tfrac{1}{2}\epsilon^{ijk}(\widehat{\delta A})_j \wedge \hat{C}_k + \tfrac{1}{2}\epsilon^{ijk}\hat{A}_j \wedge (\widehat{\delta C})_k &= 0. \end{aligned} \tag{6}$$

Comparing (5b,d), we see that $(\widehat{\delta A})^i$ and $(\widehat{\delta C})^i$ can be identified with $\hat{B}^i$ and $\hat{E}^i$, so fields satisfying the constraints represent points in the cotangent bundle $T^*(T^*\mathcal{A})$.

We now observe that only the ISO(2,1) gauge transformations generated by G^i and K^i act nontrivially on points $(\hat{A}^i, \hat{C}^i)$ in the base space of this bundle. As in pure gravity, factoring out this group gives us $T^*(\mathcal{A}/\mathcal{G})$; but this *phase space* for pure gravity is now the *configuration space* of our gravity-matter system. The full gauge group (4) is again a cotangent bundle, now with base space of ISO(2,1) transformations, and it is not hard to show that the physical *phase space* for our matter-gravity system is the cotangent bundle $T^*(T^*\mathcal{M}(\Sigma))$.

With such a cotangent bundle as a phase space, quantization is straightforward. The Hilbert space is the space of square-integrable functions of the base space, $\mathcal{H} = L^2(T^*\mathcal{M}(\Sigma))$ and we have seen that the Hamiltonian is weakly zero. The fundamental observables are functions on the configuration space $T^*\mathcal{M}$ and cotangent vectors to this configuration space.

It is also of interest to consider the path integral quantization of the system (1). If M is a closed three-manifold, the partition function $Z_M = \int [dA][dB][dC][dE]e^{iS}$ can be evaluated explicitly. The result is:

$$Z_M = \sum_{\alpha} (T(A_{(\alpha)}))^2 \ . \tag{7}$$

The quantity $T(A_{(\alpha)})$ is a topological invariant, namely the Ray-Singer analytic torsion of the flat bundle over M with connection $A_{(\alpha)}$[19],[20].

4. Physical Interpretation

In most physical theories we think of gravity as providing an arena in which matter interacts. Our model demonstrates, however, that it is not so easy to separate "gravity" from "matter."

To see this, consider the simplest topology, $\Sigma = S^2$. Note first that if $\tau^i = 0$ in (4), the action (1) is invariant, up to a total divergence, under gauge transformations with *finite* values of the remaining gauge parameters. We now take the trivial configuration $A^i = E^i = B^i = C^i = 0$, and act on it by the composition of two (finite) gauge transformations to obtain $E^i = -d\xi^i + \frac{1}{2}\epsilon^{ijk}d\rho_j\lambda_k$, $B^i = -d\rho^i$, $C^i = -d\lambda^i$. It is then possible to show that the metric obtained from the *dreibein* E^i is diffeomorphic to the most general form of the ADM metric.

In other words, given any metric on $\mathbb{R} \times S^2$, there exists some matter configuration such that the combined system is gauge-equivalent to flat empty spacetime. This provides a stark illustration of the fact that the metric cannot be treated separately from matter. Gauge transformations mix geometry and matter; it is only the combination that has physical significance. This is especially clear in the quantum theory, where states and operators depend only on gauge-invariant quantities. Indeed, for the spatial topology S^2, the moduli space $\mathcal{M}$ consists of a single point, and the Hilbert space contains only one state. Although we can produce any metric by a gauge transformation, the quantum physics remains trivial.

Acknowledgements

J. G. acknowledges the partial support of the Natural Sciences and Engineering Research Council of Canada for this research, and also thanks G. Kunstatter and G. Miller for useful conversations. Both authors thank the Canadian Association of Physicists, at whose 1990 Banff Conference on Gravitation this work was begun.

References

[1] G. 't Hooft, *Commun. Math. Phys.* **117** (1988), 685.

[2] S. Deser and R. Jackiw, *Commun. Math. Phys.* **118** (1988), 495.

[3] E. Witten, *Nucl. Phys.* **B311** (1988), 46.

[4] S. Carlip, *Nucl. Phys.* **B324** (1989), 106.

[5] S. P. Martin, *Nucl. Phys.* **B327** (1989), 178.

[6] J. Nelson and T. Regge, *Nucl. Phys.* **B328** (1989), 190.

[7] A. Ashtekar, V. Husain, C. Rovelli, J. Samuel, and L. Smolin, *Class. Quant. Grav.* **6** (1989), L185.

[8] E. Witten, *Nucl. Phys.* **B323** (1989), 113.

[9] S. Carlip, *Phys. Rev.* **D42** (1990), 2647.

[10] S. Carlip, *Class. Quant. Grav.* **8** (1991) 5.

[11] A. Ashtekar and J. D. Romano, *Phys. Lett.* **B229** (1989), 56.

[12] T. Kugo, "Renormalizability of Three-Dimensional Gravity Coupled to Matter," Kyoto preprint KUNS 1014 HE(TH) 90/05 (1990).

[13] J. Gegenberg, G. Kunstatter, and H. P. Leivo, *Phys. Lett.***B252** (1990), 381.

[14] G. T. Horowitz, *Comm. Math. Phys.* **125** (1989), 417.

[15] G. T. Horowitz and M. Srednicki, *Comm. Math. Phys.* **130** (1990), 83.

[16] R. Weitzenböck, *Invariantentheorie* (Noordhoff, Gronigen, 1923).

[17] K. Hayashi and T. Shirafuji, *Phys. Rev.* **D19** (1979), 3524.

[18] W. Goldman, *Adv. Math.* **54** (1984) 200.

[19] A. S. Schwarz, *Lett. Math. Phys.* **2** (1988), 247; *Comm. Math. Phys.* **67** (1979), 1.

[20] D. B. Ray and I. M. Singer, *Adv. Math.* **7** (1971), 145.

[21] L. Baulieu and M. Bellon, *Phys. Lett.* **B196** (1987), 142.

More colored black holes [1]

H.P. Künzle
Department of Mathematics, University of Alberta
Edmonton, Canada T6G 2G1

ABSTRACT

The discrete sequences of globally regular static spherically symmetric solutions of the $SU(2)$-Einstein-Yang-Mills equations and the similar sequences of black hole solutions are generalized to the gauge groups $SU(n)$. Spherical symmetry is defined as invariance of the Yang-Mills connection and the space-time metric under any suitable action of the group $SU(2)$ on the $SU(n)$-principal bundle. All these actions are classified. Many of them lead to equations that admit only a Reissner-Nordström like solution but there are some "irreducible" cases that appear to admit discrete sets of regular and black hole solutions.

1. Introduction

The discrete sequence of numerical solutions found by Bartnik and McKinnon[1] of globally regular static spherically symmetric and asymptotically flat solutions of the combined $SU(2)$-Einstein-Yang-Mills equations have generated a lot of interest. They were soon[2–4] extended to families of black hole solutions (for any radius of the black hole) which exhibit very similar characteristics. All these solutions are parametrized by the number of zeros of a suitable component of the Yang-Mills field and the total (Schwarzschild) masses of these solutions form a (probably infinite) monotonically increasing sequence that appears to converge towards a finite limit. Some rigorous non existence results for particular parameter values have been obtained[3], and in a number of papers[5–7] it was shown (also numerically) that these solutions are most likely unstable with respect to time dependent spherical perturbations. Only very recently[8] a rigorous existence proof was given for the "ground state" regular solution.

It is natural to expect other gauge theories, in particular those with gauge group $SU(n)$, to admit similar sets of solutions, and this note reports a derivation of the necessary equations and the first attempts to solve them numerically. Since there is no natural action of the rotation group or of $SU(2)$ on the principal bundles of a Yang-Mills gauge theory all possible actions with spherical orbits on space-time must be considered. While this problem has already been studied for theories over Minkowski space[9,10] a more systematic derivation and more convenient description is given in Ref.11 and summarized here. It has long been known that the Einstein-Yang-Mills equations for these theories are solved by trivial generalizations of the Reissner-Nordström solution. It now turns out that for most of the possible actions of $SU(2)$ this is in fact the only solution. But there remain a number of "irreducible" actions which result in coupled systems of ordinary differential equations in the static

[1]Supported in part by NSERC grant no. A8059.

case that remain to be analyzed in more detail. So far it is established that they lead to similar singular boundary value problems as the Bartnik-McKinnon one and the numerical analysis, while not complete, has shown that they admit nontrivial regular as well as black hole solutions.

2. Spherically symmetric $SU(n)$-connections

A *spherically symmetric gauge field* is by definition a connection on a principal bundle P (with structure group G, say) over space-time M that is invariant under an action of $SU(2)$ by principal bundle automorphisms on P. Such an action is given by a pair of maps $(\Phi, \bar{\Phi})$

$$\Phi : SU(2) \times P \to M : (a, p) \mapsto \Phi_a p, \qquad \bar{\Phi} : SU(2) \times M \to M$$

such that $\pi \circ \Phi = \bar{\Phi} \circ \pi$ where $\pi : P \to M$ the bundle projection and $\bar{\Phi}_a x = \bar{\Phi}_{a'} x$ if $a, a' \in SU(2)$ project onto the same element of $SO(3)$ under the covering map.

We assume that the $\bar{\Phi}_a$ are isometries on M and that the orbits of the action $\bar{\Phi}$ are 2-spheres and thus of constant curvature. Let also M be regularly foliated by these orbits. Then the actions Φ are classified by all possible actions of $SU(2)$ on a principal G-bundle over S^2. The following now follows from general results about principal bundle automorphisms proved in Ref.12 and/or Ref.13.

The equivalence classes of principal G-bundles over $S^2 \cong SU(2)/U(1)$ are in one-to-one correspondence with the conjugacy classes of homomorphisms $\lambda : U(1) \to G$. Here $U(1)$ is the isotropy group of the action $\bar{\Phi}$ at any point $x_0 \in S^2$ and two homomorphisms λ and μ are conjugate if $\mu(a) = b\lambda(a)b^{-1} \quad \forall a \in U(1)$ and for some $b \in G$. Applying this general result to $SU(n)$ we find that

> *the equivalence classes of $SU(n)$-principal bundles over S^2 with fiber transitive actions of $SU(2)$ by principal bundle automorphisms are classified by the sets of n integers whose sum is zero*

For any given bundle with an $SU(2)$-action we will choose $k_1 \geq k_2 \geq \ldots \geq k_n$ and define the matrix $D \equiv D_{\mathbf{k}} \equiv D_{(k_1 \ldots k_n)} \equiv \mathrm{diag}(k_1, \ldots, k_n)$, depending on whether the set $\{k_1, \ldots, k_n\} \subset \mathbb{Z}$ with $\sum_{i=1}^{n} k_i = 0$ needs to be specified or not.

For a fixed choice of the principal bundle P and action of $SU(2)$ we now consider the possible invariant connections. According to a theorem by Wang[13] the $SU(2)$-invariant connections on P are in one-to-one correspondence with the linear maps $\mathfrak{su}(2) \to \mathfrak{su}(n)$ satisfying $\Lambda(X) = \lambda'(X)$ for $X \in \mathfrak{u}(1)$ and $\Lambda \circ \mathrm{ad}_z = \mathrm{ad}_{\lambda(z)} \circ \Lambda \quad \forall z \in U(1)$ where λ is the map

$$\lambda : U(1) \subset SU(2) \to SU(n) : \begin{pmatrix} z & 0 \\ 0 & z^{-1} \end{pmatrix} \mapsto \mathrm{diag}(z^{k_1}, \ldots, z^{k_n})$$

and λ' the induced map of the Lie algebras. The curvature of this connection over S^2 is then fully determined. Over a spherically symmetric space-time of the form

$M = \tilde{M} \times S^2$ where $\tilde{M}$ is a Lorentz-2-manifold the $\tilde{M}$-part of the connection form must be invariant under the adjoint action of the subgroup $\lambda[U(1)]$ on $\mathfrak{su}(n)$ (cf. Ref.12). We have the result:

If P is a principal $SU(n)$-bundle over $\tilde{M} \times S^2$ with a given $SU(2)$-action the curvature is of the form

$$F = \tilde{F} + \tfrac{1}{2}C_a dx^a \wedge \bar{\Theta} - \tfrac{1}{2}C_a^H dx^a \wedge \Theta + \hat{F}\sin\theta d\theta \wedge d\phi$$

where $\hat{F} = -\frac{i}{2}([C, C^H] - D_{\mathbf{k}})$, $\tilde{F} = d\tilde{A} + \frac{1}{2}[\tilde{A} \wedge \tilde{A}]$, $\Theta := d\theta + i\sin\theta d\phi$, *and* $C_a \equiv D_a C = \nabla_a C + [\tilde{A}_a, C]$ *denotes the gauge covariant derivative of the matrix-valued scalar function C on $\tilde{M}$. The complex matrix C is strictly upper triangular and $C_{ij} \neq 0$ only if $k_i = k_j + 2$, C^H is its Hermitian conjugate, and $\tilde{A} = \tilde{A}_a dx^a$ is an $\mathfrak{su}(n)$-valued 1-form on $\tilde{M}$ whose off-diagonal entries $(\tilde{A}_a)_{ij}$ vanish unless $k_i = k_j$.*

Since $\tilde{M}$ is a twodimensional Lorentz manifold we can also write $\tilde{F}_{ab} = \tilde{\mathcal{F}}\tilde{\epsilon}_{ab}$ with $\tilde{\mathcal{F}} = -\frac{1}{2}\tilde{\epsilon}^{ab}\tilde{F}_{ab}$ where $\tilde{\epsilon}_{ab}$ are the components of the volume element on $\tilde{M}$.

3. Field equations

For simplicity we will state them only for the static case in a Schwarzschild type coordinate system for which the metric of M can be written

$$ds^2 = -NS^2dt^2 + N^{-1}dr^2 + r^2(d\theta^2 + \sin^2\theta d\phi^2)$$

where $N \equiv 1 - 2m/r$ and S depend on r only. Let also $\tilde{A} = -NSAdt + Bdr$ and write $C' := dC/dr + [B, C]$ and similarly for any scalar valued matrix function. Then the *Yang-Mills equations* are

$$\begin{aligned}
&(1) & -2r^2[A, \tilde{\mathcal{F}}] + [C, C'^H] - [C, C'^H]^H &= 0,\\
&(2) & 2(r^2\tilde{\mathcal{F}})' - [C, [A, C]^H] + [C, [A.C]^H]^H &= 0,\\
&(3) & r^2NC'' + (2m - \kappa r^3 p_\theta)C' + (C + \tfrac{1}{2}[C, [C, C^H]] - r^2N[A, [A, C]]) &= 0,
\end{aligned}$$

where $p_\theta = -\mathrm{tr}(\tilde{\mathcal{F}}^2 + r^{-4}\hat{F}^2)$ and $\tilde{\mathcal{F}} = N(-A' + [A, B]) - (2m/r^2 - \kappa r p_\theta)A$.
Einstein's equations become

$$(4) \qquad m' = \tfrac{1}{2}\kappa(NG + r^2p_\theta), \quad S^{-1}S' = \kappa r^{-1}G,$$

where $G = \mathrm{tr}([A, C][A, C]^H + C'C'^H)$, and for the mass-energy density and the radial pressure we find

$$e = N(r^{-2}G + p_\theta) \quad \text{and} \quad p_r = N(r^{-2}G - p_\theta).$$

In general, without imposing stationarity, we can prove a new variant of Birkhoff's theorem:

If the matrix C is identically zero then the space-time metric is the one of the Reissner-Nordström solution (unless r =const.). In particular, this is the case for all $SU(n)$ EYM fields if the $SU(2)$-action on P is such that $k_i \neq k_j + 2$ for all $1 \leq i, j \leq n$.

On this space-time where $N = 1 - 2m_\infty r^{-1} + \kappa k r^{-2}$ for some constant k and $S = 1$ the Yang-Mills equations admit a solution with $A = N^{-1}r^{-1}F_0, B = 0$ where F_0 is a constant antihermitian matrix commuting with D. This is a trivial generalization of the standard RN-solution. It is likely, though not fully proved, that all other solutions (with the same mass and "charge") are gauge equivalent to this one.

In this situation we really have a solution of the $U(1)$, i.e. the Einstein-Maxwell theory, trivially imbedded in the $SU(n)$-theory. There are generalizations of this result for the cases when a solution of the $SU(m)$-theory is imbedded in the $SU(n)$ theory for $m < n$. They follow from the holonomy theorem since in such a case the Yang-Mills curvature has its values in $\mathfrak{su}(m) \subset \mathfrak{su}(n)$.

Now Ershov and Gal'tsov[14] have shown that in the $SU(2)$ case a globally regular asymptotically flat solution cannot contain an "electric" component, i.e. have $A \neq 0$. Their result can be extended to the black hole case where it means that the only global solutions are the (two-parameter) Reissner-Nordström family and a discrete sequence of "colored" solutions, but that there are no combinations of these two types. One suspects that a similar situation may hold in the $SU(n)$ case. In the following we consider therefore just the special case where $A = 0$ and the matrix C has as many nonzero entries as possible, i.e. $k_i - k_j = 2$ for all $i > j$. Let $C_{j,j+1} = w_j e^{i\gamma_j}$. Then a gauge choice is possible so that γ_j and B_{ij} can be eliminated and Eq.(3) becomes

$$(5) \qquad r^2 N w_j'' + (2m - \kappa r^3 p_\theta) w_j' + [1 - w_j^2 + \tfrac{1}{2}(w_{j-1}^2 + w_{j+1}^2)] w_j = 0,$$

where now $G = \sum_{j=1}^{n-1} w_j'^2$ and $p_\theta = \frac{1}{4} r^{-4} \sum_{j=1}^{n} (w_j^2 - w_{j-1}^2 - n - 1 + 2j)^2$ (with $w_0(r) = w_n(r) = 0$).

The system of Eqs.(4) and (5) must now be analyzed together with appropriate boundary conditions at infinity and either $r = 0$ or at a regular black hole horizon. We require that $N = S = 1$ at infinity, $N = 1$ at $r = 0$ or $N = 0$ and $N' > 0$ at the horizon while S must be positive everywhere. Establishing sufficient conditions for the regularity of the gauge potential is not quite straightforward, but the necessary conditions following from regularity of pressure and mass density appear to determine the boundary value problem uniquely. They imply that

$$w_j(0)^2 = j(n-j) \qquad \text{and} \qquad \lim_{r \to \infty} w_j(r)^2 = j(n-j) \quad \text{for} \quad 1 \leq j \leq n-1$$

while at the horizon finite values of w_j can be chosen arbitrarily subject to the condition that $N' > 0$. It turns out that existence and uniqueness of local solutions near these boundaries can be proved, provided that n parameters are fixed which are just the values of w_j and S at the horizon but are complicated combinations of the first n derivatives of w_j and the values of S or m at $r = 0$ and at infinity. For a global solution of this singular two-point boundary value problem only some discrete values of these parameters will be allowed. What kind of discrete set of values will in fact parametrize all solutions is unknown at present. So far the numerical analysis has shown that at least some nontrivial globally regular as well as exterior black hole solutions exist, but much work remains to be done.

References

1. R. Bartnik and J. McKinnon, *Phys. Rev. Lett.* **61** (1988), 141–144.
2. M.S. Volkov and D.V. Gal'tsov, *Soviet J. Nuclear Phys.* **51** (1990), 747–753.
3. H.P. Künzle and A.K.M. Masood-ul-Alam, *J. Math. Phys.* **31** (1990), 928–935.
4. P. Bizon, *Phys. Rev. Lett.* **64** (1990), 2844–2847.
5. N. Straumann and Z.H. Zhou, *Phys. Lett. B* **237** (1990), 353–356.
6. N. Straumann and Z.H. Zhou, *Phys. Lett. B* **243** (1990), 33–35.
7. Z.H. Zhou and N. Straumann, *Nonlinear perturbations of Einstein-Yang-Mills solitons and non-Abelian black holes*, Preprint University of Zürich, January 1991.
8. J.A. Smoller, A.G. Wasserman, S.T. Yau, and J.B. McLeod, *Smooth static solutions of the Einstein/Yang-Mills equations*, Preprint University of Michigan.
9. C.H. Gu and H.S. Hu, *Comm. Math. Phys.* **79** (1981), 75–90.
10. N.S. Manton, *Ann. Physics* **132** (1981), 108–120.
11. H.P. Künzle, *Classical Quantum Gravity*, (in press).
12. J. Harnad, S. Shnider, and L. Vinet, *J. Math. Phys.* **21** (1980), 2719–2724.
13. S. Kobayashi and K. Nomizu, *Foundations of differential geometry*, vol. 1, John Wiley & Sons, New York, 1963.
14. A.A. Ershov and D.V. Gal'tsov, *Phys. Lett. A* **150** (1990), 159–162.

GRAVITATIONAL THETA-STATES IN 2+1 DIMENSIONAL QUANTUM COSMOLOGY

DOMENICO GIULINI
Institut für Theoretische Physik, Universität Freiburg
Hermann-Herder Strasse 3, D-7800 Freiburg, Germany

JORMA LOUKO
Theoretical Physics Institute, Department of Physics, University of Alberta
Edmonton, Alberta T6G 2J1, Canada

Abstract

Canonical quantisation of spatially homogeneous, flat and closed cosmologies in 2+1 and 3+1 spacetime dimensions exhibits the possibility of gravitational theta-states akin to those of the full theory. We illustrate this phenomenon in a 2+1 dimensional model with toroidal spatial topology, classifying the possible transformation laws of the theta-states. The consistency of the theta-states with the no-boundary proposal of Hartle and Hawking in relation to the topology of the (Euclideanised) spacetime manifold is discussed.

It is well known that quantum theories on topologically nontrivial configuration spaces display a richer structure than theories on topologically trivial configuration spaces.[1] A familiar example is provided by many-particle nonrelativistic quantum mechanics, where identical indistinguishable particles in three spatial dimensions are described either by totally symmetric or totally antisymmetric wave functions (bosons and fermions, respectively).[2] Another familiar example is provided by the theta-sectors of Yang-Mills theories.[3]

Similar theta-sectors are known to arise also in canonically quantised gravity, both for compact[4] and asymptotically flat[5] spatial surfaces. The configuration space of the theory is Riem(Σ), that is, the space of Riemannian metrics on a three-manifold Σ (with suitable fall-off conditions in the asymptotically flat case). The momentum constraint equations of the theory[6,7] imply that the wave functions must be invariant under that subset of the diffeomorphism group Diff(Σ) that can be obtained by exponentiating the Lie algebra of Diff(Σ) (with again suitable fall-off conditions in the asymptotically flat case). This is usually understood to imply that the wave function should be invariant under $\text{Diff}_{\text{id}}(\Sigma)$, the connected component of Diff(Σ). If Diff(Σ) is not connected, there however remains a freedom in choosing how the wave function is to transform under the fundamental group of Diff(Σ), $\pi_0(\text{Diff}(\Sigma)) = \text{Diff}(\Sigma)/\text{Diff}_{\text{id}}(\Sigma)$.

By the principle of coordinate invariance, one would expect to require the transformation of the wave function under $\pi_0(\text{Diff}(\Sigma))$ to be such that any physical predictions of the theory are invariant under Diff(Σ). What this means would depend on how one expects "physical predictions" to ultimately arise from canonically quantised gravity. However, experience with ordinary quantum mechanics and quantum field theory suggests that a sufficient condition would be to require that the wave function transform according to a one-dimensional complex representation of $\pi_0(\text{Diff}(\Sigma))$.[4,8] This means that the wave function need not be strictly a function on Riem(Σ)/Diff(Σ), but it can be a cross section of a flat complex line bundle. The classification of the possible theta-states then reduces to finding the one-dimensional complex representations of $\pi_0(\text{Diff}(\Sigma))$.

An arena extensively used for investigating canonical quantum gravity in the cosmological context are spatially homogeneous cosmologies.[9] The hope is that such "minisuperspace quantisation" might qualitatively preserve at least some of the global features of the full theory. (For a critical discussion, see Ref. 10.) The present article is based on the observation that there exist spatially homogeneous cosmological models with flat spatial sections such that these models possess a nontrivial analogue of the group $\pi_0(\text{Diff}(\Sigma))$.[11] In such models there arises the possibility of minisuperspace theta-states. We shall here illustrate the situation in a model with toroidal spatial topology in 2+1 spacetime dimensions. Because of the local triviality of Einstein gravity in 2+1 dimensions, this model can be argued to represent a quantisation of the full 2+1 theory with toroidal spatial topology.[12–14] The issues arising in the 3+1 dimensional spatially flat models are qualitatively similar but technically more involved.[11,15]

We consider a cosmological model in 2+1 spacetime dimensions defined by the metric ansatz

$$ds^2 = -N^2(t)dt^2 + h_{ij}(t)dx^i dx^j \tag{1}$$

where the two spatial coordinates $\{x^i\} = \{x, y\}$ are a pair of standard angle coordinates on the two-torus T^2, periodically identified by

$$(x, y) \sim (x + 2\pi, y) \sim (x, y + 2\pi) \ . \tag{2}$$

A consistent action for the model can be derived by inserting the ansatz into the modified Einstein action[16,17] and integrating over the spatial coordinates. The constrained Hamiltonian form of this action is[11]

$$I = \int dt \ (p_\alpha \dot{q}^\alpha - N\mathcal{H}) \tag{3}$$

where $\{q^\alpha\}$ are a set of three variables parametrising the symmetric positive definite matrix h_{ij}, and $\mathcal{H}$ is the super-Hamiltonian constraint quadratic in the momenta.

Upon Dirac quantisation,[18] $\mathcal{H}$ is lifted into an operator $\hat{\mathcal{H}}$ via the canonical substitution $p_\alpha \to -i\partial/\partial q^\alpha$ (modulo questions of operator ordering) and the only equation for the wave function $\Psi(h_{ij})$ is the Wheeler-DeWitt equation[6,7]

$$\hat{\mathcal{H}}\Psi = 0 \ . \tag{4}$$

The reason for the absence of supermomentum constraints from the action (3) is evident: the ansatz (1) is too restrictive to accommodate time-dependent spatial coordinate transformations. However, the ansatz does accommodate time-independent spatial coordinate transformations of the form

$$x^i = M^i{}_j \bar{x}^j \tag{5}$$

where the 2×2 matrix $M^i{}_j$ is such that the global identifications (2) look identical when expressed in the "old" coordinates $\{x^i\}$ and in the "new" coordinates $\{\bar{x}^i\}$. The largest group of matrices preserving the identifications (2) is $GL(2, \mathbb{Z})$, that is, the group of 2×2 matrices with integer entries and determinant equal to ± 1.[19] As the two-torus is orientable, it is natural to restrict the attention to the subgroup $SL(2, \mathbb{Z})$ of orientation-preserving coordinate transformations, where $\det(M) = 1$. Thus, the group of orientable minisuperspace diffeomorphisms in our model is $SL(2, \mathbb{Z})$.

The action of the $SL(2, \mathbb{Z})$ transformations (5) on the metric components is

$$\bar{h}_{ij} = M^k{}_i \, h_{kl} \, M^l{}_j \ . \tag{6}$$

The group of the transformations (6) is now the minisuperspace analogue of $\pi_0(\mathrm{Diff}(\Sigma))$. As the only degeneracy in (6) is that matrices M and $-M$ both lead to the same transformation of h_{ij}, this group is $PSL(2, \mathbb{Z}) = SL(2, \mathbb{Z})/\{E, -E\}$, where E is the 2×2 unit matrix. We therefore require the wave function to transform under the $PSL(2, \mathbb{Z})$ transformations according to

$$\Psi\left(M^k{}_i \, h_{kl} \, M^l{}_j\right) = \chi(M)\, \Psi(h_{ij}) \ , \tag{7}$$

where χ is a one-dimensional complex representation of $PSL(2, \mathbb{Z})$.

To find all the representations χ, recall[19] that $SL(2, \mathbb{Z})$ is generated by the two matrices

$$S = \begin{pmatrix} 1 & 1 \\ -1 & 0 \end{pmatrix} \ , \quad T = \begin{pmatrix} 0 & 1 \\ -1 & 0 \end{pmatrix} \tag{8}$$

whose only independent relations are

$$S^3 = T^2 = -E \ . \tag{9}$$

It follows that $PSL(2,\mathbb{Z})$ is generated by a corresponding pair of generators with relations where the right hand side of (9) is replaced by E. Hence, the only constraints for the one-dimensional representations of $PSL(2,\mathbb{Z})$ are

$$\chi^3(S) = \chi^2(T) = 1 \ . \tag{10}$$

This gives a total of six representations, uniquely characterised by the independent choices of $\chi(S) \in \{1, e^{2\pi i/3}, e^{-2\pi i/3}\}$ and $\chi(T) \in \{1, -1\}$. Note that only two of these representations are real.

Let us end by briefly discussing the compatibility of the theta-states with the no-boundary path integral prescription by Hartle and Hawking.[20,21] Consider the semiclassical contributions to the no-boundary integral, and consider for simplicity the case of a vanishing cosmological constant. The (Euclidean) actions of the classical solutions with the no-boundary data are labelled by ordered pairs (m,n) of coprime integers, modulo the overall sign of the pair.[11] The explicit formula is

$$I_{(m,n)}(h_{ij}) = -\frac{\pi}{2G}\sqrt{\frac{h}{m^2 h_{xx} + n^2 h_{yy} + 2mn h_{xy}}} \tag{11}$$

where $h = \det(h_{ij})$. The set $\{I_{(m,n)}\}$ is closed under the $PSL(2,\mathbb{Z})$ transformations (6), and conversely any $I_{(m,n)}$ can be obtained from any other by a $PSL(2,\mathbb{Z})$ transformation. The question now is what representations of $PSL(2,\mathbb{Z})$ are compatible with states written as sums of the semiclassical contributions $\exp\left[-I_{(m,n)}\right]$.

Formally, one can construct a state transforming according to any representation χ by taking one semiclassical contribution (say, $(m,n) = (0,1)$) and summing this over all of $PSL(2,\mathbb{Z})$ with weight $\chi(M)$. However, if one wishes to consider a sum that would contain each (m,n) just once, then the only surviving representation appears to be the trivial one, $\chi = 1$. The reason is that $SL(2,\mathbb{Z})$ has an alternative set of generators given by $\{TS^{-1}, T^{-1}S\}$. Each of these generators leaves exactly one semiclassical contribution invariant and must therefore be represented trivially.

It can be shown that these results concerning the no-boundary theta-states in our model transcend the semiclassical approximation. The classical solutions giving the actions (11) are defined on the three-manifold $\mathcal{M} = \bar{D} \times S^1$, where $\bar{D}$ is the closed two-dimensional disc. The countable set of classical solutions (11) is related to the fact that there are some diffeomorphisms on $\partial\mathcal{M} = T^2$ that cannot be extended into diffeomorphisms on all of $\mathcal{M}$. The same reasoning can also be used to argue, extending the work of Hartle and Witt,[8] that these results about the theta-states transcend the limitations of working within a minisuperspace model.

Acknowledgment

One of us (J.L.) was supported in part by The Natural Sciences and Engineering Research Council of Canada.

References

1. C. J. Isham, in: *Relativity, Groups and Topology II: Les Houches 1983,* ed. B. S. DeWitt and R. Stora (North-Holland, Amsterdam, 1984).
2. M. Laidlaw and C. DeWitt, *Phys. Rev.* D**3** (1971) 1375.
3. R. Jackiw, in: *Relativity, Groups and Topology II: Les Houches 1983,* ed. B. S. DeWitt and R. Stora (North-Holland, Amsterdam, 1984).
4. C. J. Isham, *Phys. Lett.* **106**B (1981) 188.
5. J. L. Friedman and R. Sorkin, *Phys. Rev. Lett.* **44** (1980) 1100.
6. B. S. DeWitt, *Phys. Rev.* **160** (1967) 1113.
7. J. A. Wheeler, in: *Battelle Rencontres 1967,* ed. C. DeWitt and J. A. Wheeler (Benjamin, New York, 1968).
8. J. B. Hartle and D. M. Witt, *Phys. Rev.* D**37** (1988) 2833.
9. M. P. Ryan and L. C. Shepley, *Homogeneous Relativistic Cosmologies* (Princeton University Press, Princeton, 1975).
10. K. V. Kuchař and M. P. Ryan, *Phys. Rev.* D**40** (1989) 3982.
11. J. Louko and P. J. Ruback, *Class. Quantum Grav.* **8** (1991) 91.
12. E. Witten, *Nucl. Phys.* B**311** (1988) 46.
13. A. Hosoya and K. Nakao, *Class. Quantum Grav.* **7** (1990) 163.
14. A. Hosoya and K. Nakao, *Prog. Theor. Phys.* **84** (1990) 739.
15. D. Giulini and J. Louko, in preparation.
16. J. W. York, *Phys. Rev. Lett.* **28** (1972) 1082.
17. G. W. Gibbons and S. W. Hawking, *Phys. Rev.* D**15** (1977) 2752.
18. P. A. M. Dirac, *Lectures in Quantum Mechanics* (Academic, New York, 1964).
19. H. M. S. Coxeter and W. O. J. Moser, *Generators and Relations for Discrete Groups,* 4th edition (Springer, Berlin, 1980).
20. J. B. Hartle and S. W. Hawking, *Phys. Rev.* D**28** (1983) 2960.
21. J. B. Hartle, in: *Gravitation in Astrophysics: Cargèse 1986,* ed. B. Carter and J. B. Hartle (Plenum, New York, 1987).

Collapse of a Spherically Symmetric Anisotropic Fluid in General Relativity

A. DAS
Department of Mathematics and Statistics,
Simon Fraser University,
Burnaby, B.C., V5A 1S6 Canada

and

T. BIECH
Department of Mathematics and Computing Science,
University of Prince Edward Island,
Charlottetown, P.E.I. C1A 4P3 Canada

ABSTRACT

The general case of a spherically symmetric collapsing anisotropic fluid is investigated. The Einstein field equations are used to find the metric coefficients in terms of the energy density and the radial pressure. The equation for the collapsing boundary is obtained by using simple jump conditions.

The notation and conventions are adopted from Synge's book[1]. Units are chosen so that $c = 8\pi G = 1$. Einstein's field equations for an anisotropic fluid are given by

$$G_{ab} = -T_{ab} := (\rho + p_\perp)u_a u_b + p_\perp g_{ab} + (p_\parallel - p_\perp)s_a s_b, \tag{1}$$

$$u^a u_a = -s^a s_a = 1, \qquad u^a s_a = 0. \tag{2}$$

For the case of spherical symmetry, we choose

$$ds^2 = e^{\alpha(r,t)}dr^2 + r^2 d\Omega^2 - e^{\gamma(r,t)}dt^2, \tag{3}$$

where $d\Omega^2 = (d\theta^2 + \sin^2\theta d\phi^2)$, and

$$u^a = (u^1,0,0,u^4), \qquad s^a = (s^1,0,0,s^4),\ u^4 > 0, \tag{4}$$

The various types of energy conditions as follows

$$\text{weak:} \begin{cases} \rho \geq 0 \\ \rho + p_\parallel \geq 0 \\ \rho + p_\perp \geq 0 \end{cases} \tag{5}$$

$$\text{dominant:} \begin{cases} \rho \geq 0 \\ \rho \geq |p_\parallel| \\ \rho \geq |p_\perp| \end{cases} \tag{6}$$

$$\text{strong:} \begin{cases} \rho + p_\parallel \geq 0 \\ \rho + p_\perp \geq |p_\parallel|. \\ \rho + p_\parallel + 2p_\perp \geq |p_\perp| \end{cases} \tag{7}$$

Eq. 1, Eq. 2, Eq. 3, and Eq. 4 yield

$$\rho = \frac{-T^4{}_4\cosh^2\beta - T^1{}_1\sinh^2\beta}{\cosh^2\beta + \sinh^2\beta}, \tag{8}$$

$$\rho + p_\parallel = \frac{T^1{}_1 - T^4{}_4}{\cosh^2\beta + \sinh^2\beta}, \tag{9}$$

for some $\beta \in \mathbb{R}$. If we prescribe $T^4{}_4 \leq 0$, $T^1{}_1 \geq 0$, then for a sufficiently small β, we have $\rho \geq 0$ and $\rho + p_\parallel \geq 0$.

Two of the field equations Eq. 1 are given by

$$\frac{(1 - e^{-\alpha})(1 + r\gamma,_r)}{r^2} = -T^1{}_1 \tag{10}$$

$$\frac{(1 - e^{-\alpha})(1 - r\alpha,_r)}{r^2} = -T^4{}_4. \tag{11}$$

We shall now give a recipe for local solutions. We restrict the domain of the solution to a subset D given by $D = \{(r,t): 0 < r < r_1, 0 < t < t_1\}$. Now choose a function M in $\mathscr{C}^2(D)$ so that the following five conditions hold

(i) $M(r,t) \geq 0$ on D;
(ii) $M(0+,t) = 0$;
(iii) $M,_r > 0$ on D;
(iv) $\lim_{r\to 0+} M,_r = 0$;
(v) $M,_t \not\equiv 0$.

We choose another $\mathscr{C}^2(D)$-function σ and define

$$T^4{}_4(r,t) := \frac{-2M,_r}{r^2} \le 0, \tag{12}$$

$$T^1{}_1(r,t) := [\sigma(r,t)]^2 \ge 0, \tag{13}$$

$$\psi(r,t) := \int_{0+}^{r} \frac{x[\sigma(x,t)]^2 + \dfrac{2M,_x(x,t)}{x}}{1 - \dfrac{2M(x,t)}{x}}\, dx \tag{14}$$

The field equations Eq. 10 and Eq. 11 can be integrated to obtain the line element

$$ds^2 = \frac{dr^2}{1 - \dfrac{2M}{r}} + r^2 d\Omega^2 - \left[1 - \frac{2M}{r}\right] e^{\psi(r,t)} dt^2. \tag{15}$$

The transverse pressure can be defined by the field equation

$G^2{}_2 = -T^2{}_2$ and $T^1{}_4 := \dfrac{2M,_t}{r^2}$. At this stage, all the field equations have been locally solved either by integration of by definition.

Let the collapsing boundary[2,3] be denoted by ∂D. We consider the equation

$$M(r,t) = b \tag{16}$$

where $b \neq 0$ is a constant. From the assumption $M,_r > 0$ and the implicit function theorem, there is a local function R_b such that

$$r = R_b(t). \tag{17}$$

We choose Eq.17 as the equation of the collapsing boundary (the reason will be clear shortly). The jump conditions at the boundary are chosen to be[1]

$$T_{ij} n^j |_{\partial D} = 0. \tag{18}$$

Eq. 18 reduces to two non-trivial equations when spherical symmetry is used. One of these equations is identically satisfied by the choice of collapsing boundary Eq. 17. The remaining jump condition reduces to the following nonlinear integral equation

$$\dot{R}_b(t) = \left. \frac{-r^2\sigma^2\left[1 - \frac{2M}{r}\right]^2 e^{\psi}}{2M,_t} \right|_{r = R_b(t)} . \tag{19}$$

Eq. 19 shows that the functions M and σ cannot be prescribed arbitrarily. In case M is prescribed, then $\sigma(r,t)\Big|_{r = R_b(t)}$ is determined by Eq. 19.

Eq. 19 is a formidable equation, however a theorem on the local existence of solutions can be proved.

References

1. J. L. Synge, *Relativity: The General Theory* (North-Holland Publishing Co., Amsterdam, 1960) p.1, 40.
2. N. O. Santos, *Phys. Lett.* **106A** (1984) 296.
3. E. N. Glass, *Phys. Lett.* **86A** (1981) 351.

A PHOENIX UNIVERSE? DISCUSSIONS ON COSMOGONY

ARNOLD E. SIKKEMA†
Department of Mathematics and Statistics, University of Guelph, Guelph, Ontario N1G 2W1, Canada

ABSTRACT

Investigations in cosmogenesis have taken many approaches throughout history. In this paper, I discuss in particular the cosmic bounce scenario which has recently been revived because of theoretical advances in the physics of black-hole interiors and in quantum gravity. The perceived difficulty associated with the thermodynamic requirement of entropy increase is removed by the phenomenon of mass inflation which accompanies the gravitational collapse of a generic astronomical body, since it both affects our view of the merging of black holes in a cosmological collapse and produces the sharp deflation in specific entropy required for the regeneration of order in a subsequent smooth big bang. Despite the physical plausibility of such a scenario, I present fundamental limitations on any cosmogonical model.

1. Introduction

"What is the origin of the Universe and the life it contains?" This question has been the subject of investigations of many kinds throughout the ages. Philosophers grapple with the issue of the first cause, Darwinian biology purports to explain the origin of species, and cosmologists weigh astronomical evidence in search of the big bang. But the oldest argument, that of a Divine Creator with a deliberate design, continues to give the most sensibility to the question of ultimate causation.[1]

In this paper I discuss the issue of cosmogenesis, focusing on the recently revived scenario of the cosmic bounce.[2]

2. The Cosmic Bounce Scenario

The idea that the Universe is passing through just one of a sequence of cycles is at least as ancient as the Hindu wheel of life. The earliest Friedmann models were also subject to the consideration of a phoenix universe, and Tolman realized that continued entropy increase would cause the universe to reach a greater size in each subsequent cycle.

2.1. The Entropy Paradoxes

Penrose[3] has discussed the improbably low entropy associated with a smooth big bang as compared to a chaotic big crunch. In terms of the cosmic bounce scenario, this raises an important issue. A plausible picture of the big crunch is given by

†Address after 1 August 1991: Department of Physics, University of British Columbia, Vancouver, British Columbia V6T 1Z1, Canada

considering the coalescence of black holes forming a universal black hole of mass $M \sim 10^{60}$ (we employ Planck units: $c = \hbar = k_B = G = 1$). The Bekenstein-Hawking formula, although strictly applicable only in asymptotically flat spacetime, gives the final entropy as $S \sim M^2 \sim 10^{120}$, giving a specific entropy of $s \sim M$. The current entropy of the Universe, found largely in the microwave background, is $S_{\rm rad} \sim 10^{27} M \sim 10^{87}$. Thus obtaining the Universe from a similar progenitor requires an entropy deflation by a factor of at least $S/S_{\rm rad} \sim 10^{33}$ at the big bounce. However, this is in contradiction to the requirement of the second law of thermodynamics, which is among the most fundamental physical laws not easily discarded even in the context of quantum gravity.

Another difficulty with a cosmic bounce also presents itself, noting that Planck levels are upper limits for physically meaningful and attainable values of intrinsic quantities[4] such as mass density, specific entropy, and temperature: the specific entropy would become completely unphysical ($s \gg 1$) long before the crunch.

2.2. Mass Inflation and Specific Entropy Deflation

The application of recent advances in the physics of black hole interiors[5] removes both of these entropy paradoxes.[2] As a generic (rotating) stellar object collapses asymptotically to its Kerr-Newman endstate, all irregularities are necessarily radiated away. The coupling between the subsequent contraction of the inner horizon and the blue-shifting of back-scattered gravitational radiation gives rise to a classically unbounded inflation of the core's mass, which is limited to $m' \sim m^3$ by imposing a Planck cutoff. Of course, the news of this mass inflation is shielded by the outer horizon, so until an observer enters the black hole no mass inflation is perceived.

Visualizing the big crunch as the coalescence of n black holes each of mass $m(\gg m_{\rm Pl})$, the following sequence of events is considered.[2]

THE COSMIC BOUNCE	time t	mass M	volume V	specific entropy s
almost merged	$-m$	nm	nm^3	m
just merged	-1	nm^3	nm^3	1
after transition	$+1$	nm^3	nm^3	1
continued expansion	$+m$	$nm^{5/2}$	$nm^{9/2}$	$m^{1/2}$

The time from merging to the "singularity" at $t = 0$ is of the order of the light-crossing time past a black hole, which is its radius m. Due to the mass inflation of the black-hole cores and the corresponding time-deflation, this time collapses to one Planck time immediately after the universe is completely contained within the horizon, when also the mass density and specific entropy become Planckian. Assuming that the onset of Planckian values triggers a quantum transition (which transports coherent states time-symmetrically[6]), we emerge in a radiation-dominated expansion at $t = +1$, which I continue here through $t = +m$ for comparison to the pre-crunch values. The key feature of this scenario is that the specific entropy s has dropped dramatically from m to $m^{1/2}$. Viewing specific entropy as the appropriate measure of disorder, it is apparent that there would be significant restoration of order as a result of the mass inflation, allowing the universe to continue in its re-expansion as it would from an "ordinary" big bang.

2.3. Incompleteness of the Cosmic Bounce Scenario

Quite apart from the speciosity of a cosmogonical model, one must address its relevance to the actual Universe and the degree to which it deals with the fundamental question of origins.

In the case of the cosmic bounce scenario, analysis of the latter is quite simple. The regeneration of order at the big crunch is crucially dependent on the existence of black holes, and so tracing the history of the universe back through successive bounces must terminate at that cycle which was sufficiently large and long-lived to permit the formation of black holes and the stellar objects from which they form.[2] Thus the first cause issue is not addressed: whence arose this "baby universe"?

It appears that this cosmic bounce scenario may resolve the horizon problem by pre-establishing homogeneity on larger-than-horizon scales.[2] Of course a full quantum theory of gravity must precede any definite predictions in connection with the highly blue-shifted gravitons which would pervade the re-expansion. Comparison with bounce scenarios not involving mass inflation[7] may also prove fruitful.

3. Fundamental Limitations on Cosmogonical Models

Both the difficulty in deducing models from astronomical observations and the tautology that the origin of the Universe occurred only in the past make cosmogony a particularly problematic field. There are, however, limitations even more fundamental which in effect remove cosmogony from the scientific realm: *it is impossible to prove anything that has happened in the past.* Clearly the data present on *any* spacelike hypersurface to our past could satisfactorily serve as a set of initial conditions.

In order to determine the history of the Universe, science can receive input only from geology and astronomy. These observations span but a tiny fraction of the four-dimensional spacetime of the Universe. Local measurements are restricted near our world line, and telescopes have scanned only a thin light-cone shell. With this in mind, how can one expect cosmology to give a reliable description of the Universe, much less a model for its origin?

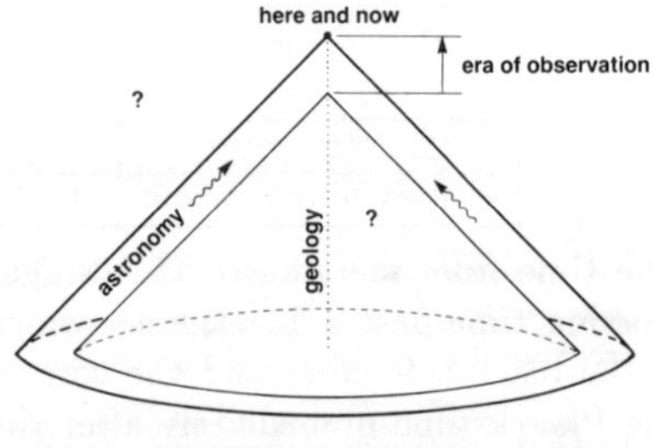

The Observational Light-Cone Shell

In fact, cosmological models are intrinsically based on completely unverifiable assumptions, two of which Ellis[8] gives as the local predictability assumption that "whenever physical laws can be applied, they correctly predict the structure of the universe" and the assumption of a spatially homogeneous universe. In addition, Burbidge[9] points out the prevalence of the presumption of the immutability of physical laws. Within the framework of a given set of assumptions, a model may be seen to perform reasonably well, but that these are built in must not be ignored.

4. Conclusions

While the cosmic bounce scenario is physically plausible, it remains completely unverifiable. But much of present-day cosmology embraces as fact cosmogonical hypotheses which are intrinsically unprovable. Objectivity must be restored, and the complete picture provided. This includes elucidating all underlying assumptions, promoting the investigation of the complete range of possible origins,[1,10] and incorporating all available astrophysical evidence.[9,10] Even so, considering the inaccessibility of the past, cosmology can rightly deal only with the present and future.

5. Acknowledgments

I should like to thank Werner Israel for his guidance and encouragement during my short stay at the University of Alberta, where this research on the cosmic bounce scenario was initiated. I also acknowledge many stimulating and informative discussions with Robb Mann and with Margaret Helder. The hospitality of George Leibbrandt and the staff at the University of Guelph during the preparation of this paper is greatly appreciated. I also thank the Universities of Waterloo and Guelph for their financial assistance in my attendance of the conference. This work was supported in part by NSERC.

6. References

1. G.F.R. Ellis, Preprint 1990/1 (Department of Applied Mathematics, University of Cape Town).
2. A.E. Sikkema & W. Israel, *Nature* **349** (1991) 45–47.
3. R. Penrose, *Ann. N.Y. Acad. Sci.* **571** (1989) 249–264.
4. J.D. Bekenstein, *Int. J. Theor. Phys.* **28** (1989) 967–981.
5. E. Poisson & W. Israel, *Phys. Rev.* **D41** (1990) 1796–1809; C. Barrabès, W. Israel & E. Poisson, *Class. Quantum Grav.* **7** (1990) L273–L278.
6. J.D. Brown & E.A. Martinez, *Phys. Rev.* **D42** (1990) 1931–1943.
7. W.C. Saslaw, *Nature* **350** (1991) 43–45.
8. G.F.R. Ellis, *Q.J.R. Astron. Soc.* **16** (1975) 245–264; reprinted in part in *Physical Cosmology and Philosophy*, ed. J. Leslie (Macmillan, New York, 1990) pp. 113–120.
9. G. Burbidge, in *New Ideas in Astronomy*, eds. F. Bertola, J.W. Sulentic & B.F. Madore (Cambridge University Press, 1988) pp. 223–238.
10. H.C. Arp, G. Burbidge, F. Hoyle, J.V. Narlikar & N.C. Wickramasinghe, *Nature* **346** (1990) 807–812.

RICCI INHERITANCE SYMMETRY IN FLUID SPACETIMES

K.L. DUGGAL

Department of Mathematics and Statistics
University of Windsor, Windsor, Ontario N9B 3P4, Canada

Abstract. This paper deals with a new symmetry called "Ricci inheritance"(RI) defined by $L_\xi R_{ab} = 2\alpha R_{ab}$, where R_{ab}, L_ξ and α are Ricci tensor, Lie-derivative with respect to Ricci inheritance vector (RIV) ξ^a and scalar function respectively. Sub case is the well-known symmetry Ricci collineation (RC) when $\alpha = 0$. We prove that a proper RI (i.e., $\alpha \neq 0$) has direct interplay with the physically significant proper conformal motions. We solve the fundamental problem of symmetry inheritance for imperfect fluid space-times. Finally, we show that, a proper RIV generates new and physically relevant solutions for a variety of fluid spacetimes.

Ricci Inheritance Symmetry. Consider a 4-dimensional spacetime, of general relativity, that admits a particular symmetry called Ricci inheritance (RI) defined by

$$L_\xi R_{ab} = 2\alpha R_{ab} \tag{1}$$

where $\alpha = \alpha(x)$ is a scalar function. R_{ab} and L_ξ are the Ricci tensor and the Lie-derivative with respect to a Ricci inheritance vector (RIV) respectively. Sub case of RI is Ricci collineation $(RC)^1$ when $\alpha = 0$. In the sequel, RIV is proper if $\alpha \neq 0$, and

$$L_\xi g_{ab} = \xi_{a;b} + \xi_{b:a} \equiv K_{ab} \tag{2}$$

where g_{ab} is the metric tensor for a Levi-civita connection.

The objective is to investigate some physical properties of RI in fluid spacetimes. Following identities hold.

$$L_\xi R^a_b = 2\alpha R^a_b - R^m_b K^a_m, \quad L_\xi r = 2\alpha r - 2R^a_b \xi^b_{:a} \tag{3}$$

where r is the scalar curvature.

Theorem 1. If a spacetime admits an RIV ξ^a then

$$(R^a_m \xi^m)_{;a} = \alpha r \tag{4}$$

The proof follows by explicitly writing the Eq.(1) and then contracting b with a and using the contracted Bianchi identity. Above theorem is a generalization of an earlier result proved by Collinson[2] for an RCV when $\alpha = 0$. If the Einstein's field equations are in the form

$$R_{ab} - \frac{1}{2} g_{ab} r = T_{ab} \tag{5}$$

then dynamics is introduced by using (5) in (4) which provides

$$\left[(T^{ab} - \frac{1}{2} T g^{ab}) \xi_b\right]_{;a} = \alpha r, \qquad T = -r. \tag{6}$$

The equation (6) is the basis for generating new solutions for a variety of fluid spacetimes under consideration. First we deal with the fundamental problem of finding conditions under which any matter tensor T_{ab} inherits the symmetry with respect to an RIV ξ^a . That is. find conditions so that

$$L_\xi T_{ab} = 2\alpha T_{ab} \tag{7}$$

holds. For this we have the following theorem.

Theorem 2. If a spacetime with Einstein's field equations (5) admits an RIV ξ^a, given by (1), then its energy momentum tensor T_{ab} inherits symmetry, given by (7), if ξ^a is also a conformal Killing vector (CKV) satisfying:

$$L_\xi g_{ab} = 2\psi g_{ab}, \qquad r\psi = R^a_b \xi^b_{;a}. \qquad r \neq 0 \tag{8}$$

For a CKV ξ^a, the following identities hold[3].

$$L_\xi R_{ab} = -2\psi_{;ab} - g_{ab} \Box\psi, \, . \quad L_\xi r = 2\psi r - 6\Box\psi \tag{9}$$

where $\Box\psi \equiv g^{ab}\psi_{;ab}$. Comparing (1) and (3) with (9), we state:

Corollary 1. An RIV is also a CKV iff

$$\psi_{;ab} = \frac{\alpha}{2}\left[\frac{r}{3} g_{ab} - 2R_{ab}\right], \quad \Box\psi + \frac{\alpha r}{3} = 0. \quad L_\xi r = 2(\alpha - \psi) r \tag{10}$$

Above result is a generalization of an earlier result of Katzin et al [4], proving that an RCV ($\alpha = 0$) is also a CKV iff $\psi_{;ab} = 0$ i.e.. ξ^a is a <u>*special*</u> CKV.

Remark 1. Recent study[5] indicates very limited use of <u>*special*</u> CKV's due to the existence of a covariant constant vector $\psi_{.a}$. On the other hand, proper CKV's are physically significant. in particular. reference to their existence in FRW spaces[6]. Thus. our theorem 2 is important as it establishes an interplay between the new symmetry RI and proper CKV's. This suggests that there is a need for further study on RI.

Imperfect Fluids and Symmetry Inheritance. Consider a class of spacetimes that admit an RIV ξ^a. given by the Eq.(1). and Einstein's field equations (5) with imperfect fluid in the form:

$$T_{ab} = \mu u_a u_b + p h_{ab} - 2\eta\sigma_{ab} + 2q_{(a}u_{b)} \tag{11}$$

when $\mu, p, q^a, \eta(\geq 0), \sigma_{ab}$ and h_{ab} are the energy density, the isotropic pressure, the energy flux vector relative to the four velocity vector u^a, the shear viscosity coefficient, the shear tensor and the projection tensor ($h_{ab} = g_{ab} + u_a u_b$) respectively. Here, $\sigma_{ab}u^a = 0. \sigma^a_a = 0$ and $q_a u^a = 0$. We are interested in finding conditions under which all the physical and geometric quantities of the Einstein's field equations. with T_{ab} satisfying (11), inherit the spacetime symmetry with respect to an RIV ξ^a. It follows from the theorem 2, that, T_{ab} and r will inherit symmetry if g_{ab} inherits symmetry i.e., if ξ^a is also a CKV. given by the Eq.(8). Thus, it is logical to assume that an RIV ξ^a is also a CKV. Observe that this condition is consistent with Coley and Tupper's work[5] on symmetry inheritance <u>*of special*</u> CKV which obviously is also an RCV ($\alpha = 0$) . Based on above, we prove the following theorem:

Theorem 3. Let an imperfect fluid spacetime admit an RIV ξ^a which is also a CKV. If the fluid flow lines are mapped conformally by ξ^a, then, the following symmetry inheritance equations hold.

$$L_\xi \mu = 2(\alpha - \psi)\mu, \quad L_\xi p = 2(\alpha - \psi)p. \quad L_\xi u_a = -\psi u_a \tag{12a}$$

$$L_\xi \sigma_{ab} = \psi\sigma_{ab}, \quad L_\xi \eta = -(2\alpha + \psi)\eta. \quad L_\xi q_a = (2\alpha - \psi)q_a \tag{12b}$$

Proof. For a CKV ξ^a, if the fluid flow lines are mapped conformally by ξ^a, then we know[5] that $L_\xi u_a = -\psi u_a$ and $L_\xi \sigma_{ab} = \psi\sigma_{ab}$ will hold. Also, in general, for a CKV

$$L_\xi q_a = (Q^{-1}L_\xi Q + \psi)q_a + w_a \tag{13}$$

where $Q^2 = q_a q^a$ and $w_a q^a = 0$. By taking the Lie-derivative of the field equations (5), with T_{ab} defined by the Eq.(11), and contracting this equation in turn with $u^a u^b$, h^{ab}, $u^a h^b_c$, $h^a_c h^b_d - \frac{1}{3}h^{ab}h_{cd}$, we get

$$L_\xi \mu = -2\psi\mu - 2\Box\psi - 2\psi_{;ab}u^a u^b \tag{14a}$$

$$L_\xi p = -2\psi p + \frac{4}{3}\Box\psi - \frac{2}{3}\psi_{:ab}u^a u^b \tag{14b}$$

$$(Q^{-1}L_\xi Q + 2\psi)q_a + w_a = 2\psi_{:cb}u^b h^c_a + 2(\psi_{:cd}u^c u^d)u_a \tag{14c}$$

$$\begin{aligned} L_\xi(\eta\sigma_{ab}) = \psi_{;ab} + \psi_{;cb}u^c u_a + \psi_{;ac}u^c u_a - \frac{1}{3}g_{ab}(\Box\psi + \psi_{:cd}u^c u^d) \\ -\frac{1}{3}u_a u_b(\Box\psi - 2\psi_{:cd}u^c u^d) \end{aligned} \tag{14d}$$

Using the field equations (5) with T_{ab} given by (11) and (10) in (14a) and (14b) we recover first two equations of (12a). Again using above mentioned in (14d), we obtain

$$\begin{aligned} L_\xi(\eta\sigma_{ab}) &= (L_\xi\eta + \psi\eta)\sigma_{ab} \\ &= \frac{\alpha}{6}(r + 2\mu)g_{ab} - \alpha R_{ab} + \frac{\alpha}{3}(4\mu - r)u_a u_b + 2q_{(a}u_{b)} \end{aligned} \tag{15}$$

Contracting (15) with σ^{ab} and then using (5) and (11), we get $(L_\xi\eta + \psi\eta)(2\sigma^2) = -4\alpha\eta\sigma^2$, where $2\sigma^2 \equiv \sigma_{ab}\,\sigma^{ab}$. For $\sigma \neq 0$, we recover $L_\xi\eta = -(2\alpha + \psi)\eta$. Finally, using (10) in (14c) we get $Q^{-1}L_\xi Q + 2\psi - 2\alpha)q_a + w_a = 0$, which when substituted in (13) recovers $L_\xi q_a = (2\alpha - \psi)q_a$.

Remark. Theorem 3 is a generalization of a recent work by Coley and Tupper[5] on symmetry inheritance of <u>*special*</u> CKV which is also an RCV (i.e., $\alpha = 0$).

Equations of State. Assume a spacetime satisfying Einstein's field equations (5) which admits an RIV ξ^a. Let ξ^a be also a CKV so that theorem 3 holds. For a CKV ξ^a satisfying (8), following dynamic result is known[7].

$$\left[(T^{ab} - \frac{1}{2}Tg^{ab})\xi_b\right]_{:a} = -3\Box\psi \tag{16}$$

The equation (16) is consistent with the Eq.(6) if ξ^a is also an RIV as it follows from (10) that $3\Box\psi = -\alpha r$. Thus, under the hypothesis of theorem 3, the equation (6) is the basis for generating new equations of state for a variety of fluid spacetimes. To illustrate this point, consider a special case of perfect fluid ($\sigma_{ab} = 0$ and $q_a = 0$) with $\xi^a u_a = 0$. Then, using this in (6) and substituting (12a) provides

$$\psi(\mu - p) = -2\alpha p \tag{17}$$

If ξ^a is parallel to u^a, proceeding as above, we get

$$\psi(\mu + 3p) = -2\alpha\mu \tag{18}$$

The equation (18) also holds for an imperfect fluid with $q_a = 0$. For physically meaningful equations of state we need a relation between α and ψ which can come from the Eq.(10) for a suitable prescription of $\Box\psi$. As a special case, we consider eigenfunction equation

$$\Box\psi = -k\psi; \qquad k = constant. \tag{19}$$

Using (19) in (10), we obtain

$$\alpha = \frac{3k\psi}{\mu - 3p}, \qquad r = \mu - 3p \neq 0 \tag{20}$$

Substituting (20) in (17) and (18) provides the following equations of state:

$$\mu^2 + 3p^2 - 4\mu p + 6kp = 0 \quad : \qquad \xi^a u_a = 0 \tag{21}$$

$$\mu^2 - 9p^2 + 6k\mu = 0 \quad ; \qquad \xi^a 11 u^a \tag{22}$$

where the range of k is subject to satisfying energy conditions.

Remark. Observe that for $k = 0$, $\Box\psi = 0$ and we recover the known[7] stiff ($u = p, \xi^a u_a = 0$) and unphysical ($u + 3p = 0$, $\xi^a 11 u^a$) equations of state. For this special case ($k = 0$), $\alpha = 0$ and ξ^a reduces to an RCV. In conclusion, we have shown that a proper RIV ($\alpha \neq 0$) can generate new and physically relevant solutions for a variety of fluid spacetimes

References.

(1) D.R. Oliver and W.R. Davis, Gen. Relat. Gravit.8, (1977)905.

(2) C.D. Collinson, Gen. Relat. Gravit. 1, (1970)137.

(3) K. Yano, The theory of Lie derivatives and its applications (North-Holland, Amsterdam, 1955).

(4) G.H. Katzin, J. Levine and W.R. Davis, J. Math. Phys.10 (1969)617.

(5) A.A. Coley and B.O.J. Tupper, J. Math. Phys.30 (1989)2616.

(6) R. Maartens and S.D. Maharaj, Class. Quantum. Grav.3 (1985)1005.

(7) R. Maartens, D.P. Mason and M. Tsamparlis, J. Math. Phys.27 (1986)2987.

DETERMINATION OF DARK MATTER FROM GRAVITATIONAL LENSING

Rachel L. Webster
Canadian Institute for Theoretical Astrophysics,
60 St. George St., Toronto, Ontario, M5S 1A1, Canada.

ABSTRACT

The use of gravitational lensing to determine the constituents of dark matter, the large scale distribution of dark matter and its contribution to Ω, is reviewed.

1. Introduction

Since Zwicky[1] first used the velocity dispersion in the Coma Cluster of galaxies to deduce the mass of that cluster, there has been evidence to suggest that not all matter in the universe emits observable radiation, and that a significant fraction of the total mass might therefore be dark. Dark matter (DM) will be defined as matter whose existence is known only through its gravitational effects. In this review, a fundamentally new technique for detecting DM, viz. gravitational lensing, will be described. Although the gravitational lensing effect has been known for many decades, it is only in the last 10 years, since the effect was first observed, that serious consideration has been given to utilising it to measure mass concentrations and their distribution.

Gravitational lensing provides an alternative method for the observational cosmologist to investigate the distribution of matter. In particular, since one is unravelling the effects of gravitational potentials along the line-of-sight to a background object, dark matter concentrations containing little luminous matter might be detected, if they exist. The mass scales probed are $\gtrsim \theta^2_{source} D_{os}$, where θ_{source} is the angular size of the source and D_{os} is the angular diameter distance from the observer to the source, and range from $\sim 10^{-2} M_\odot$ when the background sources are quasars to $\gtrsim 10^{12} M_\odot$ when the background sources are galaxies. The primary limitations of using gravitational lensing to probe the mass distribution, are (i) only the two-dimensional mass distribution can be determined, and (ii) lensing effects are only observable for surface densities which have $\Sigma \gtrsim 0.1\Sigma_{crit}$. Σ_{crit} is the minimal surface density required before the observer 'sees' two images of a background source; in other words it is the surface mass density required to focas the light rays at the observer. Algebraically

$$\Sigma_{crit} = \frac{c^2}{4\pi G} \frac{D_{os}}{D_{ol} D_{ls}} \sim 1\, gm\, cm^{-2}$$

where the distances are angular diameter distances and the subscripts refer to observer, lens and source.

The effects of gravitational lensing may be characterised by five types of observation:
(i) multiple-imaging of background sources, where the multiple images are resolvable. There are about a dozen known instances of multiply imaged quasars and galaxies[2]. Some putative cases of multiple imaging, where only two similar quasar images are seen, have yet to be confirmed as cases of gravitational lensing . Confirmation might come, for example, from measuring the timedelay between the two images[3].
(ii) arcs and rings of background galaxies. There are four known radio rings and perhaps ten instances where rich clusters of galaxies have been seen to produce arcs or rings of background galaxies[2].
(iii) microlensing variability, where one might observe a characteristic variation in the lightcurve of a quasar, and thereby deduce lensing by foreground stars. Instances of microlensing have only been observed in quasars which are already gravitationally lensed on resolvable scales[4].
(iv) statistical associations, for example where foreground galaxies are shown to be statistically associated with background quasars on the sky[5]. The most natural explanation of the associations is that the background objects are being gravitationally lensed, magnified and therefore preferentially included in magnitude-limited samples. Another example of statistical associations might be the observation of correlated ellipticities in background objects, such as galaxies, due to foreground mass concentrations[6].
(iv) luminosity bias, where a source is differentially gravitationally lensed, thereby producing a population of bright objects which might look very different from the parent population. For example it has been conjectured that some BL Lacs may be gravitationally lensed quasars[7] whose central continuum region has been greatly magnified, but whose more extended emission region has not.

In the first section of this review, the use of gravitational lensing to measure the free parameters in FRW models of the universe will be discussed, since ultimately the validity and scale of the cosmological model have a significant impact on the measurement of DM. In the sections which follow, specific techniques for the determination of DM constituents, the distribution of DM and the contribution of DM to Ω, the ratio of the mean mass density in the universe to the critical density, will be presented.

2. Determination of the Cosmological Model

Cosmological examples of gravitational lensing are a poor test of General Relativity; indeed the observable effects of gravitational lensing might be seen in a universe governed by Newtonian gravity. The bending of light predicted by Newtonian gravity is half that predicted by General Relativity; however the sense of the effect is the same. Therefore unless the masses of the deflecting objects are known to better than a factor of two, one cannot distinguish between these two models of gravity. In the case of light rays being bent in the gravitational field of the sun however, where we do know the mass accurately, we are able to differentiate between the two models of gravity. In a strong gravitational field the gravitational lensing differences between the two models are more marked; however the weak field limit of General Relativity is applicable to all observed instances of gravitational lensing .

Gravitational lensing has had modest success in determining or at least constraining the three parameters in the standard FRW model, Ω, Λ and H_0. Perhaps the most difficult of these parameters to determine from lensing will be H_0. The differential timedelay between two images comprises two components; one is the geometrical difference in the path lengths, and the second is the difference in the gravitational potential along the two light paths. Both terms are $\propto H_0$. There is only one instance of multiple imaging where the time delay has been measured, though there is still some controversy about the actual value obtained from the statistical analysis of the data. A timedelay of $\Delta t \sim 510\, days$ has been measured for 0957+561[8]. A simple calculation which models the galaxy as an isothermal lens, and the cluster of galaxies as a constant density sheet, predicts that

$$H_0 = 90 \pm 10 \left(\frac{\sigma}{390\, kms} \right)^2 \left(\frac{\Delta t}{1\, yr} \right)^{-1} km\, s^{-1}\, Mpc^{-1}$$

$$\sim 45\, km\, s^{-1}\, Mpc^{-1}$$

with the recently measured value[9] of $\sigma = 320 \pm 50\, km\, s^{-1}$. However if the mass model is seriously in error, and the contribution to the potential term in the timedelay has been overestimated, then the value of H_0 may have a large systematic error.

Turner and Fukugita[10] have shown that intersting limits can be set on $\lambda = \Lambda/3H_0^2$ since the predicted distribution of double images from gravitational lensing is rather sensitive to this parameter. For cosmologies with $k = 0$ and $\Omega + \lambda = 1$, using galaxies as lenses, these authors show that $\Omega = 1.0$ and $\lambda = 0$ is a good fit to the available data; however the available data is poor. High values of λ are almost certainly excluded, but a modest range of values is allowed.

A variety of techniques have been used to measure both $\Omega = \rho_0/\rho_{crit}$ and the total mass identified with a specific concentration of light. The most primitive of these techniques is the summation technique, whereby one simply adds all visible components, and then by assuming a value for the M/L (mass-to-light ratio), one can obtain the total mass. This technique is now rarely used, but will give a minimal estimate of mass[11]. The most commonly used technique is some form of an equilibrium equation, where the kinetic energy is estimated using a specific statistic and then the mass is derived from an assumed form of the gravitational potential, vis $M/R \propto V^2$. Not only does one need to know the form of the potential but also the structure of the velocity field. This technique is used on all length scales, from our galaxy to the radius of the visible universe. Applications of this technique require a population of luminous tracers, which react to the gravitational potential and the population is assumed to be uniformly distributed through the underlying mass distribution. It is suggested from these calculations that M/L is an increasing function of scale, when the regions being considered are centred on luminous concentrations[12]. Recent measurements on the largest scales suggest that $\Omega \sim 0.8 - 1.0$[13].

Completely independent techniques for determining Ω include nucleosynthesis, which gives an estimate for the mass density in baryons of $\Omega_b h_{100}^2 \sim 0.015 - 0.15$[14], where $h_{100} = H_0/100$, and inflation scenarios for the early universe which require that $\Omega = 1$[15]. In addition, classical kinematic tests can be used to measure q_0, although these techniques are bedevilled with the uncertainties associated with the evolution of 'standard candles'.

3. Constituents of Dark Matter

Measurements of microlensing events will provide the possibility of determining whether or not the constituents of DM are compact objects. There are several observational possibilities. Paczyński[16] has suggested that stars or other compact objects in the halo of the Milky Way might be detected since they would microlense background sources, such as stars in other galaxies like the Large Magellanic Cloud. The range of mass of objects in the halo which might be detected, is $10^{-6} - 10^{2} M_\odot$. Several groups are currently implementing observational programs which might detect such events; however because of the low probability of these events ($\sim 10^{-6}$), these progams will acquire a huge amount of data before making a positive detection. The analysis of this data is requiring new methodologies in the on-line processing, analysis and storage of astronomical data.

Microlensing will also provide a method for detecting the DM constituents of clusters of galaxies, if they are compact[17,18]. Recent measurements of microlensing events in the quasar 2237+0305[4] has enabled the size of the quasar continuum region to be estimated. The value obtained is $\sim 4 \times 10^{-4} h_{100}^{-1}\, pc$. If this source size is generic, and can be applied to other quasars, then in the case of 0957+561 it will be possible to place limits on the DM constituents of the cluster mass distribution. For this lens, the optical depth in cluster mass is ~ 0.5 at both image locations. The duration of a microlensing event by a solar mass object is

$$\Delta t_{M_\odot} \sim 30 \left(\frac{500\, km\, s^{-1}}{v_t} \right) yrs$$

where v_t is the transverse velocity of the source relative to the observer-deflector line. Currently ~ 10 years of observations have been made. It is certain that a microlensing event during this period would have been observed, since the light curve has been sampled frequently. Eventually microlensing events due to the stars in the galaxy must also be observed. If microlensing from the cluster DM is not observed, then $M_{DM} \lesssim 10^{-1} M_\odot$ for clusters of galaxies.

4. Large-scale Distribution of DM

If a background luminous source is seen along a light path which is near a massive object, the bundle of light rays from the source will be sheared, and an ellipticity will be induced in the image, with the principle axis perpendicular to the impact parameter to the deflecting mass. A large-scale mass will therefore induce correlations in the ellipticity of the background sources. A number of authors have calculated these effects for various cosmolgical distributions of matter. For example, for the CDM model of the universe, a root-mean-square distortion of a few percent is predicted over scales of about one degree[5,19,20]. An upper limit on the measured correlation in faint galaxy ellipticities will put an upper limit on the amplitude of density fluctuations in the universe. However interesting limits can probably only be obtained by observations which are beyond our current technical abilities.

Another possibility which has been considered, is the use of luminous arcs to constrain the DM distribution in clusters of galaxies. A typical observation of these arcs shows a rich cluster of galaxies at a redshift of a few tenths, with an arc subtending up to $60\,degrees$ around the centre of the cluster. The arcs are the gravitationally lensed images of faint background galaxies which partially cover the caustic lines in the source plane. Using Birkoff's theorem, and the radius of the arc, one can obtain a good estimate of the mass interior to the arc. Essentially the arc is tracing out the contour enclosing that part of the cluster where the mean surface density is critical. A critical line in the image plane corresponds to a caustic in the source plane, the latter being the locations whose images appear (formally) to be infinitely magnified.

The width of the arc can be used to determine the slope of the surface mass distribution at the location of the arc. Hammer[21] has derived a simple relationship between the arc thickness (θ_{arc}) and the source size (θ_{source}) as,

$$\theta_{arc}/\theta_{source} = 0.5/(1-\kappa) + O(A^{-1})$$

where κ is a measure of the mean surface density at the arc and A is the amplification. For a fixed source size, the more concentrated the deflecting mass distribution, the smaller κ at the arc position and the thinner the arc. A point mass or a mass distribution with a de Vaucoleurs profile will produce arcs which are half as thick as those produced by an isothermal mass distribution, and a quarter as thick as those produce with an isothermal mass distribution which has a large core radius. Observationally, the thickness of most arcs is unresolved ($\lesssim 0.5\,arcsec$), but there are a few exceptions. The latter predict small core radii of $\lesssim 30\,arcsecs$, while X-ray determinations of the core radii of clusters give $60-100\,arcsecs$ and the distribution of luminous galaxies typically gives $100-250\,arcsecs$. If one were able to confidently predict the size of the source galaxies, for example as 'normal' spiral galaxies, then the DM in clusters is more concentrated than either the X-ray or optical light distributions.

5. Determination of Ω

Gravitational lensing provides a direct method for the determination of Ω_{clump} where only matter in clumps with $\Sigma \gtrsim \Sigma_{crit}$ and $\theta_E \gtrsim \theta_{source}$ can be detected, where θ_E is the radius of the Einstein ring associated with the deflecting mass. Many authors have used gravitational lensing to place limits on Ω_{clump}. The standard technique is to use the fraction of multiple images observed to set a limit on $\Omega_{compact}^{22-26}$. In general, these calculations are confined to the regimes where multiple images of background sources are observable, which at optical wavelengths is about an arcsecond, but may be sub-arcsecond at radio wavelengths. The primary limitation of these calculations is that they are only sensitive to masses corresponding to relatively large angular scales vis $\sim 10^{10} M_\odot$. All published results are summarised in Table 1. A more sophisticated proposal is to look for variations in the emission lines and continua of quasars, which might be caused by microlensing. For example, the root-mean-square scatter in the ratio of the line flux to the optical or X-ray continua would be expected to increase as a function of redshift if gravitational lensing were important[27].

Table 1. Limits on $\Omega_{compact}$ from different gravitational lensing studies

$\Omega_{compact}$	$M_L/M_\odot$	Reference
< 0.1	$0.1 - 200$	Canizares[27]
< 1.0	$200 - 10^5$	Canizares[27]
< 0.4	$10^{11} - 10^{13}$	Hewitt *et al.*[24]
< 1.0	$10^{9.9}$	Neimroff[25]
< 0.25	$10^{10.3}$	Neimroff[25]
< 0.4	$10^7 - 10^9$	Kassiaola *et al.*[26]

If we are able to determine the fraction of quasars which are gravitationally lensed, then we can set a limit on $\Omega_{compact}$. It is by no means easy to determine this fraction. We already know that only a fraction of the percent of quasars are multiply imaged on arcsecond angular scales; however many more might be lensed on smaller scales by objects such as microlenses, and we would have no independent way of detecting them. Observations of statistical associations, such as quasar-galaxy associations provide another measure of the number of quasars which might be effected by gravitational lensing. Again though, this observation provides a lower limit on the number of quasars which might be effected by gravitational lensing.

Webster[28] has considered the effect that gravitational lensing might have on the shape of the quasar luminosity function. We know that the observed quasar luminosity function is the convolution of the intrinsic quasar luminosity function and the amplification function. From the physics of gravitational lensing it is possible to construct a generic amplification function. It is straightforward to show that when the intrinsic quasar luminosity function is fairly flat, that the observed quasar luminosity function has a similar slope. However when the instrinsic slope is steeper, then it might be considerably changed by gravitational lensing . Ideally one would deconvolve the observed luminosity function, to obtain the intrinsic one, however this process is difficult when the full form of the observed luminosity function is not known. Thus a more tractable method is to (iteratively) convolve a 'best guess', until the observed luminosity function is obtained. The calculation of both the observed quasar luminosity function and the amplification function must be performed with the same cosmological model.

For faint quasars, the observed luminosity function is fairly flat. Therefore we know that the shape of the intrinsic luminosity function must also be fairly flat, though the normalisation may be slightly changed. The bright end of the quasar luminosity function, on the other hand is quite steep, steep enough that it could be seriously effected by gravitational lensing .

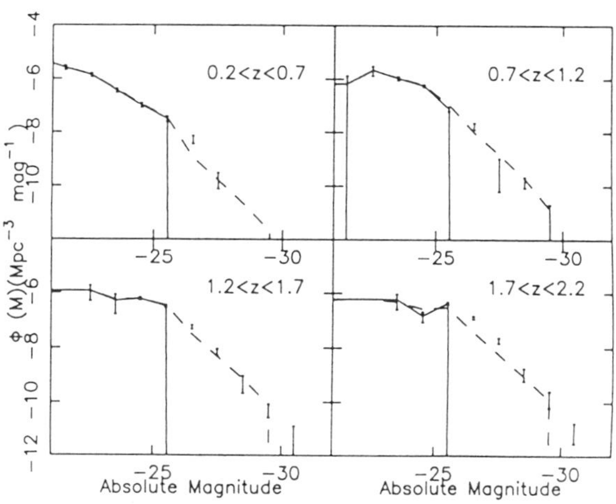

Fig. 1. The effects of gravitational lensing on the bright end of the quasar luminosity function are shown for a population of objects with $\Omega_{compact} = 1$. Each plot shows a different redshift bin. The intrinsic quasar luminosity function (solid line) has been truncated at $M_B = -26$. The resultant quasar luminosity function after lensing is shown as a dashed line, and the observed quasar luminosity function as the data points. Small changes in the resultant quasar luminosity function an be introduced by changing the parameters of the amplification function. For these plots, $H_0 = 50\,\mathrm{km\,s^{-1}\,Mpc^{-1}}$ and $\Omega = 1$.

It is interesting to ask whether or not the entire bright end of the quasar luminosity function might be due to gravitational lensing . Figure 1 shows the effect of lensing by a population of objects with $\Omega_{compact} = 1$. Clearly it is possible to create the entire bright end by gravitational lensing . Is there any evidence that the entire bright bright end of the quasar luminosity function is due to gravitational lensing ? The studies of quasar-galaxy associations by Webster *et al.*[29], show that more than 50% of bright quasars have a nearby galaxy in excess of those expected by random chance. If the convolution calculation is repeated, one is able to show that $\Omega_{compact} \sim 0.1$ is required to obtain this result. Thus in order to obtain the observed number of quasar-galaxy associations, we require that $\Omega_{compact} \gtrsim 0.1$ in objects with $M \gtrsim 10^{-1} M_{\odot}$.

6. Conclusions

As a tool in observational cosmology, gravitational lenses are still in their infancy. It is likely that gravitational lensing will provide one of the most signifcant probes of DM. The preliminary results we have already obtained suggest that $\Omega_{clump} \gtrsim 0.1$. It seems that DM in clusters may be quite centrally concentrated, and there is a weak suggestion that DM in clusters may not be clumped on scales $\gtrsim 10^{-1} M_{\odot}$.

7. Acknowledgements

RLW thanks NSERC for financial support.

8. References

1. F. Zwicky, *Helv. Phys. Acta*, **6** (1933) 110.
2. B. Fort,in *A Toulouse Workshop on Gravitational Lensing*, eds. Y. Mellier, G. Soucail, B. Fort, (Springer-Verlag, 1990), p221.
3. R. L. Webster and M. J. Fitchett, *Nature*, **324** (1986) 617.
4. R. T. Corrigan *et al.*, *A.J.* , **102** (1991) 34.
5. R. L. Webster, P. C. Hewett, M. E. Harding and G. A. Wegner, *Nature*, **336** 358.
6. R. D. Blandford, A. B. Saust, T. G. Brainerd and J. V. Villumson, *M.N.R.A.S.*, **251** (1991) 600.
7. J. P. Ostriker and M. Vietri, *Nature*, **344** (1990) 45.
8. D. H. Roberts, J. Lehár, J. N. Hewitt and B. F. Burke *Nature*, **352** (1991) 43.
9. G. Rhee, *Nature*, **350** (1991) 211.
10. M. Fukugita and E. L. Turner, *M.N.R.A.S.*, **253** (1991) 99.

11. P. J. E. Peebles, *Physical Cosmology*, Princeton Univerity Press, Princeton, N. J. (1971).
12. V. Trimble, *Ann. Rev. Astron. Astrophys.*, **25** (1987) 425.
13. N. Kaiser, G. Efstathiou, W. Saunders, A. Lawrence, M. Rowan-Robinson, R. S. Ellis and C. S. Frenk, *M.N.R.A.S.*, (1991) in press.
14. A. M. Boesgaard and G. Steigman, *Ann. Rev. Astron. Astrophys.*, **23** 319.
15. A. H. Guth, *Ann.NY Acad.Sci.*, **422** (1984) 1.
16. B. Paczyński, *Ap. J. (Letters)*, **308** (1986) L43.
17. J. R. Gott, *Ap. J.*, **243** (1981) 140.
18. R. L. Webster, A. N. M. Ferguson, R. T. Corrigan and M. J. Irwin, *A.J.* , (1991) in press.
19. J. Miralda-Escudé, *Ap. J.*, **380** (1991) 1.
20. N. Kaiser, (1991) preprint.
21. F. Hammer, (1991) preprint.
22. W. H. Press and J. E. Gunn, *Ap. J.*, **185** (1973) 397.
23. R. D. Blandford and M. Jaroszynski, *Ap. J.*, **246** (1981) 1.
24. J. N. Hewitt, B. F. Burke, E. L. Turner, D. P. Schneider, C. R. Lawrence, G. I. Langston and J. P. Brody, in *Gravitational Lenses*, ed. J. M. Moran, J. N. Hewitt and K. Y. Lo (Springer-Verlag 1989), p147.
25. R. J. Neimroff, *Phys. Rev. Lett.*, **66** (1991) 538.
26. A. Kassiola, I. Kovner and R. D. Blandford, *Ap. J.*, **381** (1991) 6.
27. C. R. Canizares, *Ap. J.*, **263** (1982) 508.
28. R. L. Webster, *A.J.* , submitted.
29. R. L. Webster, M. J. Drinkwater and P. A .Thomas, *A.J.* , submitted.

QUANTUM MECHANICS FROM THE PERSPECTIVE OF QUANTUM COSMOLOGY

JAMES B. HARTLE
Department of Physics, University of California
Santa Barbara, CA 93106

ABSTRACT

In the search for a theory of the initial condition of the universe, quantum mechanics must be applied to the universe as a whole. For this the "Copenhagen interpretations" of quantum mechanics are insufficiently general. Characteristically these interpretations assumed that there was a division of the universe into "observers" and "observed", that the outcomes of "measurements" were the primary focus of scientific statement, and, in effect, posited the existence outside of quantum mechanics of the quasiclassical domain of familiar experience. However, in a theory of the whole thing there can be no fundamental division into observers and observed. Measurements and observers cannot be fundamental notions in a theory that seeks to describe the early universe where neither existed. In a unified theory of cosmology there is no fundamental basis for a separate classical physics. Copenhagen quantum mechanics must, therefore, be generalized to apply to cosmology.

This talk sketched a quantum mechanics for closed systems adequate for cosmology developed in joint work with Murray Gell-Mann.[1,2,3] This framework is an extension and clarification of that of Everett[4] and builds on several aspects of the post-Everett development. It builds especially the work of Zeh[5], Zurek[6], Joos and Zeh[7], and others on the interactions of quantum system with the larger universe and on the ideas of Griffiths[9], Omnès[10], and others on the requirements for consistent probabilities for histories.

Three forms of information are necessary for prediction in the quantum mechanics of a closed system. In an approximation in which quantum spacetime is ignored, these are the Hamiltonian, the initial quantum state of the universe, and the information specifying particular histories. The most general objective of quantum theory is the prediction of the probabilities of the individual histories in a set of alternative histories for the universe. However, the characteristic feature of a quantum theory is that not every set of histories that may be described can be assigned probabilities because of quantum interference. Probabilities can be assigned only to sets of histories for which there is negligible interference between the individual histories in the set as a consequence of the initial density matrix of the universe and the Hamiltonian governing its dynamics. Such sets of histories are said to decohere.

Histories described at an arbitrarily fine-grained level do not decohere; some coarse graining is necessary for decoherence. Coarse-graining was described in the talk and the decoherence functional that measures the level of decoherence for sets of alternative coarse-grained histories was introduced. Mechanisms for decoherence were reviewed in simple models. Habitual decoherence was argued to be widespread in the universe for coarse-grained histories defined by certain quasiclassical variables.

A quasiclassical domain is roughly a set of alternative coarse-grained histories that is as refined as possible consistent with decoherence and has individual branches that are defined by quantities that are similar from one time to the next correlated in time mostly according to classical deterministic laws. The problem of precisely defining quasiclassical domains was discussed. The question of whether or not the universe exhibits a quasiclassical domain like the one of familiar experience is a calculable one in quantum cosmology given a suitably precise definition, the Hamiltonian of the elementary particles and the initial density matrix of the universe. In particular, the variables that describe classical physics and the form of its phenomenological equations of motion should be derivable from that Hamiltonian and the initial condition.[10]

A measurement situation is one in which a variable becomes correlated with a quasiclassical operator of the "quasiclassical domain". The theory of measurements in quantum mechanics was discussed from this point of view. The recovery of the Copenhagen formulation of quantum mechanics as an approximation to the more general framework appropriate in measurement situations was described. An "observer" (or information gathering and utilizing system, IGUS) was treated as a complex adaptive system that evolves to utilize the relative predictability of a "quasiclassical domain".

The talk concluded that resolution of many of the problems of interpretation presented by quantum mechanics is not to be found within the theory *in general* but rather through an examination of the universe's initial condition and the emergent features that it, together with the Hamiltonian of the elementary particles, implies. Quantum mechanics may be best and most fundamentally understood in the context of quantum cosmology.

Acknowledgement

Preparation of this abstract was supported in part by the National Science Foundation under NSF grant PHY90-08502.

References

1. M. Gell-Mann and J.B. Hartle, "Quantum Mechanics in the Light of Quantum Cosmology" in *Complexity, Entropy and the Physics of Information, Santa Fe Institute Studies in the Sciences of Complexity*, Volume VIII, ed. W. Zurek (Addison Wesley, Reading, 1990) or in *Proceedings of the 3rd International Symposium on the Foundations of Quantum Mechanics in the Light of New Technology*, eds. S. Kobayashi, H. Ezawa, Y. Murayama, S. Nomura (Physical Society of Japan, Tokyo, 1991).
2. See, e.g. J.B. Hartle, "The Quantum Mechanics of Cosmology" in *Quantum Cosmology and Baby Universes: Proceedings of the 1989 Jerusalem Winter School*, eds. S. Coleman, J.B. Hartle, T. Piran, and S. Weinberg (World Scientific, Singapore, 1991).
3. M. Gell-Mann and J.B. Hartle, "Alternative Decohering Histories in Quantum Mechanics", to appear in the *Proceedings of the International Conference on High Energy Physics*, Singapore, August 1990.
4. H. Everett, *Rev. Mod. Phys.* **29**, 454, (1957).
5. H.D. Zeh, *Found. Phys.* **1**, 69, (1971).
6. W. Zurek, *Phys. Rev. D* **24**, 1516, (1981); *ibid.* **26**, 1862, (1982).
7. E. Joos and H.D. Zeh, *Zeit. Phys.* **B59**, 223, (1985).
8. R. Griffiths, *J. Stat. Phys.* **36**, 219, (1984).
9. R. Omnès, *J. Stat. Phys.* **53**, 893, (1988); *ibid.* **53**, 933, (1988); *ibid.* **53**, 957, (1988); *ibid.* **57**, 357, (1989).
10. M. Gell-Mann and J.B. Hartle, "Classical Equations for Quantum Systems". Preprint UCSBTH91-15. .

Energy Localization and the Kerr-Newman Metric

F. I. COOPERSTOCK and S. A. RICHARDSON

Department of Physics and Astronomy, University of Victoria,
P. O. Box 3055, Victoria, B.C., Canada V8W 3P6

ABSTRACT

The localization of energy in general relativity continues to be a subject of interest. Recently, Virbhadra showed that three different pseudotensors, to third order in the rotation parameter, give the same result for the localized Kerr-Newman field energy. We have extended the Virbhadra pseudotensor calculations to higher order. In addition, we present an exact expression for the Kerr-Newman energy localization. This expression is consistent with the pseudotensor results and the weak field limit, but differs from that of Cohen and de Felice, which does not have the correct weak field limit.

1. Introduction

In general relativity, efforts to formulate a meaningful, generally covariant, expression for energy density have proven inconclusive. The canonical energy-momentum, derived from variational formulations of general relativity, leads to expressions which are neither unique nor tensorial. Consequently, various authors, such as Komar[1] and Penrose[2], have developed more sophisticated approaches. Nevertheless, the canonical energy-momentum pseudotensors have been useful in the computation of integrated energies, provided suitable care is taken in their application. The metric must be asymptotically Minkowskian and all spatial integrals must include the entire field. This prevents any attempt, in general, to assign the energy to a particular distribution within the field.

The situation is different, however, if the field is static and spherically symmetric. In this special case, one can derive an energy distribution that is physically reasonable[3–5]. Specifically, the total energy $m(r)$ within a sphere of radius r is given by

$$m(r) = 4\pi \int_0^r T_0{}^0\, r^2 dr = \int T_0{}^0\, dV, \tag{1}$$

where $T_i{}^k$ is the source of the gravitational field. Note that in this presciption the energy is localized in regions where $T_0{}^0$ is non-vanishing.

It is natural to ask if there is any middle ground between the general and spherically symmetric cases. Consider the axially symmetric, Kerr family of metrics. These fields reduce to spherically symmetric solutions as their rotation parameter approaches zero. For small deviations from spherical symmetry, can one still select physically meaningful localizations of energy?

To this end, Virbhadra[6–7] expanded the Kerr-Newman metric to third order in the rotation parameter and calculated a number of the energy-momentum pseudotensors. He found that the energy complexes of Einstein, Tolman and Landau-Lifshitz all gave the same exterior energy value while that of Møller gave twice this value. In this paper, we extend the pseudotensor approximations to seventh order and develop an exact expression for the energy localization in the Kerr-Newman field.

2. Kerr-Newman Metric

The Kerr-Newman metric describes the exterior, axially symmetric, gravitational field of a charged, rotating object.* It belongs to the class of algebraically special metrics of the form

$$g_{ik} = \eta_{ik} - l_i l_k \tag{2}$$

where η_{ik} is the Minkowski metric and l_i is a null vector field. Two properties of these special solutions are particularily useful. First, the inverse metric components are easily obtained,

$$g^{ik} = \eta^{ik} + l^i l^k \quad \text{where} \quad l^i = g^{ik} l_k = \eta^{ik} l_k, \tag{3}$$

and, second, the metric determinant satisfies $\sqrt{-g} = 1$. In addition, provided the vector fields l_i vanish appropriately at spatial infinity, these metrics are clearly asymptotically Minkowskian.

Specifically, the Kerr-Newman line element in cartesian coordinates is[8,6],

$$ds^2 = d\eta^2 - \frac{(2mr - e^2)r^2}{r^4 + a^2 z^2}\left[dt + \frac{z\,dz}{r} + \frac{r(x\,dx + y\,dy) + a(x\,dy - y\,dx)}{r^2 + a^2}\right]^2, \tag{4}$$

where a is the angular momentum per unit mass and e is the charge of the source. The radial parameter r is related to the cartesian radius R by

$$r^4 - (R^2 - a^2)r^2 - a^2 z^2 = 0 \quad \text{with} \quad R^2 = x^2 + y^2 + z^2. \tag{5}$$

Note that the Kerr-Newman metric reduces to the uncharged Kerr solution, as $e \to 0$, or reduces to the spherical Reissner-Nordström solution, as $a \to 0$. If both a and e vanish then the Schwarzchild metric results, in Eddington-Finkelstein form.

3. Pseudotensor Expansions

The various canonical energy complexes lead to integral conservation laws of the form

$$m = \int_{\infty} (T_0{}^0 + t_0{}^0)\sqrt{-g}\,dV, \tag{6}$$

where $t_i{}^k$ is the particular pseudotensor. As mentioned, the integrands in Eq. 6 do not describe a generally covariant energy density. For the special case of the Kerr-Newman field, following Virbhadra, we examined four different pseudotensors. Using computer-aided algebra (Maple version 4.3), the metric was expanded in

---

*Geometrical units are used, $c = G = 1$. Latin indices run from 0-3. Metric signature is $+ - - -$.

powers of a/R to seventh order. Then Eq. 6 was evaluated for the region outside a sphere of radius R, i.e., the total energy $m(R)$ is given by

$$m(R) = m - E(R) \quad \text{where} \quad E(R) = \int_R^\infty (T_0{}^0 + t_0{}^0)\sqrt{-g}\, dV. \tag{7}$$

The results confirm the Virbhadra relationships. The complexes of Einstein, Tolman, and Landau and Lifshitz all agreed yielding

$$E(R) = \frac{e^2}{2R}\left[1 + \frac{2}{3}\left(\frac{a}{R}\right)^2 + \frac{3}{5}\left(\frac{a}{R}\right)^4 + \frac{4}{7}\left(\frac{a}{R}\right)^6\right], \tag{8}$$

while that of Møller was twice this value.

4. Energy Localizations

Several candidate energy expressions were examined in order to obtain an exact energy localization. Using Komar's conserved quantity, Cohen and de Felice[9] deduced a K-N mass localization,

$$m(r) = m - \frac{e^2}{2r} - \frac{e^2(r^2 + a^2)}{2ar^2}\arctan\frac{a}{r}, \tag{9}$$

using Boyer-Lindquist coordinates. Note that in the absence of a gravitational field, the Kerr-Newman energy must reduce to $e^2/2R$, the flat space electrostatic energy density. However, in the limit $a \to 0$, Eq. 9 approaches $m - e^2/r$. Since the radial coordinates r and R merge in this limit, the Cohen and de Felice expression fails to give the correct flat space limit.

We also considered the Tolman total energy integral[10],

$$m = \int_\infty (T_0{}^0 - T_\alpha{}^\alpha)\sqrt{-g}\, dV, \tag{10}$$

as a candidate for localization by interpreting the integrand as an energy density. However, it has been shown[11,12], for a variety of spherically symmetric distributions, that the Tolman integral extended over only a portion of the system, disagrees with the corresponding localized energy computed according to Eq. 1. In fact, for the exterior Reissner-Nordström field, where $T_i{}^k$ is traceless, Eq. 10 reduces to[11]

$$m(R) = m - 2\int_R^\infty T_0{}^0\, dV = m - e^2/R. \tag{11}$$

giving again the wrong flat space limit. In contrast, the integral of $T_0{}^0$ alone for $E(R)$ gives the correct limit. Consequently, Eq. 1 was selected as the means to an exact Kerr-Newman energy localization.

5. Exact Expression

We now compute the integral of $T_0{}^0$ over the volume dV for the exterior Kerr-Newman field. To simplify the calculation, the tensor $T_i{}^k$ is first determined using the original Kerr-Newman coordinates. The result is then transformed to the more complicated, asymptotically Minkowski coordinates, given in Eq. 4, and the integration is performed.

In original Kerr-Newman coordinates, the line element has the form[8,13],

$$\begin{aligned} ds^2 &= (1-\rho^{-2}M)\,du^2 + 2\,dr\,du + 2a\rho^{-2}M\sin^2\theta\,d\phi\,du \\ &\quad -\rho^2 d\theta^2 - 2a\sin^2\theta\,dr\,d\phi - \sin^2\theta(r^2+a^2+a^2\rho^{-2}M\sin^2\theta)\,d\phi^2. \end{aligned} \tag{12}$$

where

$$\rho^2 = r^2 + a^2\cos^2\theta \quad \text{and} \quad M = 2mr - e^2. \tag{13}$$

In these coordinates, the electromagnetic field tensor F is known[8],

$$\begin{aligned} F &= \frac{e}{\rho^4}[(r^2-a^2\cos^2\theta)\,dr\wedge du - 2a^2r\cos\theta\sin\theta\,d\theta\wedge du \\ &\quad -a\sin^2\theta(r^2-a^2\cos^2\theta)\,dr\wedge d\phi + 2ar(r^2+a^2)\cos\theta\sin\theta\,d\theta\wedge d\phi]. \end{aligned} \tag{14}$$

Recall that the energy-momentum tensor, in terms of the Maxwell tensor, is

$$T_i{}^k = \frac{1}{4\pi}F_i{}^m F_m{}^k + \frac{1}{16\pi}\delta_i{}^k F_{lm}F^{lm}. \tag{15}$$

Thus substitution of the Maxwell tensor and the Kerr-Newman metric into Eq. 15 yields the components of $T_i{}^k$. With help from Maple, one finds

$$T_i{}^k = \frac{e^2}{8\pi\rho^6}\begin{bmatrix} r^2+a^2+a^2\sin^2\theta & 0 & 0 & 2a \\ 0 & \rho^2 & 0 & 0 \\ 0 & 0 & -\rho^2 & 0 \\ -2a(r^2+a^2)\sin^2\theta & 0 & 0 & -(r^2+a^2+a^2\sin^2\theta) \end{bmatrix} \tag{16}$$

The energy-momentum tensor $T_i{}^k$ is now transformed to the cartesian coordinate system. The coordinates (u, r, θ, ϕ) are related to the cartesian set by

$$t = -u - r, \quad x + iy = (r - ia)e^{i\phi}\sin\theta, \quad \text{and} \quad z = r\cos\theta. \tag{17}$$

The result is

$$T_i{}^k = \frac{e^2}{8\pi\rho^6}\begin{bmatrix} R^2+a^2 & 2ay & -2ax & 0 \\ -2ay & -(R^2+a^2-2x^2) & 2xy & 2xz \\ 2ax & 2xy & -(R^2+a^2-2y^2) & 2yz \\ 0 & 2zx & 2yz & -(R^2-a^2-2z^2) \end{bmatrix} \tag{18}$$

where now

$$\rho^4 = (R^2-a^2)^2 + 4a^2z^2. \tag{19}$$

The integration region is the volume outside a sphere of radius R. Thus,

$$E(R) = 2\pi \int_R^\infty \int_0^\pi \frac{e^2}{8\pi} \frac{R^2 + a^2}{[(R^2 - a^2)^2 + 4a^2 R^2 \cos^2\theta)]^{3/2}} R^2 \sin\theta \, d\theta \, dR. \tag{20}$$

Completing the integral gives

$$E(R) = \frac{e^2}{8}\left(\frac{2R}{R^2 - a^2} + \frac{1}{a} \ln \frac{R + a}{R - a}\right), \qquad R > a, \tag{21}$$

which is our new exact expression for the Kerr-Newman energy distribution.

The approximate pseudotensor calculations were based upon expanding the Kerr-Newman metric in powers of a/R. Thus of particular interest is the power series expansion of Eq. 21. One finds

$$E(R) = \frac{e^2}{2R} \sum_{n=0}^{\infty} \frac{n+1}{2n+1} \left(\frac{a}{R}\right)^{2n}. \tag{22}$$

This series agrees with the previous expansions. That is, the Einstein, Tolman and Landau-Lifshitz pseudotensor expansions agree exactly while the Møller pseudotensor expansion is equal to twice Eq. 22.

6. Summary

In this paper, we investigated the energy distribution of the Kerr-Newman metric. Approximate distributions using different pseudotensors were evaluated to seventh order in the rotation parameter a. These results agree with an exact expression we derived by an integration of $T_0{}^0$ over euclidean volume dV as opposed to proper volume. This exact expression is also consistent with the weak field limit. This approach to energy localization is a generalization of the result of Misner and Sharp for the case of spherical symmetry.

References

1. A. Komar, *Phys. Rev.* **113**, 934 (1959).
2. R. Penrose, *Proc. R. Soc. London A* **381**, 53 (1982).
3. C. W. Misner and D. H. Sharp, *Phys. Rev.* **136**, 571 (1964).
4. F. I. Cooperstock and R. S. Sarracino, *J. Phys. A: Math. Gen.* **11**, 877 (1978).
5. E. Nahmad-Achar, *Mon. Not. R. Astr. Soc.* **228**, 51p (1987)
6. K. S. Virbhadra, *Phys. Rev. D* **41**, 1086 (1990).
7. K. S. Virbhadra, *Phys. Rev. D* **42**; 2919 (1990).
8. B. Carter, *Phys. Rev.* **174**, 1559 (1968).
9. J. M. Cohen and F. de Felice, *J. Math. Phys.* **25**, 992 (1984).
10. R. C. Tolman, *Phys. Rev.* **35**, 875 (1930).
11. F. I. Cooperstock, R. S. Sarracino and S. S. Bayin, *J. Phys. A: Math. Gen.* **14**, 181 (1981).
12. J. Devitt and P. S. Florides, *Gen. Rel. Grav.* **21**, 6 (1989).
13. E. T. Newman, et. al., *J. Math. Phys.* **6**, 918 (1965)

SINGULARITIES IN COLLIDING PLANE-WAVE SPACETIMES

D. A. KONKOWSKI
Department of Mathematics, United States Naval Academy
Annapolis, MD 21402, U.S.A.

T. M. HELLIWELL
Department of Physics, Harvey Mudd College
Claremont, CA 91711, U.S.A.

ABSTRACT

Singularities in colliding plane-wave spacetimes are studied. Colliding impulsive gravitational plane-wave spacetimes and colliding sandwich gravitational plane-wave spacetimes are shown to possess quasiregular singularities while colliding thick gravitational plane-wave spacetimes have nonscalar curvature singularities. It is argued that these singularities are generally unstable. Scalar and electromagnetic wave perturbations support this result. However, a special subset of electromagnetic modes does not lead us to predict a scalar curvature singularity. A quasiregular singularity is predicted instead, and it is confirmed in the colliding gravitational- and electromagnetic-wave spacetime of Chandrasekhar and Xanthopoulos.

1. Introduction

As the singularity theorems have shown, singular spacetimes are generic[1] . The nature of these singularities, however, is less well understood[1]. In this paper we shall examine the singularity structure of maximally extended colliding-plane gravitational wave spacetimes. We shall look at colliding impulsive waves[2], colliding sandwich waves[3], and colliding thick waves[4]. In all three cases, a scalar curvature singularity results after the waves collide[5,6]. However, in all three cases, singularities also form in other regions of the spacetime. We shall show that these singularities are not scalar curvature singularities. In the colliding impulsive wave spacetime and the colliding sandwich wave spacetime, they are quasiregular singularities and in the colliding thick gravitational wave spacetime they are nonscalar curvature singularities. We do not, however, expect these singularities to be stable to the addition of matter or fields to the spacetime. It is argued in the case of colliding impulsive waves that the behavior of the stress-energy tensor for most scalar and electromagnetic wave modes makes the quasiregular singularity unstable, converting it into a scalar curvature singularity. However, a special subset of electromagnetic modes does not lead to the divergence of stress-stress-energy scalars at the singularities. In the presence of such modes the singularity should remain quasiregular in an exact backreaction calculation, as confirmed in the colliding gravitational- and electromagnetic-wave spacetime of Chandrasekhar and Xanthopoulos[7].

2. Colliding impulsive waves

In 1971, Khan and Penrose[2] described the spacetime for two colliding impulsive (δ-function profile) gravitational waves. The global structure of this colliding wave spacetime was further considered by Matzner and Tipler[5] and Yurtsever[8].

In the Khan-Penrose spacetime, δ-function gravitational waves propagate into an initially flat region. A two-dimensional slice is diamond shaped. Region I is at the bottom, Regions II and III are on the sides, separated from region I by impulsive gravitational waves. All three regions are flat. Region IV is at the top, on the future of the collision of the waves. The metrics in regions I, II, and III are

$$\text{I:} \quad ds^2 = 2dudv - dx^2 - dy^2 \tag{1a}$$

$$\text{II:} \quad ds^2 = 2dudv - (1+u)^2 dx^2 - (1-u)^2 dy \tag{1b}$$

$$\text{III:} \quad ds^2 = 2dudv - (1+v)^2 dx^2 - (1-v)^2 dy^2 . \tag{1c}$$

Region IV is curved and is described by a more complicated metric.

Khan and Penrose[2] showed through an analysis of the Green function for the solution that a curvature singularity forms in the future of the collision at $u^2 + v^2 = 1$. No curvature singularity, however, forms at the points $u = 0, v = 1$ or $v = 0, u = 1$. If one evaluates the Riemann tensor in a PPON frame along timelike geodesics approaching these points, it is obviously always zero[9]. These points are nevertheless singular, since they possess no neighborhood which is isolated from the singularity[8]; they are therefore quasiregular singularities. As Matzner and Tipler[5] point out, the point $u = 0, v = 1$ is identified under the Killing symmetries with the entire surface $v = 1$, and similarly the surface $u = 1$ is identified with $u = 1, v = 0$. Therefore, these surfaces are also quasiregular singularities.

3. Colliding sandwich gravitational waves

In 1972, Szekeres[3] described the spacetime for two colliding sandwich gravitational waves. Regions I, II, and III are flat. Between regions II and I is region A, and between regions III and I is region B. Regions A and B are sandwich gravitational waves. Region IV is the curved spacetime resulting after the waves collide; it is bounded by a scalar curvature singularity. As in the colliding impulsive case only timelike geodesics from flat regions reach $u = 0, v = 1$ and $v = 0, u = 1$. The same arguments apply, so these are quasiregular singularities. Likewise, one can argue that these singular surfaces are identified with the entire $u = 1$ or $v = 1$ surfaces, which must therefore be quasiregular singularities.

4. Colliding thick gravitational waves

The spacetime which models the collision of two thick gravitational waves was given by Szekeres[4]. In this spacetime regions II and III are thick gravitational waves. Each is curved and there is of course a scalar curvature singularity to the future of the collision of the waves. However, the nature of the singularity in the plane-wave regions II and III is less obvious.

Consider region II. To determine the nature of the singularity at $u = 1$, we must evaluate the Riemann tensor in a PPON frame. In a simple x,y-constant PPON frame, the nonzero components of the Riemann tensor are

$$R_{0202} = R_{0212} = -R_{0303} = -R_{0313} = -R_{1313} = R_{1212} = \sqrt{6}c_1^2(1-u^4)^{-5/2} \tag{2}$$

where c_1 is a constant. Clearly as u approaches one, these components diverge and a curvature singularity results. It is straightforward to show that all 14 C^0 scalars in g_{ab}, η_{abcd}, and R_{abcd} are zero. The singularity in Region II at $u = 1$ (and similarly in region III at $v = 1$) is therefore a C^0 nonscalar curvature singularity.

5. Stability of the singularity structure

In the two cases of colliding impulsive waves and colliding sandwich waves, quasiregular singularities result. Are these mild, topological singularities stable? One indication as to their stability is given by the behavior of test waves. For definiteness we will only consider the colliding impulsive wave spacetime; an analysis in the colliding sandwich wave spacetime would be similar but more complicated mathematically.

Consider minimally coupled scalar x,y-independent waves. The wave equation $\Box\Phi=0$ is trivially solved to yield the solution $\Phi = f(v)$ as the waves approach the $u = 0$ gravitational wave. In region II the solution $F = f(v)(1-u^2)^{-1/2}$ matches the one in region I; it moves from $u = 0$ towards $u = 1$. The wave is focused and its amplitude diverges as u approaches one unless $f(v) = 0$. One can similarly show that the stress-energy tensor of the scalar wave generically diverges in both coordinate and PPON frames, and the stress-energy scalars T^a_a and $T_{ab}T^{ab}$ generically diverge as well.

Likewise, the stress-energy for an electromagnetic wave generically diverges in a coordinate frame. If one requires $T_{\mu\nu}T^{\mu\nu}$ to be finite and then calculates the stress-energy tensor in a PPON frame, one finds

$$T_{ab} = \left(a^1\right)^2 T_{00}\begin{pmatrix} 1 & 0 & 0 & 1 \\ 0 & 0 & 0 & 0 \\ 0 & 0 & 0 & 0 \\ 1 & 0 & 0 & 1 \end{pmatrix} \tag{3}$$

where a^1 is a nonzero constant, and

$$T_{00} = \frac{1}{4\pi}\left\{(1+u)^{-2}\left[\frac{c(v)}{(1-u)^{3/2}(1+u)^{1/2}} + 2(1+u)f_2 + (1+u)^2 f_2'\right]^2\right.$$

$$\left. +(1-u)^{-2}\left[\frac{-d(v)}{(1+u)^{3/2}(1-u)^{1/2}} - 2(1-u)f_3 + (1-u)^2 f_3'\right]\right\} \tag{4}$$

where c, d, f_2, f_3 are arbitrary functions. Thus generically $T_{ab} \to \infty$ as $u \to 1$ unless $c = d = 0$ and f_3 is chosen to make T_{00} finite.

Both T_{ab} and stress-energy scalars diverge as $u \to 1$ for generic scalar and electromagnetic fields. The colliding impulsive wave spacetime plus generic fields therefore "mimics" the behavior of a spacetime which reacts to the presence of the fields by converting the quasiregular singularity into a scalar curvature singularity. Although no back-reaction calculation has been carried out in the generic case, we expect that these fields will convert the quasiregular singularities at $u = 1$ in region II and at $v = 1$ in region III into scalar curvature singularities. There is no proof that this conversion takes place. However, in the few cases where a back-reaction calculation on a quasisingular spacetime has been made, the mimicking of scalar curvature, nonscalar curvature, and quasiregular singularities by the behavior of test fields is a completely reliable guide.[9-11]

6. An Exact Backreaction Solution

It is interesting that in a special case of the general electromagnetic field an exact back-reaction calculation has already been carried out by Chandrasekhar and Xanthopoulos[7]. Their spacetime features the collision between two plane impulsive gravitational waves, each supporting an electromagnetic shock wave. In our region II, their fields depend only upon u and not v, corresponding to waves moving in the v direction toward the boundary with the interaction region IV. These are also the waves for which $T_{\mu\nu}T^{\mu\nu}$ and T_{ab} are finite as $u \to 1$; they therefore mimic the behavior of a quasiregular singularity. Consistency with our conjecture then requires that the Chandrasekhar-Xanthopoulos spacetime retains the quasiregular singularities exhibited by the colliding impulsive wave spacetime at $u = 1$ in region II and at $v = 1$ in region III. And, in fact, this is the case.[12]

In region II the Chandrasekhar-Xanthopoulos metric is

$$ds^2 = -2dudv + X^2(u)(1+u)^2dx^2 + X^{-2}(u)(1-u)^2dy^2. \tag{5}$$

In the coordinate frame, the only nonzero components of the Riemann tensor are R^{v}_{xux} and R^{v}_{yuy} and others obtained from symmetries of $R_{\mu\nu\lambda\sigma}$. All components are finite as $u \to 1$. However, to discern the singularity type, it is necessary to evaluate R_{abcd} in PPON frames. In a PPON frame the nonzero components of R_{abcd} converge and are continuous as $u \to 1$. The $u = 1$ singularity in region II (and similarly the $v = 1$ singularity in region III) are therefore quasiregular as expected.

7. Acknowledgments

This research is partially supported by NSF Grant PHY-891160. The authors also wish to thank the Aspen Center for Physics, where some of this research was carried out.

8. References

1. G. F. R. Ellis and B. G. Schmidt, *Gen. Relativ. Gravit.* **8** (1977) 915.
2. K. A. Khan and R. Penrose, *Nature* **229** (1971) 185.
3. P. Szekeres, *J. Math. Phys.* **13** (1972) 286.
4. P. Szekeres, *Nature* **228** (1970) 1183
5. R. A. Matzner and F. J. Tipler, *Phys. Rev. D* **29** (1984) 1575.
6. F. J. Tipler, *Phys. Rev. D* **22** (1980) 2929.

7. S. Chandrasekhar and B. C. Xanthopoulos, *Proc. R. Soc. London* **A398** (1985) 223.
8. U. Yurtsever, *Phys. Rev. D* **37** (1988) 2803.
9. D. A. Konkowski and T. M. Helliwell, *Proc. 11th Int. Conf. on General Relativity and Gravitation*, ed B. Laurent and K. Rosquist (Stockholm: Int. Soc. on G. R. G., 1986) Abstracts p. 173.
10 D. A. Konkowski and T. M. Helliwell, *Phys. Rev. D* **31** (1985) 1195 .
11. D. A. Konkowski and T. M. Helliwell, *Class. Quantum Grav.* **6** (1989) 1847.
12. T. M. Helliwell and D. A. Konkowski, *Phys. Rev. D* **41** (1990) 2507.

NON-GAUSSIAN MICROWAVE BACKGROUND FLUCTUATIONS FROM NONLINEAR GRAVITATIONAL EFFECTS

D.S. Salopek
NASA/ Fermilab Astrophysics Center
P.O. Box 500 MS-209
Batavia, Illinois, USA 60510

After October 1, 1991
Department of Applied Mathematics and Theoretical Physics
Silver Street, Cambridge, England CB3 9EW

ABSTRACT

Whether the statistics of primordial fluctuations for structure formation are Gaussian or otherwise may be determined if the Cosmic Background Explorer satellite (COBE) makes a detection of the cosmic microwave-background temperature anisotropy $\Delta T_{CMB}/T_{CMB}$. Non-Gaussian fluctuations may be generated in the chaotic inflationary model if two scalar fields interact nonlinearly with gravity. Theoretical contour maps are calculated for the resulting Sachs-Wolfe temperature fluctuations at large angular scales ($> 3^0$). In the long-wavelength approximation, one can confidently determine the nonlinear evolution of quantum noise with gravity during the inflationary epoch because (1) different spatial points are no longer in causal contact, and (2) quantum gravity corrections are typically small— it is sufficient to model the system using classical random fields. If the potential for two scalar fields $V(\phi_1, \phi_2)$ possesses a sharp feature, then non-Gaussian fluctuations may arise. An explicit model is given where cold spots in $\Delta T_{CMB}/T_{CMB}$ maps are suppressed as compared to the Gaussian case. The fluctuations are essentially scale-invariant.

1. INTRODUCTION

With the decline of the Cold Dark Matter scenario, it is imperative that theorists propose alternative models which can be compared with the observations. Here, I describe a variation of the inflation model which yields non-Gaussian primordial fluctuations. This model may be tested in the near future if the Cosmic Background Explorer satellite[1] (COBE) measures the cosmic temperature anisotropy.

Even with the difficulties in accounting for large scale structure, it is still reasonable to retain the inflationary scenario. Redshifts of IRAS galaxies[2] and their inferred peculiar velocities indicate that the Universe is at critical density,[3,4] $\Omega = \rho/\rho_{crit} = 0.8 \pm 0.3$. This result gives support to inflation whose most outstanding prediction was that $\Omega = 1$. However, one must attempt to modify or improve the scenario so as to produce a set of initial conditions that are richer than scale-invariant Gaussian. Here, I will describe a chaotic inflation model that gives non-Gaussian fluctuations which are basically scale-invariant.[5,6]

There are three essential ingredients to the inflationary scenario. Firstly, a scalar field with potential $V(\phi)$ models the decay of the cosmological constant. Secondly, gravity is crucial in order to account for the expansion of the Universe. Finally, scalar field quantum fluctuations are necessary to produce inhomogeneities that will eventually produce structure in our Universe. One should view the inflation model as a microscope that magnifies quantum fluctuations at the smallest imaginable distance scales (less than the Planck length) to scales that are cosmologically observable. One of the problems with linear perturbation theory was that there was no short distance cutoff. This gave the illusion that one could extrapolate to arbitrarily small distances. However, nonlinearities must be important at some scale.

Non-Gaussian fluctuations would be the signature of nonlinearities[7] in the inflationary scenario. Their calculation is problematic because one requires a formalism that governs the evolution of quantum noise with gravity. Ideally, one needs a quantum theory of the gravitational field. In order to bypass this very severe difficulty, I will use three tricks:

(1) The classical nonlinear evolution of long-wavelength scalar fields and gravity is tractable.[8,] When the wavelength of a fluctuation exceeds the Hubble radius, different spatial points are no longer in causal contact, and they evolve as independent homogeneous universes. One may safely neglect second order spatial gradients in the action for scalar fields and gravity. Nonetheless, one must carefully join the independent spatial points to make one Universe.

(2) Long-wavelength quantum noise behaves classically, and it may be described using classical random fields in a process termed stochastic inflation.[10,8] Using the Wheeler-DeWitt equation, one may show that quantum gravity corrections are typically small.[8]

(3) The long-wavelength equations may be solved completely when the logarithm of the scalar field potential is linear,[9] $\ln V(\phi_j) = \sum_k a_k \phi_k$, where the a_k are constants. More complicated potentials may be approximated by joining various linear $\ln V$ potentials together. In this way, one may produce models that yield non-Gaussian fluctuations that are consistent with current microwave background anisotropy limits.

2. LONG-WAVELENGTH FIELDS: GENERALIZATION OF HOMOGENEOUS MINISUPERSPACE

The long-wavelength system is an elegant extension to inhomogeneous fields of homogeneous minisuperspace.[8,9] They enable one to construct non-Gaussian models for structure formation.

Given some initial conditions, I will outline how to solve for the classical long-wavelength evolution of scalar fields $\phi_j(t,x)$ with potential $V(\phi_j)$ interacting through gravity. It will be assumed that the metric has the form,

$$ds^2 = -N^2(t,x)dt^2 + e^{2\alpha(t,x)}((dx^1)^2 + (dx^2)^2 + (dx^3)^2), \tag{2.1}$$

which describes an isotropic Universe with inhomogeneous scale factor $e^{\alpha(t,x)}$. The lapse function N is determined when one decides the time hypersurface, although an explicit choice is not necessary. In what follows, $H(t,x) \equiv \dot{\alpha}/N$ is the Hubble parameter and $\pi^{\phi_i}(t,x) = e^{3\alpha}\dot{\phi}_j/N$ are the momentum densities of the scalar fields.

All second order spatial gradients in the Lagrangian of Einstein gravity with scalar fields will be neglected. It is necessary to retain first order spatial gradients otherwise one returns to homogeneous minisuperspace. The energy constraint,

$$H^2 = \frac{8\pi}{3m_{\mathcal{P}}^2}\Big[\frac{e^{-6\alpha}}{2}\sum_j {\pi^{\phi_j}}^2 + V(\phi_j)\Big], \tag{2.2a}$$

and the evolution equations are valid at each comoving spatial point, and they are the same as those for homogeneous flat cosmologies. The new ingredient is the momentum constraint, which joins together the independent spatial points to make one Universe:

$$H_{,i} = -\frac{4\pi}{m_{\mathcal{P}}^2}\, e^{-3\alpha}\pi^{\phi_j}\phi_{j,i}. \tag{2.2b}$$

The solution of this equation is familiar to those who study fluid mechanics. The Hubble parameter is assumed to be a function of the scalar fields, and the momentum densities are given by partial derivatives with respect to the scalar fields,

$$H \equiv H(\phi_j), \qquad \pi^{\phi_j} = -\frac{m_{\mathcal{P}}^2}{4\pi} e^{3\alpha}\frac{\partial H}{\partial \phi_j}. \tag{2.3}$$

When these are substituted into the energy constraint, one obtains the separated Hamilton-Jacobi equation,

$$H^2 = \frac{m_{\mathcal{P}}^2}{12\pi}\sum_j\Big(\frac{\partial H}{\partial \phi_j}\Big)^2 + \frac{8\pi V(\phi_j)}{3m_{\mathcal{P}}^2}. \tag{2.4}$$

This self-contained equation for the Hubble parameter governs the nonlinear dynamics of the long-wavelength gravitational system. It is covariant in that is does not refer either to the time hypersurface nor to the spatial coordinates.

In a significant improvement for calculations based on Hamilton-Jacobi methods,[9,5] I gave an complete analytic solution of the SHJE for two scalar fields (ϕ_1, ϕ_2) interacting with gravity through an exponential potential,[11]

$$V(\phi_1,\phi_2;p,\theta) = V_0 \exp\Big[-\sqrt{\frac{16\pi}{p}}\frac{(-\phi_1\sin\theta + \phi_2\cos\theta)}{m_{\mathcal{P}}}\Big]. \tag{2.6}$$

The coupling parameter p controls the steepness of the potential whereas θ is the angle that surfaces of uniform potential make with the ϕ_1 axis. The complete solution for the Hubble parameter depends on two arbitrary parameters, b and m,

$$H \equiv H(\phi_1,\phi_2;p,\theta;b,m), \tag{2.7}$$

and it is shown in Fig.(1). Surfaces of constant Hubble parameter are plotted as solid curves in Fig.(1) for the case $m = 1$, $\theta = 0$. The family of orthogonal lines are the physical trajectories. All solutions of the SHJE with potential (2.6) may be derived from this solution.

The complete solution (2.7) generates a transformation to new canonical variables, b, m, with conjugate momenta, π^b and π^m, given by differentiation of the Hubble parameter,

$$\pi^b = \frac{m_P^2}{4\pi} e^{3\alpha} \frac{\partial H}{\partial b}, \qquad \pi^m = \frac{m_P^2}{4\pi} e^{3\alpha} \frac{\partial H}{\partial m}. \tag{2.8}$$

The new canonical variables (b, m, π^b, π^m) are constants in time, although they may be spatially dependent. They completely govern the evolution of the system. However, they are not all independent because they are constrained through the new version of the momentum constraint,

$$0 = \pi^b b_{,i} + \pi^m m_{,i}, \tag{2.9}$$

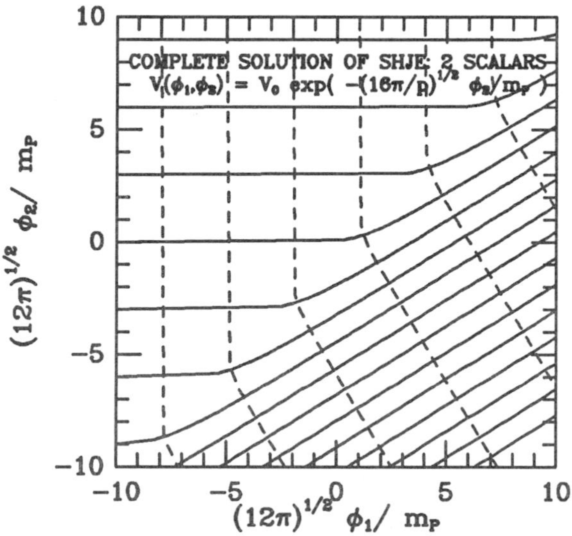

Fig.(1): The complete solution, $H(\phi_j; b, m; p, \theta)$, of the separated Hamilton-Jacobi equation is shown for two scalar fields interacting through a linear $\ln V$ potential, eq.(2.6). Here, the mixing angle θ vanishes, and the constant parameters are chosen to be $b = 0$ and $m = 1$. Surfaces of constant potential are just horizontal lines. The broken lines are trajectories of the fields which move up the page; they are orthogonal to the surfaces of constant Hubble parameter (solid lines).

For a single scalar field,[12,8] when the wavelength of the perturbation is larger than the Hubble radius, the system is characterized by a constant of integration, $\zeta \equiv \zeta(\alpha, \phi)$. It is the quantity of primary interest for adiabatic models of structure formation. For example, in the Cold Dark Matter Model, microwave background anisotropies at angular scales greater than $\sim 3^0$ are directly proportional to $-\zeta$,

$$\Delta T_{CMB}/T_{CMB} = -\zeta/15. \tag{2.10}$$

This is just the Sachs-Wolfe relation. For multiple scalar fields, eq.(2.10) may be taken as the definition of ζ, and using eqs.(2.7,2.8), one may write down an exact expression which is a function of the initial values of the fields,[9]

$$\zeta \equiv \zeta(\alpha_0, \phi_{j0}, \pi_0^{\phi_j}). \tag{2.11}$$

The initial conditions of the long-wavelength problem are generated from short wavelength quantum fluctuations that are assumed to begin in the ground state (Bunch-Davies vacuum). The fluctuations expand beyond the Hubble radius, and they become a part of the long-wavelength background.

3. NON-GAUSSIAN MODEL CALCULATIONS

One can obtain non-Gaussian fluctuations on cosmologically observable scales from a potential created by joining three linear $\ln V$ regions as shown in Fig.(2). It is assumed that our patch of the Universe began homogeneously in the lower half-plane, region 1, where the potential parameters are given by $p_1 = 20$, $\theta_1 = -50^0$. (For a justification of the homogeneous starting point, see Salopek and Bond.[8]) Short wavelength quantum noise then generates Gaussian initial conditions for the various fields. The broken curves in Fig.(2) depict the subsequent evolution of the scalar fields at several spatial points in a 64^3 lattice calculation. When they roll over the various interfaces in the potential, non-Gaussian fluctuations are generated.

The Hubble function, $H(\phi_j)$, in region 1 is taken to be the attractor solution,

$$H_{att}(\phi_j) = \sqrt{\frac{8\pi}{3m_{\mathcal{P}}^2}\frac{V_0}{1-1/(3p_1)}}\exp\left[-\sqrt{\frac{4\pi}{p_1}}\frac{(-\phi_1 \sin\theta_1 + \phi_2 \cos\theta_1)}{m_{\mathcal{P}}}\right] \tag{3.1}$$

corresponding to $b = -\infty$ and $m = 0$ in (2.7) having homogeneous values. The new momentum constraint (2.9) is then satisfied at early times, and the evolution equations guarantee that it will be satisfied at late times. In region 1, the fields then evolve in time α according to,

$$\phi_1(x, \alpha) = -\frac{m_{\mathcal{P}}}{\sqrt{4\pi p_1}}\alpha \sin\theta_1 + \phi_{10}(x), \quad \phi_2(x, \alpha) = \frac{m_{\mathcal{P}}}{\sqrt{4\pi p_1}}\alpha \cos\theta_1 + \phi_{20}(x), \tag{3.2a}$$

$$\pi^{\phi_1}(x, \alpha) = -\frac{m_{\mathcal{P}}}{\sqrt{4\pi p_1}}e^{3\alpha}H_{att}(\phi_j)\sin\theta_1, \quad \pi^{\phi_2}(x, \alpha) = \frac{m_{\mathcal{P}}}{\sqrt{4\pi p_1}}e^{3\alpha}H_{att}(\phi_j)\cos\theta_1. \tag{3.2b}$$

The initial values of the scalar fields, $\phi_{i0}(x)$, are classical Gaussian random fields with power spectrum

$$\mathcal{P}_{\phi_{i0}}(k) \equiv \frac{k^3}{2\pi^2} < |\phi_{i0}(k)|^2 >= \left(\frac{H_0}{2\pi}\right)^2 \left(\frac{k}{H_0 e^{\alpha_0}}\right)^{-2/(p_1-1)},$$

whose amplitude is determined by the value of the Hubble parameter H_0 when the lattice size exceeded the Hubble radius. Here, $H_0 = 10^{-6} m_P$ is chosen to give results consistent with current temperature anisotropy limits. The amplitudes of the homogeneous $k = 0$ Fourier modes are arbitrary.

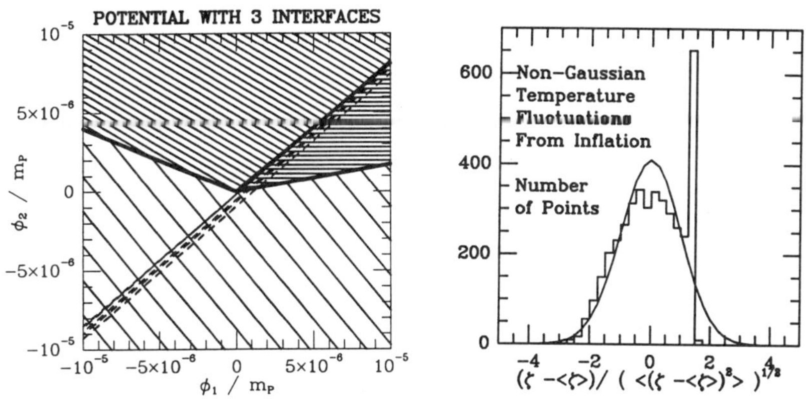

Fig.(2): (a) Non-Gaussian fluctuations consistent with CMB limits may be generated when the scalar fields pass over several interfaces in their potential. The light solid curves are lines of uniform potential, whereas the heavy lines are the interfaces which continuously join the 3 regions with linear $\ln V(\phi_1, \phi_2)$. If the scalar field trajectories (broken lines) beginning at the bottom of the diagram pass sufficiently near the origin, then nonlinear effects at long wavelengths become important. For a 64^3 lattice simulation, the histogram of the resulting fluctuations in ζ are shown in Fig.(b). The distribution of microwave background fluctuations at large angular scales is found by reflecting the distribution about the vertical axis: $\Delta T_{CMB}/T_{CMB} = -\zeta/15$. For comparison, a Gaussian distribution (smooth curve) with the same mean and dispersion as the histogram is also shown.

If the trajectories pass into the upper right hand area, region 2 ($\theta_2 = 0^0$), they receive an upward kick from the potential, which later forces them into region 3 ($\theta_3 = -30^0$). (If this diagram were extended, one would find that trajectories actually cross into region 3.) The angles of the interfaces starting with the lower right and proceeding counterclockwise are, $\chi_{12} = 10^0$, $\chi_{23} = 39^0$ and $\chi_{31} = 158^0$. The resulting distribution of ζ is plotted in Fig.(2b). For the parameters shown, it was found that non-Gaussian fluctuations can arise if the fields passed sufficiently near the origin, which can be arranged through the arbitrary choice of the homogeneous mode amplitudes in eq.(3.2). The calculations are performed using the analytic expression for ζ eq.(2.11).

A 2-D slice of the 64^3 lattice simulation is shown in Fig.(3). Large positive excursions of ζ are heavily suppressed (*i.e.* negative excursions of $\Delta T_{CMB}/T_{CMB}$ are suppressed). With some further analysis, this plot could be interpreted as a large angle microwave background map. The power spectrum for the full 3-D simulation is given in Fig.(4). It is essentially flat.

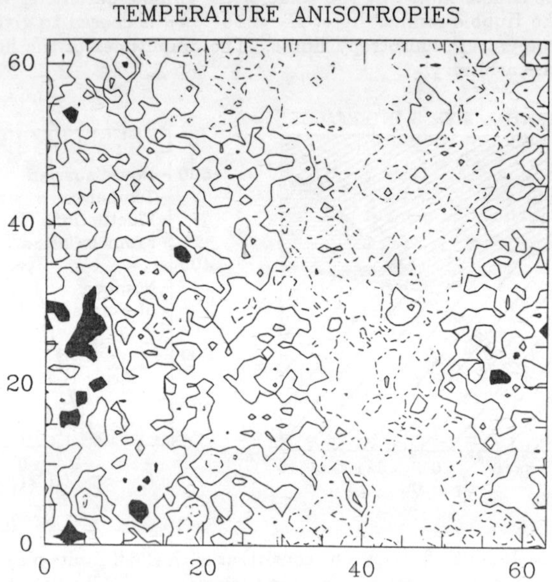

Fig.(3): Sachs-Wolfe temperature fluctuations are shown on a 2-D planar slice for the non-Gaussian model considered in Fig.(2). The shaded areas are 2σ deviations from the mean for $\Delta T_{CMB}/T_{CMB}$. The solid lines correspond to $1, 0\sigma$ contours, whereas the broken lines are -1σ deviations. The most significant feature of this plot is that cold spots in the temperature anisotropy are suppressed over the usual Cold Dark Matter model with Gaussian primordial fluctuations. In fact, there are no -2σ fluctuations of $\Delta T_{CMB}/T_{CMB}$ in this 64^3 lattice calculation, a result which is expected from the distribution function of Fig.(2b).

4. SUMMARY AND DISCUSSION

The distribution of temperature anisotropies could serve as a valuable discriminator of various models of the early Universe. For example, it could indicate that nonlinearities in inflation model were important. It could even determine what was the initial quantum state of the Universe (see, for example, Hartle[13]).

The resulting structure formation scenario for the proposed non-Gaussian model can be described qualitatively. Since the nonlinearities did not change the shape of the primordial fluctuation spectrum in the Newtonian potential (see Fig.(4)), it is natural to assume (at least in the first approximation) that the normalization of the spectrum using the two-point correlation function of galaxies would be the same as the standard Cold Dark Matter model[14] with Gaussian fluctuations. For example, the variance of temperature fluctuations $<(\Delta T_{CMB}/T_{CMB})^2>$ would be the same. (In fact, the initial value of the Hubble parameter, $H_0 = 10^{-6} m_{\mathcal{P}}$, given in Sec. 4 was chosen for this reason.) Differences from standard CDM would appear in the distribution of the fluctuations. Cold spots in the temperature fluctuations would be suppressed as shown in Fig.(3). In addition, I expect that high positive density excursions in the density field will be suppressed.

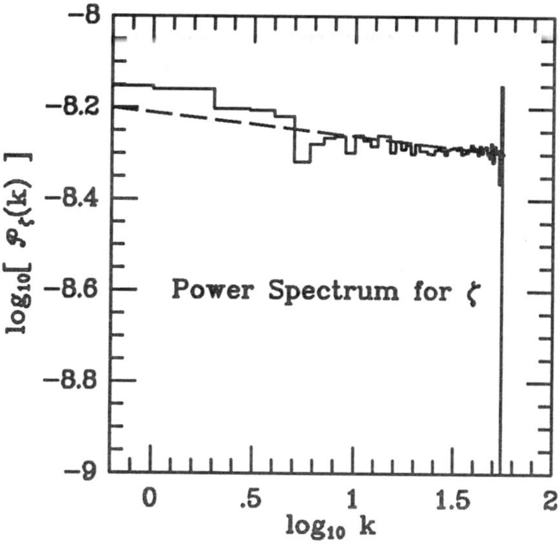

Fig.(4): The power spectrum, $\mathcal{P}_\zeta(k) \equiv k^3 < |\zeta|^2 > /(2\pi^2)$, for the variable ζ is calculated by taking the Fourier transform of the 64^3 lattice simulation described in in Fig.(2). Nonlinear effects do not change the shape of the flat spectrum whose amplitude is approximately $P_\zeta(k) \sim 10^{-8.3}$.The comoving wavenumber $k = 1$ is the largest mode that can fit in the lattice. The slow monotonic decrease for increasing values of k is a consequence of the Hubble parameter decreasing in region 1. The larger the potential parameter p_1 of region 1 is chosen, the flatter is the fluctuation spectrum. The spike in the last bin is not significant; it arises because the number of modes in the last bin is small, leading to large shot noise.

There are problems with non-Gaussian scale-invariant fluctuations, or at least with the simple-minded picture adopted above. It is difficult to account for the excess power in the two-point correlation function of galaxies as indicated by the APM survey.[15] However, one should not be overly eager to reject this model because there are observational uncertainties.

I would like to thank N. Kaiser for some interesting discussions. This work was supported by the U.S. Department of Energy and NASA at Fermilab (Grant No. NAGW-1340).

[1] Smoot, G.F. et al, COBE Preprint (1991).

[2] Strauss, M.A. and Davis, M., *Large Scale Structure in the Universe, IAU Symposium 130*, eds. J. Audouze and A. Szalay, Reidel, Dordrecht (1991).

[3] Kaiser, N. and Lahav, O., Mon. Not. R. Astr. Soc. **237** 129 (1989).

[4] Kaiser, N., in this proceedings (1991).

[5] Salopek, D.S., NATO Advanced Research Workshop: *Observational Tests of Inflation*, Dec. 10-14, 1990, ed. T. Shanks, Kluwer Academic Publishers.

[6] Salopek, D.S., to be submitted to the Ap.J. (1991).

[7] Allen, T.J., Grinstein, B. and Wise, M., Phys. Lett. **B197**, 66 (1987).

[8] Salopek, D.S. and Bond, J.R., Phys. Rev. **D42**, 3936 (1990), **D43**, 1005 (1991).

[9] Salopek, D.S., Phys. Rev. **D43**, 3214 (1991).

[10] Vilenkin, A., Phys. Rev. **D27**, 2848 (1983);

Starobinski, A.A., in *Current Topics in Field Theory, Quantum Gravity, and Strings*, Proc. Meudon and Paris VI, ed. H.T. de Vega and N. Sanchez **246** 107 (Springer-Verlag, 1986);

Linde, A.D., Phys. Lett. **175B**, 395 (1986);

Bardeen, J.M. and Bublik, G.J., Class. Quant. Grav. **4**, 573 (1987).

[11] Lucchin, F. and Matarrese, S., Phys. Rev. **D32**, 1316 (1985).

[12] Bardeen, J.M., Steinhardt, P.J. and Turner, M.S., Phys. Rev. **D28**, 670 (1983).

[13] Hartle, J.B., *Gravitation in Astrophysics*, eds. J.B. Hartle and B. Carter, Plenum Press, New York (1986).

[14] Bond, J.R. and Efstathiou, G., Ap.J. **285**, L45 (1984);

Bond, J.R., in Proc. of *The Early Universe*, ed. W. Unruh and G. Seminoff, Dordrecht: Kluwer (1988).

[15] Maddox, S.J., Efstathiou, G., Sutherland, W.J. and Loveday, J., Mon. Not. R. Astr. Soc. **242**, 43P (1990).

SYMMETRY OPERATORS FOR MAXWELL'S EQUATIONS ON CURVED SPACE-TIME.

E. G. KALNINS,
Department of Mathematics and Statistics, University of Waikato, Hamilton, New Zealand.

R. G. MCLENAGHAN and G. C. WILLIAMS
Department of Applied Mathematics, University of Waterloo, Waterloo, Ontario, Canada.

ABSTRACT

We derive the necessary and sufficient conditions that a second order covariant differential operator be a symmetry operator of Maxwell's equations in a general curved space-time background. It is found that such operators are naturally formulated in terms of conformal Killing vectors, tensors and spinors. Operators of this type play a role in the solution of Maxwell's equations via separation of variables in the Kerr background space-time.

1. Introduction

Recently Kalnins, Miller and Williams [1] found a second order covariant differential operator which is a symmetry operator of Maxwell's equations in the Kerr background space-time and which also characterizes the separable solution of Maxwell's equations in that background. In particular they found that the separable solution was an eigensolution of the operator with the eigenvalue being the Teukolsky separation constant appearing in the separated solution. This result is the primary motivation behind this work which is aimed at providing a means whereby all the second order differential symmetry operators of Maxwell's equations in any given space-time may be found.

We shall take the source free Maxwell equations

$$\nabla^a F_{ab} = 0 \quad , \quad \nabla_{[a} F_{bc]} = 0, \qquad\qquad (F_{ab} = F_{[ab]}) \tag{1.1}$$

in an arbitrary background space-time and find the conditions under which a second order covariant differential operator is a symmetry operator of those equations. Throughout we use the conventions and notation of Penrose and Rindler [2].

Similar results on the existence conditions for symmetry operators of the Dirac and Weyl neutrino equations in curved space-times have been obtained by McLenaghan and Spindel [3] and Kamran and McLenaghan [4].

2. Symmetry operators

A symmetry operator of Maxwell's equations is an operator which maps solutions of Maxwell's equations into themselves. Thus $Q_{ab}{}^{cd}$ is a symmetry operator if given any field F_{ab} satisfying Maxwell's equations the new field $\widetilde{F}_{ab}$ formed by

$$\widetilde{F}_{ab} = Q_{ab}{}^{cd}F_{cd} \tag{2.1}$$

also satisfies Maxwell's equations. Note that $Q_{ab}{}^{cd} = Q_{[ab]}{}^{[cd]}$. We shall in general consider a complex Maxwell field, the analysis when the field is real being essentially similar in all respects. It is also computationally convienient to use a spinor description rather than a tensor one. We have then

$$F_{ab} = \varepsilon_{A'B'}\phi_{AB} + \varepsilon_{AB}\chi_{A'B'}, \tag{2.2}$$

$$\begin{aligned} Q_{ab}{}^{cd} &= \varepsilon_{A'B'}\varepsilon^{C'D'}\,{}_1Q_{AB}{}^{CD} + \varepsilon_{AB}\varepsilon^{C'D'}\,{}_2Q_{A'B'}{}^{CD} \\ &\quad + \varepsilon_{A'B'}\varepsilon^{CD}\,{}_3Q_{AB}{}^{C'D'} + \varepsilon_{AB}\varepsilon^{CD}\,{}_4Q_{A'B'}{}^{C'D'}, \end{aligned} \tag{2.3}$$

with Maxwell's equations being

$$\begin{aligned} \nabla^A{}_{A'}\phi_{AB} &= 0, \\ \nabla_A{}^{A'}\chi_{A'B'} &= 0. \end{aligned} \tag{2.4}$$

In order that $Q_{ab}{}^{cd}$ be a symmetry operator we require that the two spinor pieces of $\widetilde{F}_{ab}$ each satisfy the spinor form of Maxwell's equations Eq. (2.4), that is

$$\nabla^A{}_{A'}\,{}_1Q_{AB}{}^{CD}\phi_{CD} = 0, \tag{2.5}$$

$$\nabla_A{}^{A'}\,{}_2Q_{A'B'}{}^{CD}\phi_{CD} = 0, \tag{2.6}$$

for every solution ϕ_{CD} of Maxwell's equations. The conditions on ${}_3Q_{AB}{}^{C'D'}$ and ${}_4Q_{A'B'}{}^{C'D'}$ are similar and we shan't consider them further.

We consider then two types of operators, those described by ${}_1Q$ which map solutions ϕ_{AB} into solutions $\widetilde{\phi}_{AB}$ and those described by ${}_2Q$ which map solutions ϕ_{AB} into solutions $\widetilde{\chi}_{A'B'}$. The most general possible second order covariant differential operators ${}_1Q_{AB}{}^{CD}$ and ${}_2Q_{A'B'}{}^{CD}$ are

$${}_1Q_{AB}{}^{CD} = K_{AB}{}^{CDEE'FF'}\nabla_{EE'}\nabla_{FF'} + L_{AB}{}^{CDEE'}\nabla_{EE'} + M_{AB}{}^{CD}, \tag{2.7}$$

$${}_2Q_{A'B'}{}^{CD} = K_{A'B'}{}^{CDEE'FF'}\nabla_{EE'}\nabla_{FF'} + L_{A'B'}{}^{CDEE'}\nabla_{EE'} + M_{A'B'}{}^{CD}. \tag{2.8}$$

Using Maxwell's equations and the Ricci identities and taking account of the symmetries of $Q_{ab}{}^{cd}$ we can impose without loss of generality a number of symmetry conditions on the various parts of these operators. For example, we will require

$$K_{ABCDEFE'F'} = K_{(AB)(CDEF)(E'F')}. \tag{2.9}$$

3. Operators of the first type

Condition (2.5) contains derivatives of ϕ_{CD} of up to third order. In order that it be satisfied the coefficents of the various symmetric derivatives of ϕ_{CD} must vanish.

Requiring that the coefficent of the symmetric third order derivatives of ϕ_{CD} vanish gives

$$K_{ABCDEFE'F'} = \varepsilon_{A(C}\varepsilon_{|B|D}K_{EF)E'F'}, \tag{3.1}$$

where

$$K_{EFE'F'} = \frac{3}{5}K_{KL}{}^{KL}{}_{(EF)(E'F')}. \tag{3.2}$$

Requiring that the coefficent of the symmetric second order derivatives of ϕ_{CD} vanish gives

$$\nabla_{(A(A'}K_{EF)E'F')} = 0, \tag{3.3}$$

$$L_{ABCDEE'} = -\frac{4}{3}\varepsilon_{(C|(A}G_{B))|DE)E'} + \varepsilon_{A(C}\varepsilon_{|B|D}L_{E)E'}, \tag{3.4}$$

where

$$G_{ABCC'} = \nabla_{(AK'}K_{BC)}{}^{K'}{}_{C'} \tag{3.5}$$

and

$$L_{EE'} = \frac{1}{2}L_{KL}{}^{KL}{}_{EE'}. \tag{3.6}$$

Eq. (3.3) is precisely the condition that $K_{ab} = K_{ABA'B'}$ be a valence 2 conformal Killing tensor.

Requiring that the coefficent of the symmetric first order derivative of ϕ_{CD} vanish gives

$$M_{(BCDE)} = \frac{1}{3}\nabla_{(BK'}\nabla_{CL'}K_{DE)}{}^{K'L'} - \frac{1}{2}\Phi_{(BCK'L'}K_{DE)}{}^{K'L'}, \tag{3.7}$$

$$\begin{aligned}\nabla_{(A(A'}L_{E)E')} = &-\frac{5}{6}\nabla^{K}{}_{(A'}\nabla_{(K|K'|}K_{AE)}{}^{K'}{}_{E')} - \frac{3}{2}\Psi_{KLAE}K^{KL}{}_{A'E'}\\ &-\overline{\Psi}_{K'L'A'E'}K_{AE}{}^{K'L'} - 5\Phi_{(A|K|(A'|K'|}K_{E)}{}^{K}{}_{E')}{}^{K'}\\ &-10\Lambda K_{AEA'E'},\end{aligned} \tag{3.8}$$

$$\begin{aligned}M_{(A|K|}{}^{K}{}_{E)} = &\frac{5}{6}\nabla^{K}{}_{K'}\nabla_{(KL'}K_{AE)}{}^{K'L'} + \nabla_{(AK'}L_{E)}{}^{K'}\\ &+2\Phi_{K(AK'L'}K_{E)}{}^{KK'L'}.\end{aligned} \tag{3.9}$$

Note that the condition on $L_a = L_{AA'}$ Eq. (3.8) is invariant under the addition of a conformal Killing vector to L_a.

Finally, we require the coefficent of ϕ_{CD} to vanish giving

$$\begin{aligned}
&\frac{14}{15} K_{(B}{}^{FE'}{}_{A'} \nabla_{|K|E'} \Psi_{CM)F}{}^{K} + K^{EF}{}_{A'}{}^{E'} \nabla_{(EE'} \Psi_{BCFM)} \\
&+ \frac{1}{3} K_{B}{}^{FE'F'} \nabla_{(C(E'} \Phi_{MF)A'F')} + \frac{1}{3} K_{C}{}^{FE'F'} \nabla_{(M(E'} \Phi_{BF)A'F')} \\
&+ \frac{1}{3} K_{M}{}^{FE'F'} \nabla_{(B(E'} \Phi_{CF)A'F')} + \frac{1}{6} K_{(BC}{}^{E'F'} \nabla_{M)K'} \overline{\Psi}_{E'A'F'}{}^{K'} \\
&- \frac{5}{3} K_{(BC}{}^{E'}{}_{A'} \nabla_{M)E'} \Lambda + \frac{1}{3} G_{DF(BA'} \Psi_{CM)}{}^{DF} + \frac{5}{3} G_{K(BCF'} \Phi_{M)}{}^{K}{}_{A'}{}^{F'} \\
&- \frac{10}{3} \Lambda G_{BCMA'} - L_{FA'} \Psi_{BCM}{}^{F} - L_{(BF'} \Phi_{CM)A'}{}^{F'} + \nabla^{A}{}_{A'} M_{A(BCM)} = 0 \quad (3.10)
\end{aligned}$$

and

$$\begin{aligned}
&\frac{7}{6} K^{EFE'}{}_{A'} \nabla_{KE'} \Psi_{CEF}{}^{K} + \frac{5}{4} K^{EFE'F'} \nabla_{(E(E'} \Phi_{CF)A'F')} \\
&+ \frac{2}{3} K_{C}{}^{FE'F'} \nabla_{FK'} \overline{\Psi}_{A'E'F'}{}^{K'} - \frac{20}{3} K_{C}{}^{EE'}{}_{A'} \nabla_{EE'} \Lambda + \frac{4}{3} G_{DEFA'} \Psi_{C}{}^{DEF} \\
&+ \frac{5}{6} G_{CDEE'} \Phi^{DEE'}{}_{A'} + 2 L_{FF'} \Phi_{C}{}^{F}{}_{A'}{}^{F'} - 6 \Lambda L_{CA'} + \nabla^{A}{}_{A'} M_{AK}{}^{K}{}_{C} = 0. \quad (3.11)
\end{aligned}$$

We observe that the above two conditions will impose consistency conditions on $K_{ABA'B'}$ and $L_{AA'}$ when the terms in the derivative of M_{ABCD} are substituted for using Eqs. (3.7) and (3.9).

The operator discovered by Kalnins et al. [1] in the Kerr space-time is an operator of this first type.

4. Operators of the second type

We now examine condition (2.6). Requiring that the coefficent of the symmetric third order derivatives of ϕ_{CD} vanish gives

$$K_{A'B'E'F'CDEF} = \varepsilon_{A'(E'} \varepsilon_{|B'|F')} K_{CDEF}, \tag{4.1}$$

where

$$K_{CDEF} = \frac{1}{3} K_{K'L'}{}^{K'L'}{}_{(CDEF)}. \tag{4.2}$$

Requiring that the coefficent of the symmetric second order derivatives of ϕ_{CD} vanish gives

$$\nabla_{(AA'} K_{CDEF)} = 0, \tag{4.3}$$

$$L_{A'B'C'ABC} = \frac{8}{5} \varepsilon_{(A'|C'|} \nabla_{KB')} K_{ABC}{}^{K}. \tag{4.4}$$

We observe that Eq. (4.3) is the equation of a valence four Killing spinor.

Requiring that the coefficent of the symmetric derivative of ϕ_{CD} vanish gives

$$M_{B'E'CD} = \frac{3}{5}\nabla_{L(E'}\nabla_{KB')}K_{CD}{}^{KL}. \tag{4.5}$$

Finally, we examine the coefficent of ϕ_{CD}. Using Eqs. (4.1) - (4.5) it can be shown that this coefficent vanishes identically.

Note, both $L_{A'B'C'ABC}$ and $M_{A'B'AB}$ are completely determined by K_{ABCD}. As a consequence symmetry operators of the second type must be differential operators of at least second order.

Eq. (4.3) has the algebraic integrability condition

$$K_{K(ABC}\Psi_{DEF)}{}^{K} = 0. \tag{4.6}$$

For a non-zero Weyl spinor this condition implies

$$K_{ABCD} = \xi\Psi_{ABCD}, \tag{4.7}$$

where ξ is an arbitrary complex scalar function. In particular it follows immediately that the principal spinors of K_{ABCD} and Ψ_{ABCD} are aligned.

The results presented here will be published elsewhere in greater detail.

5. Acknowledgements

This work was supported in part by an Operating Grant from the Natural Sciences and Engineering Research Council of Canada.

6. References

1. E. G. Kalnins, W. Miller Jr. and G. C. Williams. Killing-Yano tensors and variable separation in the Kerr geometry. *J. Math. Phys.* **30**, (1989), p. 2360.
2. R. Penrose and W. Rindler. *Spinors and space-time Vol. 1. Two spinor calculus and relativistic fields.* (C. U. P., 1984).
3. R. G. McLenaghan and Ph. Spindel. Quantum numbers for Dirac spinor fields on a curved space-time. *Phys. Rev.* **D20**, (1979), p. 409.
4. N. Kamran and R. G. McLenaghan. Symmetry operators for neutrino and Dirac fields on curved space-time. *Phys. Rev.* **D30**, (1984), p. 357.

SEPARATION OF VARIABLES AND CONSTANTS OF MOTION FOR RELATIVISTIC WAVE EQUATIONS ON CURVED SPACE-TIME

R.G. McLENAGHAN

Department of Applied Mathematics, University of Waterloo
Waterloo, Ontario N2L 3G1 Canada

ABSTRACT

The lecture reviewed the known separability results for the Hamilton-Jacobi, Klein-Gordon, Weyl neutrino, Maxwell, gravitational perturbation and Dirac equations on the Kerr space-time[1], and more generally on vacuum space-times of Petrov type D[2]. The relation between the separability properties and the existence of underlying symmetry operators on the space-time was described[3]. The characterization of these symmetric operators in terms of the Yano-Killing tensor existing on the Kerr space-time and on a large subclass of the vacuum D metrics was explained[4]. Some partial results on the intrinsic characterization of the separability of the Dirac equation were presented[5]. Recently obtained results on the symmetry operators underlying the separability of Maxwell's equation on the Kerr solution were discussed[6,7] .

References

1. B. Carter and R.G. McLenaghan, in *Proceedings of the second Marcel Grossmann Meeting on General Relativity*, ed. R. Ruffini (North-Holland, Amsterdam, 1982).

2. N. Kamran and R.G. McLenaghan, in *Gravitation and Geometry*, ed. W. Rindler and A. Trautman (Bibliopolis, Naples, 1987).

3. B. Carter and R.G. McLenaghan, Phys. Rev. D**19** (1979) 1093.

4. N. Kamran, in *Proceedings of the second Canadian Conference on General Relativity and Relativistic Astrophysics*, ed. A. Coley, C. Dyer and B.O.J. Tupper (World Scientific, Singapore, 1988).

5. N. Kamran and R.G. McLenaghan, Bull. C. Sci. Acad. R. Belgique **LXX** (1984) 596.

6. E.G. Kalnins, W. Miller Jr. and G.C. Williams, J. Math. Phys. **30** (1989) 2360.

7. E.G. Kalnins, R.G. McLenaghan and G.C. Williams, article in this volume.

TWO-DIMENSIONAL QUANTUM GRAVITY FROM MATRIX MODELS

ROBERT C. MYERS

Physics Department, McGill University, Ernest Rutherford Building

Montréal, Québec, CANADA H3A 2T8

ABSTRACT

The simplicial description of two-dimensional surfaces via random matrix models is explained. Some of the recent developments in non-critical string theory arising from this approach are outlined.

The most promising avenue of research in physicists' pursuit of the Theory of Everything remains (super)string theory, but our understanding of these theories is largely limited to a perturbative construction of string scattering amplitudes. One approach is through the Polyakov path integral.[1] As a one-dimensional loop or string moves through space-time, it sweeps out a two-dimensional surface or *world-sheet.* The Polyakov path integral constructs string scattering amplitudes, by integrating over all world-sheets connecting a particular set of fixed loops at asymptotic infinity.

$$\mathcal{S} = \sum_{\chi} g^{2-\chi} \int \mathcal{D}g_{ab}\mathcal{D}X^{\mu}\, e^{-S_{Poly}} \tag{1}$$

Here g_{ab} is the intrinsic two-dimensional metric on the world-sheet, and X^{μ} are the embedding variables of the surfaces into a background space-time with D dimensions. In Eq. 1, g is the string coupling constant, or the strength of the fundamental interaction in which a single loop divides into two. The Euler character χ is a topological invariant of the two-dimensional surfaces, which may be calculated as $\chi = 2-2\,(\#\,handles)-(\#\,holes)$. For a fixed number of asymptotic strings (or holes), the perturbative expansion in Eq. 1 is the sum over the number of handles in the intervening two-dimensional geometry. This sum is similar to the sum over loops in the Feynman diagrams describing a particular transition in quantum field theory.

The action used to weight the path integral in Eq. 1 is[1]

$$S_{Poly} = \int \mathrm{d}^2\sigma \sqrt{g} g^{ab} \partial_a X^{\mu} \partial_b X^{\mu}\ . \tag{2}$$

This action could be described as the Klein-Gordon action for a set of massless scalar fields labelled X^μ, in a curved two-dimensional space. Such a point-of-view leads to an interpretation of Eq. 1 as a path integral for a set of quantum scalar field theories coupled to two-dimensional quantum gravity. In other string theories, the action in Eq. 2 may be modified by the inclusion of two-dimensional fermions and/or other fields coupled to (super)gravity on the world-sheet.[2]

Diffeomorphism invariance on the world-sheet implies that two of the three components of the metric g_{ab} are spurious gauge degrees of freedom. One may fix the diffeomorphism freedom by choosing *conformal gauge*, $g_{ab} = e^\phi \hat{g}_{ab}$ where $\hat{g}_{ab}$ is a fixed constant curvature metric. With this choice, the remaining physical degree of freedom is given by the conformal mode ϕ, and substituting into Eq. 2 yields

$$S_{Poly} = \int \mathrm{d}^2\sigma \sqrt{\hat{g}}\hat{g}^{ab}\partial_a X^\mu \partial_b X^\mu \ . \tag{3}$$

Thus the conformal mode actually does not appear in the *classical* two-dimensional action. Hence the entire metric appears to decouple in the path integral in Eq. 1. Not only does the integral $\mathcal{D}g_{ab}$ overcount the physically distinct world-sheets by integrating over surfaces which are equivalent under diffeomorphisms, but also over those related by conformal or Weyl transformations. The path integral measure must be more carefully defined though, and one must ensure that this decoupling survives in the quantum string theory. We will see that this leads to the widely known aspect of string-lore that the theory requires twenty-six dimensions (for bosonic strings, or ten dimensions for superstrings). Actually $D-26=0$ is only the leading term in one of a series of constraints required for conformal invariance in the quantum theory.[3] For example for strings propagating in a curved background, one must also require that the space-time metric satisfy Einstein's equations $R_{\mu\nu}=0$ (to leading order).

A careful regularization of the quantum path integral leads to the following gravity action on the world-sheet (for strings propagating in a flat space-time)[1]

$$\frac{26-D}{48\pi}\int \mathrm{d}^2\sigma\, \mathrm{d}^2\sigma'\, \sqrt{g}\mathcal{R}\, G(\sigma,\sigma')\, \sqrt{g'}\mathcal{R}' + \int \mathrm{d}^2\sigma \sqrt{g}\mu \ . \tag{4}$$

In the second term, μ is a two-dimensional cosmological constant, which gives the quantum shift in the world-sheet vacuum energy. This local part of Eq. 4 could be eliminated by adding an appropriate bare cosmological constant term to Eq. 2. The first term involves convoluting the two-dimensional Ricci scalar $\mathcal{R}$ with a Green's function satisfying $\nabla^2 G(\sigma,\sigma') = \delta^2(\sigma-\sigma')$. This nonlocal term receives two separate contributions: $-D$ from regularizing the embedding part of the measure, and 26 from the gravity part of the measure. In conformal gauge, Eq. 4 becomes local

$$\frac{26-D}{48\pi}\int \mathrm{d}^2\sigma\sqrt{\hat{g}}\left(\hat{g}^{ab}\partial_a\phi\partial_b\phi + 2\hat{\mathcal{R}}\phi\right) + \int \mathrm{d}^2\sigma\sqrt{\hat{g}}\mu e^\phi \tag{5}$$

which is called the Liouville action. If $D - 26 = 0$, the first term of the action vanishes, and the conformal mode has no dynamics in the quantum theory. So we see how this constraint arises to ensure conformal invariance in the Polyakov path integral. Eq.'s 4 and 5 are the two-dimensional theory of gravity which we will be considering below for $D \neq 26$ from a very different point-of-view.

With our present understanding, some of the paramount problems in *critical* string theories (*i.e.*, string theories with quantum conformal invariance) are: It is technically extremely laborious to calculate amplitudes for world-sheets with large Euler number. There are indications that the perturbative expansion produces an asymptotic series,[4] so that the series in Eq. 1 could not be summed even if all of the terms were calculable. Finally, it is expected that an understanding of non-perturbative aspects of string theory will be required to address various important issues such as selection of the correct vacuum, or supersymmetry breaking.

Recently much progress has been made in understanding *non-critical* string theories. For such theories, there is no conformal invariance in the quantum theory, and the conformal mode becomes a dynamical field. Most importantly, and contrary to earlier expectations, it was realized that these are actually sensible quantum theories, at least for certain ranges of the space-time dimension.[5] More recently, it has been found that some simple examples of such theories are completely soluble[6,7] – all physical amplitudes are readily calculable to all orders in the topological expansion. It must be emphasized that these soluble cases are very simple models (in the sense of a string theory) and cannot hope to describe our universe. On the other hand, they may serve as toy models from which we might gain insight into the problems of more complicated string theories which may actually provide the theory of everything. The soluble models might also serve in the same role for the theory of quantum gravity. For instance, the topological expansion of two-dimensional surfaces may give new insight into the role of topology fluctuations in higher dimensions. These have been conjectured to affect the observed values of all physical coupling constants,[8] and in particular to explain the vanishing of the cosmological constant.[9]

Of these toy models, I wish to discuss in more detail the simplest case of a string theory without any world-sheet matter. This example can be considered to be the theory of two-dimensional quantum gravity, since all that remains is a path integral over world-sheet metrics. The partition function reduces to

$$\mathcal{Z} = \sum_{\chi} \int \mathcal{D}g_{ab} \exp\left[-\int d^2\sigma\sqrt{g}(\mu_B + \frac{\kappa}{4\pi}\mathcal{R})\right] \tag{6}$$

where the classical action now includes a cosmological constant μ_B, and an Einstein term. Due to the peculiarities of two dimensions, this second term yields the Euler character of the surface, a topological invariant, rather than providing dynamics for the metric. Eq. 6 becomes

$$\mathcal{Z} = \sum_{\chi} \int \mathcal{D}g_{ab} \, \exp\left[-\mu_B A - \kappa\chi\right] \tag{7}$$

where A is the area of the world-sheet. Thus the bare action only contributes a term which suppresses world-sheets with large area (for positive μ_B), and a topological term from which one can extract $e^{\kappa} = g_B$ as the *bare* string coupling constant (compare to Eq. 1). The interesting dynamics of the metric will arise from the path integral measure, $\mathcal{D}g_{ab}$.

The problem is that $\mathcal{D}g_{ab}$ is a measure for an infinite number of degrees of freedom at small scales. One approach is to regulate the path integral by introducing a discrete model, which will approximate the continuum theory in the appropriate limit. The approach we will pursue is *dynamical triangulation.*[10] This method differs from the perhaps more familiar Regge calculus[11] and was specifically invented for numerical studies of noncritical string theories. The idea is to construct the surfaces from flat plaquettes, each with a fixed finite area. In the present case, we will introduce squares whose edge lengths correspond to a regulator length α. Given some fixed number of squares n, integrating over metrics corresponds to connecting the squares in all possible ways which correspond to different geometries. In the following, we refer to this as making all possible *tilings.* In these discrete surfaces, all the curvature is concentrated at vertices where the corners of the squares meet. A vertex where four squares meet corresponds to a flat element of surface. If fewer than four corners meet, an angular deficit corresponding to a concentration of positive curvature is produced. On the other hand, five or more corners meeting at a given vertex yields a concentration of negative curvature. On small scales near the regulator length α, the resulting surfaces will appear jagged, but for much larger length scales they provide a good approximation to smoothly curving geometries. For large values of n, one expects then that the summation over all tilings gives a good approximation to the continuum theory. One has

$$\mathcal{F} = \sum_{\chi} \sum_{n} \sum_{tilings} g_B^{-\chi} \, e^{-\mu_B n \alpha^2} \tag{8}$$

where the total area is given by $A = n\alpha^2$. Here we have introduced the free energy or effective action, $\mathcal{F} = -\log\mathcal{Z}$, to indicate that we will restrict the tilings in the last sum to include only connected surfaces. This relation between the sum over all surfaces and only connected surfaces should be familiar from studying Feynman diagrams in quantum field theory. For the present, we have also implicitly restricted our attention to closed surfaces without boundaries or punctures.

This approach of dynamical triangulation was first applied in numerical investigations.[12] Further progress was made analytically by observing that the tilings can be generated by random matrix models. Consider

$$\mathcal{Z} = \int \mathrm{d}\Phi \; e^{-S} \tag{9}$$

where Φ_{ij} is an $N \times N$ hermitian matrix, and the matrix action takes the form

$$S = N \,\mathrm{Tr}\left[\frac{1}{2}\Phi^2 + \frac{b}{4}\Phi^4\right] . \tag{10}$$

In Eq. 9, one simply integrates over the matrix entries with the flat measure

$$\mathrm{d}\Phi = \prod_{i=1}^{N} \mathrm{d}\Phi_{ii} \prod_{1\leq i<j\leq N} \mathrm{d}(\mathrm{Re}\,\Phi_{ij})\,\mathrm{d}(\mathrm{Im}\,\Phi_{ij}) . \tag{11}$$

Regarding Eq. 9 as a zero-dimensional field theory, it may be evaluated perturbatively in the coupling of the quartic interaction b, via Feynman diagrams. Pairs of matrix insertions are contracted (*à la* Wick's theorem) using the propagator $\langle\Phi_{ij}\Phi_{kl}\rangle = \frac{1}{N}\delta_{il}\delta_{jk}$. This propagator is represented diagramatically with a double line to indicate the contractions of the two pairs of indices. The quartic interactions are represented by a cross of double lines (see Fig. (1)). In the cross, each pair of lines represents a single matrix, with the connections of the lines indicating the contractions of the indices of consecutive matrices under the matrix trace. This is a very brief description of the Feynman diagrams, and I refer the interested reader to the complete discussion appearing in Ref. [13]. I would like to emphasize that evaluating Eq. 9 at each order in b, one performs N^2 ordinary gaussian integrals, and the Feynman diagrams simply provide a useful aid in organizing the calculation.

One should ask now how the matrix model produces the tilings of two-dimensional surfaces. Fig. 1 shows a portion of a Feynman diagram, which might be a contribution at some large order in the perturbative expansion. One sees that it begins to form some sort of two-dimensional network, but the tilings of the surfaces actually arise as the dual diagrams to the Feynman diagrams. The dual diagrams are formed by replacing each of the propagators by an edge which runs orthogonal to it, as is shown in Fig. 2. The reason that one takes the dual graph can be seen by comparing the Feynman diagram in Fig. 1 with its dual in Fig. 3. While the Feynman diagrams are composed of polygons which may have any number of edges, the dual graphs contain only quadrilaterals. While these quadrilaterals are irregularly shaped in Fig. 3, this is an artifact of the fact that the surface is being compressed flat onto the page. One should actually think of this as a projection of (a portion of) a curved surface with faces which are all regular squares. The translation between the Feynman diagrams and the tilings is completed as follows: propagators are replaced by edges; matrix interactions, by faces; and closed index loops, by vertices. Clearly the fact that the dual is

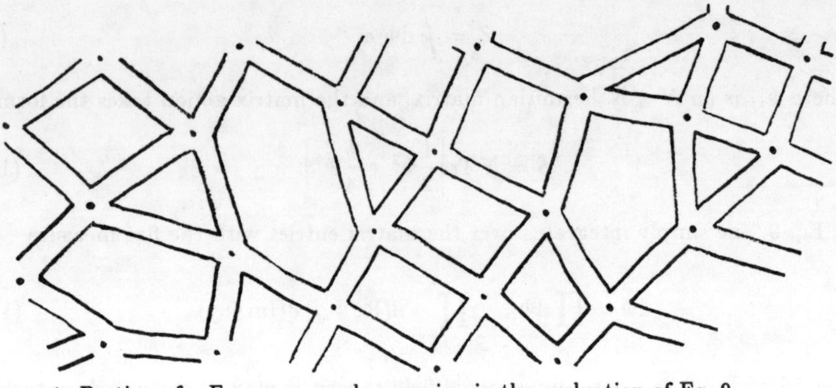

Figure 1: Portion of a Feynman graph occurring in the evaluation of Eq. 9.

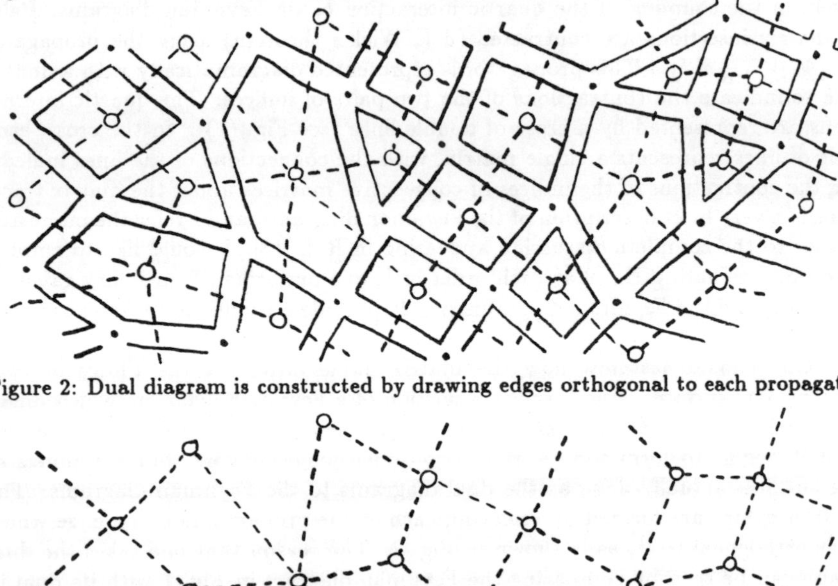

Figure 2: Dual diagram is constructed by drawing edges orthogonal to each propagator.

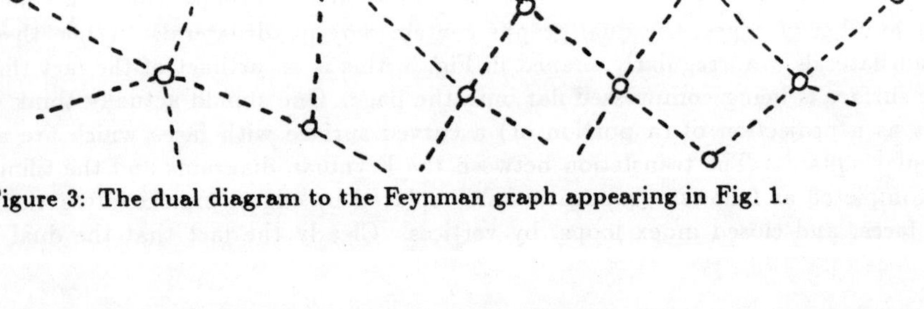

Figure 3: The dual diagram to the Feynman graph appearing in Fig. 1.

composed completely from squares arises because the interaction in the matrix action given in Eq. 10 is quartic. In general, an interaction in the matrix action involving a trace over k matrices, would contribute faces with k edges in the dual diagrams.

At this point, one may wonder why we are dealing with matrices? The reason is that the matrices organize the topological expansion of the dual graphs, as was first noted by 't Hooft.[14] To see this, one must examine where factors of N enter into the evaluation of the Feynman graphs. First every insertion of a matrix interaction is accompanied by a factor N, because of the overall factor of N appearing in the action given in Eq. 10. This factor also gives rise to the $1/N$ in the matrix propagators. Finally since N is the dimension of the matrices, each closed index loop contributes a factor of N to the Feynman diagrams. Thus each Feynman graph includes an overall factor of

$$(N)^{\#\,interactions}(\frac{1}{N})^{\#\,propagators}(N)^{\#\,loops} = N^{\#\,faces-\#\,edges+\#\,vertices} = N^{\chi} \tag{12}$$

where we have used the formula for the Euler character of a tiled two-dimensional surface: $\chi = \#\,faces - \#\,edges + \#\,vertices$. Thus the factor of N classifies the Feynman diagrams by the topology of the corresponding dual graph. So N^2 arises from the graphs corresponding to a spherical topology (these are commonly called the planar graphs), N^0 accompanies the torus graphs, et cetera. Therefore in the partition function generated by the matrix model, $1/N$ plays the role of the bare string coupling, g_B.

Now we have an equivalence between the connected Feynman graphs from Eq. 9 and the tilings in Eq. 8. Recalling that insertions of the quartic interaction carry a factor of the coupling constant b, the perturbative evaluation of the matrix model yields

$$\mathcal{F} = \sum_{\chi}\sum_{n}\left(\frac{1}{N}\right)^{-\chi} b^n\ [\#\,tilings]_{n,\chi}\ . \tag{13}$$

Eq. 13 corresponds precisely to Eq. 8 when we identify $g_B = 1/N$ as above, and also associate b with the bare cosmological constant $\mu_B = -\log(b)/\alpha^2$. Using the graphical techniques associated with the matrix model, Ref. [13] was able to evaluate the last factor in Eq. 13 for some low genus surfaces. The result to leading order in n takes the form

$$[\#\,tilings]_{n,\chi} = (b_c)^{-n} n^{-\lambda\chi-1} C_{\chi}\ . \tag{14}$$

There are corrections to Eq. 14 which are down by a factor $1/n$, but recall that we are only really interested in the results for large values of n, where we expect that Eq. 8 gives a good approximation to the continuum theory. Three new constants have appeared in Eq. 14, all of which are simply dimensionless numbers, but they are very different in character. The first, b_c, is a non-universal number, meaning that its value depends on the particular regularization scheme that was chosen. For instance, choosing to regulate with triangular plaquettes would produce a different value of b_c. The exponent λ is a universal number, which is independent of the choice of regularization. The numbers C_χ are independent of n, and carry information about the genus expansion of the theory. These numbers are not universal by themselves, but appropriate ratios yield universal results.[15]

Using the result given in Eq. 14, to leading order Eq. 13 becomes

$$\begin{aligned} \mathcal{F} = & \quad \sum_\chi \sum_n \left(\tfrac{1}{N}\right)^{-\chi} \left(\tfrac{b}{b_c}\right)^n n^{-\lambda\chi-1} C_\chi \\ \simeq & \quad \sum_\chi \int \tfrac{dA}{A} \left(\tfrac{1}{N}\right)^{-\chi} \left(\tfrac{b}{b_c}\right)^{A/\alpha^2} \left(\tfrac{A}{\alpha^2}\right)^{-\lambda\chi} C_\chi \end{aligned} \tag{15}$$

In the second line, the summation over n is approximated by a continuum area integral using $n = A/\alpha^2$. Now the task at hand is to make sense of this result in the limit that the regulator length α is taken to zero. First we note that potential divergence due to the exponent in the second factor of the integrand can be absorbed by taking the limit, $b/b_c \to 1$, while fixing $A_o \equiv \frac{\alpha^2}{\log(b_c/b)}$. This new finite (unphysical) constant may be regarded as a renormalization point for the measure in the area integral.[15] With this definition, Eq. 15 becomes

$$\mathcal{F} = \sum_\chi \int \frac{dA}{A} \left(\frac{1}{N} \frac{1}{(\log b_c/b)^\lambda} \right)^{-\chi} e^{-A/A_o} \left(\frac{A}{A_o} \right)^{-\lambda\chi} C_\chi \tag{16}$$

In this expression a renormalized string coupling constant has appeared, namely $g_R = N^{-1}(\log b_c/b)^{-\lambda}$. One can maintain a finite g_R by taking the *double scaling limit*, where N diverges at the same time that $b/b_c \to 1$. In this limit not only do we recover information about the continuum theory, but we are also able to retain information about the entire topological expansion. This is to be contrasted with earlier results with $N \to \infty$ alone, the so-called planar limit, where only graphs of spherical topology contributed (see for example Ref. [16]). Near the critical point one has

$$g_R = \frac{1}{N} \frac{1}{(\log b_c/b)^\lambda} \simeq \frac{1}{N} (b_c/b - 1)^{-1+\gamma/2} \tag{17}$$

where we have introduced $\lambda = 1 - \gamma/2$, which is the conventional parameterization of this universal exponent. γ is commonly referred to as the critical index or the string susceptibility, and is found in the present case to be $-1/2$ (*i.e.*, $\lambda = 5/4$).

Now I would like to add a detail to the above analysis. The astute reader may have noted there is actually an error which runs through the analysis from Eq. 13 to Eq. 17. The insertion of a quartic matrix interaction is in fact accompanied by a factor of $-b$, and not simply b as stated above Eq. 13. Therefore one should replace b by $-b$ in the above. This means the double scaling limit involves taking $b \to -b_c$. Now b_c was a number which came out of counting graphs, and so must be positive. Thus the critical or limiting value of the matrix coupling constant is actually a negative number. One might find this disturbing since in this case Eq. 9 appears to be divergent. Of course the gaussian integrals involved in the perturbative evaluation of Eq. 9 remain finite, but a nonperturbative definition of Eq. 9 requires a careful analysis of the contours of integration involved.[17,18]

Before considering the solution of the matrix model in the double scaling limit, I would like to discuss an interesting result due to Kazakov.[19] Consider introducing a slightly more complicated matrix action into Eq. 9 of the form

$$S = N \,\mathrm{Tr}\left[\frac{1}{2}\Phi^2 + \frac{b}{4}\Phi^4 + \frac{c}{6}\Phi^6\right] . \tag{18}$$

Now as before in a perturbative expansion, the quartic interaction generates squares with area α^2, and now the new sextic interaction generates regular hexagons with the same edge length α, and hence with area $x\alpha^2 = (3\sqrt{3}/2)\,\alpha^2$. With this new action, Eq. 13 is replaced by

$$\mathcal{F} = \sum_{\chi}\sum_{n_4}\sum_{n_6}\left(\frac{1}{N}\right)^{-\chi} b^{n_4} c^{n_6} [\#\, tilings]_{n_4,n_6,\chi} \, . \tag{19}$$

where the perturbative expansion now involves separate summations over the numbers of quartic and sextic interactions. Now the total area of the surfaces is given by $A = n_4\alpha^2 + n_6 x\alpha^2 \equiv m\alpha^2$. Defining a ratio of the matrix couplings $\delta = c/b^x$, one may write

$$\mathcal{F} = \sum_{\chi}\sum_{m}\left(\frac{1}{N}\right)^{-\chi} b^m \left[\sum_{n_6} \delta^{n_6}\, [\#\, tilings]_{m,n_6,\chi}\right] . \tag{20}$$

Now to leading order for large m, the result for the the last factor takes a form analogous to Eq. 14

$$\widehat{b_c}^{\,-m} m^{-\lambda\chi-1} C_\chi \, .$$

Generically one finds that as above $\lambda = 5/4$, but for a one specific value of δ, one finds $\lambda = 7/6$. At this point, the universal part of the C_χ also changes. These results for the specially tuned point of the matrix action in Eq. 18 is referred to as *multicritical* behavior.

What is the source of this new multicritical behavior? In some sense, it is a new realization of an old idea in general relativity, namely producing matter from geometry. Heuristically, given a region of surface (which I imagine is large on the regulator scale, α), one has the option of tiling it either with squares, or with hexagons. So the new model is like having two-dimensional gravity coupled to a two state statistical mechanics system. One should object that generically a finite region will be tiled by a jumbled conglomeration of both squares and hexagons, and so this two state behavior would never be realized. This is in keeping with the stated results that generically one recovers only the behavior of two-dimensional gravity. It is only at specially tuned values of the matrix couplings that the behavior of the system changes. This is in accord with the expectation that large correlated tilings of squares or hexagons would only make an important contributions to Eq. 9 with a specially tuned matrix action. This is a heuristic interpretation for the results, but a precise relation can be found between the matrix model with the action in Eq. 18, and a two-dimensional field theory called the Lee-Yang edge singularity, which corresponds to a certain phase of an Ising spin system in an (imaginary) magnetic field.[20] One last comment is that by adding further higher order interactions to the matrix action, which generate plaquettes with more edges, one can find higher multicritical behavior with extra tuning of the new matrix couplings.[19] These higher multicritical models are expected to correspond to two-dimensional gravity coupled to more complicated statistical mechanics systems.

We now turn to the double scaling solution of the hermitian matrix model. Our discussion will not be complete, and the full analysis may be found in Ref.'s [6, 7]. The first point to note is that the matrix action in Eq. 10 has a $U(N)$ symmetry. It remains invariant if the matrix Φ is shifted by a similarity transformation $V^\dagger \Phi V$, where V is any $N \times N$ unitary matrix. Such a unitary transformation may be used to diagonalize the hermitian matrix, $\Phi = U^\dagger \mathrm{diag}(x_1, x_2, \ldots, x_N) U$, and then the partition function in Eq. 9 may be written as

$$\mathcal{Z}_N = \int \mathrm{d}U \int \prod_{i=1}^{N} \left(\mathrm{d}x_i e^{-S(x_i)} \right) \Delta_N^2(x) \tag{21}$$

In the following we will drop the integral over the unitary group since it is simply an overall normalization constant, which does not effect the critical behavior. The term $\Delta_N^2(x)$ is the Jacobian for the change of variables from the original measure given in Eq. 11 to the integral over eigenvalues and unitary transformations. Δ_N is the Vandermonde determinant

$$\begin{aligned} \Delta_N(x) &= \prod_{1 \le j < i \le N} (x_i - x_j) \\ &= \det \begin{bmatrix} 1 & x_1 & x_1^2 & \ldots & x_1^{N-1} \\ 1 & x_2 & x_2^2 & \ldots & x_2^{N-1} \\ \vdots & & & & \vdots \\ 1 & x_N & x_N^2 & \ldots & x_N^{N-1} \end{bmatrix} . \end{aligned} \tag{22}$$

Using the freedom to add multiples of columns in the determinant, we may rewrite this expression as

$$\Delta_N(x) = \det \begin{bmatrix} P_0(x_1) & P_1(x_1) & P_2(x_1) & \dots & P_{N-1}(x_1) \\ P_0(x_2) & P_1(x_2) & P_2(x_2) & \dots & P_{N-1}(x_2) \\ \vdots & & & & \vdots \\ P_0(x_N) & P_1(x_N) & P_2(x_N) & \dots & P_{N-1}(x_N) \end{bmatrix} \tag{23}$$

where $P_k(x) = x^k + \sum_{l=0}^{k-1} p_{k,l} x^l$ where $p_{k,l}$ are arbitrary constants. A convenient choice for the polynomials is to select polynomials orthogonal with respect to the natural measure provided in Eq. 21[13]

$$\int dx\, e^{-S(x)}\, P_k(x) P_l(x) = \delta_{k,l}\, h_k \quad . \tag{24}$$

The partition function then reduces to simply

$$\mathcal{Z}_N = N! \prod_{i=0}^{N-1} h_i \; . \tag{25}$$

This is a compact expression, but actually solving for $\mathcal{Z}_N$ would still require an infinite amount of work to explicitly construct the polynomials and calculate all of the weights h_i. We can reduce the amount of work to a manageable amount by considering ratios of partition functions.

$$\begin{aligned} \frac{h_N}{h_{N-1}}\left(1+\frac{1}{N}\right) = \frac{\mathcal{Z}_{N+1}\mathcal{Z}_{N-1}}{\mathcal{Z}_N^2} &= \exp\left[-(\mathcal{F}_{N+1} - 2\mathcal{F}_N + \mathcal{F}_{N-1})\right] \\ &\simeq \lim_{N\to\infty} \exp\left[-\frac{\partial^2 \mathcal{F}}{\partial N^2}\right] \end{aligned} \tag{26}$$

For large N, then $\partial_N^2 \mathcal{F} \simeq -\log(h_N/h_{N-1})$. Thus to calculate this variation of the free energy with N, we are only interested in the ratio of the last two weights, and in fact we never need to calculate any individual weights explicitly. This then gives a flavor for the analysis that goes into solving these models. I refer the reader to the original references[6,7] for more of the details and will move ahead to the final solution.

The results are most simply described in terms of a scaling parameter, $t = N^{4/5}|1 + b_c/b|$. (Recall that the double scaling limit involves taking $N \to \infty$ and $b \to -b_c$.) The second derivative of the free energy $\partial_t^2 \mathcal{F} = f$ satisfies a nonlinear differential equation, $t = f^2 - \frac{1}{3}\partial_t^2 f$, known as the Painleve I equation.[21] For large t, one can construct a series solution for this equation

$$f = t^{1/2}\left(1 - \frac{1}{24}t^{-5/2} - \frac{49}{1152}t^{-5} - \dots\right) \tag{27}$$

which integrated twice yields

$$\mathcal{F} = \frac{4}{15}t^{5/2} + \frac{1}{24}\log t - \frac{7}{1440}t^{-5/2} - \ldots \tag{28}$$

Thus we see an expansion naturally emerges in $t^{-5/2} = N^{-2}|1 + b_c/b|^{-2+\gamma} = g_R^2$ with $\gamma = -1/2$, which is precisely the result which arose in the previous discussion.

One can also find the double scaling solution of the multicritical models in a similar way.[6,7] In the case of Eq. 18 with proper tuning of the matrix coupling constants, the specific heat is $\partial_t^2 \mathcal{F} = f$ where f satisfies a more complicated nonlinear equation, $t = f^3 - f\partial_t^2 f - \frac{1}{2}(\partial_t f)^2 + \frac{1}{10}\partial_t^4 f$. One constructs a series solution for $\mathcal{F}$ in $g_R^2 = t^{-(2+\frac{1}{3})}$. Adding further interactions to the matrix action leads to a system of multicritical solutions for which the specific heat satisfies an equation of the form $t = f^k + \ldots + \partial_t^{(2k-2)} f$ and has a series expansion in $g_R^2 = t^{-(2+\frac{1}{k})}$. This hierarchy of equations has been studied in relation with the Korteweg-deVries equation.[22]

Originally it was stated that the models under discussion were exactly soluble, but thus far I have only indicated that one can calculate the free energy. In fact though correlation functions for a large class of operators are calculated in terms of the free energy.[7] These scaling operators are special linear combinations of monomials $\mathrm{Tr}\,\Phi^k$, and are in one-to-one correspondence with the tuned multicritical matrix actions. The full results use some rather complex mathematical formalism, known as the Gelfand-Dikii equation for the resolvent of the Shrödinger operator.[23] While these scaling operators correspond to having punctures in the world-sheet, one can also calculate correlation functions for operators which introduce boundaries of a finite length.[22,24]

After the hermitian matrix model was solved, a great deal of effort was put into solving new models. Some of the models which were successfully solved involve different ensembles of matrices, but still reproduce two-dimensional gravity and the Korteweg-deVries hierarchy as in the hermitian case.[25] Yet other models correspond to new physical systems including: unoriented surfaces,[26] open strings,[27] strings in one-dimension,[28] and gravity coupled to new world-sheet matter systems.[29] Double scaling solutions were also found for models where the physical interpretation is unknown,[22] and where branched polymers are generated rather than two-dimensional surfaces.[31] Given this array of solved models, the problem remains to determine what insights are to be gained for more realistic string theories or theories of quantum gravity. One interesting result is that nonperturbative effects appear to be much stronger in string theories than would be expected from our experience with field theory.[32] There is also a less specific lesson to be learned for gravity. The double scaling approach to matrix models provides a multitude of results with relatively little effort. Beginning with the continuum theory in Eq. 5, it is a far more arduous exercise to reproduce a small fraction of these results.[33] The lesson then is that regarding quantum gravity as simply a quantum field theory of the metric is unlikely to be the most successful approach to understanding this problem. The question remains how best to understand quantum gravity in four dimensions?

This work was supported by NSERC of Canada, and Fonds FCAR du Québec. I would like to thank the organizers of the 4th Canadian Conference on General Relativity and Relativistic Astrophysics for the opportunity to present this talk. I would also like to thank the Aspen Center for Physics, where this manuscript was completed. I am also grateful to A. Anderson and V. Periwal for their comments on earlier drafts of the manuscript.

1. A. Polyakov, *Phys. Lett.* **103B** (1981) 207

2. A. Polyakov, *Phys. Lett.* **103B** (1981) 211

3. C. Callan, D. Friedan, E. Martinec and M.J. Perry, *Nucl. Phys.* **B262** (1985) 593.

4. D.J. Gross and V. Periwal, *Phys. Rev. Lett.* **60** (1988) 2105; **61** (1988) 1517

5. V.G. Knizhnik, A.M. Polyakov and A.B. Zamolodchikov, *Mod. Phys. Lett.* **A3** (1988) 819; A.M. Polyakov and A.B. Zamolodchikov, *Mod. Phys. Lett.* **A3** (1988) 1213; A.M. Polyakov, *Mod. Phys. Lett.* **A2** (1987) 893

6. E. Brézin and V.A. Kazakov, *Phys. Lett.* **236B** (1990) 144; M.R. Douglas and S.H. Shenker, *Nucl. Phys.* **B335** (1990) 635; D.J. Gross and A.A. Migdal, *Phys. Rev. Lett.* **64** (1990) 127

7. D.J. Gross and A.A. Migdal, *Nucl. Phys.* **B340** (1990) 333

8. S.W. Hawking, *Phys. Rev.* **D37** (1988) 904; *Phys. Lett.* **195B** (1987) 337; S.W. Hawking and R. Laflamme, *Phys. Lett.* **209B** (1988) 39; S. Giddings and A. Strominger, *Nucl. Phys.* **B306** (1988) 890; **B307** (1988) 854; G.V. Lavrelashvili, V.A. Rubakov and P.G. Tinyakov, *JETP Lett.* **46** (1987) 167; *Nucl. Phys.* **B299** (1988) 757; *Mod. Phys. Lett.* **A3** (1988) 1231

9. S. Coleman, *Nucl. Phys.* **B310** (1988) 643

10. F. David, *Nucl. Phys.* **B257** (1985) 45; V. Kazakov, *Phys. Lett.* **150B** (1985) 282; J. Ambjørn, B. Durhuus and J. Fröhlich, *Nucl. Phys.* **B257** (1985) 433

11. T. Regge, *Nuo. Cim.* **19** (1961) 558

12. See following and the references therein: V.A. Kazakov and A.A. Migdal, *Nucl. Phys.* **B311** (1988) 171

13. D. Bessis, C. Itzykson, and J.-B. Zuber, *Adv. Appl. Math.* **1** (1980) 109

14. G. 't Hooft, *Nucl. Phys.* **B72** (1974) 461

15. E. Witten, preprint IASSNS-HEP-90/45 (1990)

16. E. Brézin, C. Itzykson, G. Parisi, and J.-B. Zuber, *Comm. Math. Phys.* **59** (1978) 35

17. F. David, *Nucl. Phys.* **B348** (1991) 507

18. J. Anglin and R.C. Myers, in preparation

19. V.A. Kazakov, *Mod. Phys. Lett.* **A4** (1989) 2125

20. M. Staudacher, *Nucl. Phys.* **B336** (1990) 349

21. E.L. Ince, *Ordinary Differential Equations*, Dover (New York, 1956)

22. T. Banks, M. Douglas, N. Seiberg and S. Shenker, *Phys. Lett.* **238B** (1990) 279

23. I. Gelfand and L. Dikii, *Usp. Matem. Nauk.* **30** (1975) 5

24. G. Moore, N. Seiberg and M. Staudacher, preprint Rutgers RU-91-11 (1991); E. Martinec, G. Moore and N. Seiberg, preprint Rutgers RU-91-14 (1991)

25. R.C. Myers and V. Periwal, *Phys. Rev.* **D42** (1990) 3600; T.R. Morris, preprint Fermilab-Pub-90/121-T (1990)

26. E. Brézin and H. Neuberger, *Phys. Rev. Lett.* **65** (1990) 2098; *Nucl. Phys.* **B350** (1991) 513; G.R. Harris and E.J. Martinec, *Phys. Lett.* **245B** (1990) 384; R.C. Myers and V. Periwal, *Phys. Rev. Lett.* **64** (1990) 3111

27. I.K. Kostov, *Phys. Lett.* **238B** (1990) 181; J.A. Minahan, preprint U. Virginia 91-0105 (1991)

28. D.J. Gross and N. Miljković, *Phys. Lett.* **238B** (1990) 217; E. Brezin, V.A. Kazakov and Al.B. Zamolodchikov, *Nucl. Phys.* **B338** (1990) 673; P. Ginsparg and J. Zinn-Justin, *Phys. Lett.* **240B** (1990) 333

29. D.J. Gross and A. Migdal, *Phys. Rev. Lett.* **64** (1990) 717; E. Brezin, M. Douglas, V. Kazakov and S. Shenker, *Phys. Lett.* **237B** (1990) 43; C. Crnkovic, P. Ginsparg and G. Moore, *Phys. Lett.* **237B** (1990) 190; M. Douglas, *Phys. Lett.* **238B** (1990) 176

30. V. Periwal and D. Shevitz, *Phys. Rev. Lett.* **64** (1990) 1326; *Nucl. Phys.* **B344** (1990) 731; R.C. Myers and V. Periwal, *Phys. Rev. Lett.* **65** (1990) 1088; M.M. Robinson, preprint U. Chicago EFI-90-71 (1990)

31. A. Anderson, R.C. Myers and V. Periwal, *Phys. Lett.* **254B** (1991) 89; *Nucl. Phys.* **B360** (1991) 463; S. Nishigaki and T. Yoneya, *Nucl. Phys.* **B348** (1991) 787; P. Di Vecchia, M. Kato and N. Ohta, *Nucl. Phys.* **B357** (1991) 495

32. S.H. Shenker, preprint Rutgers RU-90-47 (1990)

33. M. Goulian and M. Li, *Phys. Rev. Lett.* **66** (1991) 2051

BROKEN WEYL INVARIANCE AND GEOMETRIC PARTICLES

W.R. WOOD and G. PAPINI

Department of Physics, University of Regina
Regina, Saskatchewan, S4S 0A2, Canada

ABSTRACT

The basis on which Weyl's unified theory of gravitation and electromagnetism was rejected is reconsidered from a new perspective. It is argued that while Weyl's theory, as indeed any classical theory, is incapable of explaining atomic phenomena, this does not nullify the geometric interpretation for the exterior electromagnetic field; it simply reflects the fact that some form of quantization is needed to account for atomic standards of length. To support this argument a model of extended particles represented by bubbles in Weyl space with the conformal invariance broken in the interior anti-de Sitter space is presented. In the present work, an atomic standard of length is introduced by assumption rather than as a consequence of a quantization mechanism.

1. Weyl-Dirac Theory

Weyl's proposal[1] to unify gravitation and electromagnetism geometrically was based on the suggestion that not only the direction, but also the length of a parallel transported vector could change. To describe the change mathematically Weyl introduced a vector field κ^μ, which together with the metric tensor $g_{\mu\nu}$ comprised the fundamental fields of the new geometry. In regions where the field tensor $f_{\mu\nu} = \kappa_{\nu,\mu} - \kappa_{\mu,\nu}$ is nonzero, Weyl's geometry does not support an absolute standard of length. Under the local scaling of length $\tilde{\ell} = \sigma(x)\ell$, the resulting transformations $\tilde{\kappa}_\mu = \kappa_\mu + (\ln\sigma)_{,\mu}$ and $\tilde{f}_{\mu\nu} = f_{\mu\nu}$ suggest a geometric interpretation of the electromagnetic *field*. The metric, which transforms as $\tilde{g}_{\mu\nu} = \sigma^2 g_{\mu\nu}$, is said to have a conformal weight of two: $w(g_{\mu\nu}) = 2$. It follows that $w(\sqrt{-g}) = 4$.

To develop a physical theory in Weyl space it is convenient to introduce a gauge-covariant calculus[2] based on the gauge-covariant derivative $\bar{\Box}$ and the semi-metric connection

$$\bar{\Gamma}^\alpha{}_{\mu\nu} = \Gamma^\alpha{}_{\mu\nu} + g_{\mu\nu}\kappa^\alpha - g^\alpha{}_\mu\kappa_\nu - g^\alpha{}_\nu\kappa_\mu, \tag{1}$$

where an overbar is used to distinguish objects from their Riemannian counterparts (in this case the covariant derivative $\Box$ and the Christoffel symbol of the second kind $\Gamma^{\alpha}{}_{\mu\nu}$). The curvature tensor $\bar{R}^{\alpha}{}_{\beta\mu\nu}$ that is defined in terms of $\bar{\Gamma}^{\alpha}{}_{\mu\nu}$ (the conventions of MTW[3] are adopted throughout the present work) is not antisymmetric in its first two indices as in the Riemannian case. Since the conformal weight of the scalar curvature $\bar{R} \equiv g^{\mu}{}_{\nu}\bar{R}^{[\alpha\nu]}{}_{\alpha\mu}$ is -2, a direct generalization of the Einstein-Hilbert action is not conformally invariant. Weyl's $\bar{R}^2$ theory, which did not contain an absolute standard of length, was strongly criticized since fixed standards are realized in the physical world in the form of atomic clocks. It was argued (see Ref.4) that the application of Weyl's formula for the change in length of a parallel transported vector $\delta\ell = \ell\kappa_{\mu}\delta x^{\mu}$ to the period of an atomic clock in a Coulomb potential $\varphi(r)$ yields the unphysical variable period $\tau = \tau_0 e^{c\varphi(r)t}$. However, notice that this criticism assumes that the formula for the parallel transport of a vector can be used to describe not only the properties of the geometry, but also the properties of atoms. The validity of this assumption is very questionable since a proper description of atoms requires some form of quantization. Viewed in this way, the criticism of Weyl's theory applies to any classical theory of electromagnetism, including Maxwell's theory, which predicts that a bound electron will spiral into the nucleus.

In the present work, the possibility of breaking the conformal invariance in a manner such that atomic standards can coexist with a geometric interpretation of the exterior electromagnetic field is considered as a precursor to the issue of quantization. This symmetry breaking approach to rehabilitate Weyl's theory differs fundamentally from Dirac's two metric approach[5]. In the present work a spacetime interval will be described by one and only one metric. However, Dirac's method of introducing a real scalar field β to achieve a conformally invariant action that is linear in $\bar{R}$ will be used in the present analysis of geometric particles or bubbles in Weyl space, where the conformal invariance is broken in the interior space.

2. The Particle Model

The analysis of geometric particles treated as thin shells is most naturally given in terms of the Gauss-Mainardi-Codazzi (GMC) formalism where a timelike hypersurface Σ, which represents the history of the thin shell, divides spacetime into two regions (V^I, V^E). The intrinsic metric on Σ is given by $h_{\mu\nu} = g_{\mu\nu} - n_{\mu}n_{\nu}$, where n^{μ} is well-defined as a unit spacelike vector field normal to Σ since $w(n_{\mu}n^{\mu}) = 0$. The (three-dimensional) extrinsic curvature tensor is defined by

$$\bar{K}_{\mu\nu} = -h_{\mu}{}^{\alpha}\,\bar{\Box}_{\alpha}\,n_{\nu} = K_{\mu\nu} + h_{\mu\nu}n^{\alpha}\kappa_{\alpha}, \tag{2}$$

where the last equality is obtained by using $w(\bar{K}_{\mu\nu}) = 1$. The development of the GMC formalism in Weyl geometry ultimately yields the equations[6]

$$n_\mu n^\nu G^\mu{}_\nu = -\frac{1}{2}({}^3R + K_{\mu\nu}K^{\mu\nu} - K^2) + D_\mu \kappa^\mu + 2h_\mu{}^\nu \kappa^\mu \kappa_\nu + 2Kn^\mu \kappa_\mu \tag{3}$$

$$n_\mu h_\alpha{}^\nu G^\mu{}_\nu = D_\alpha K - D_\mu K^\mu{}_\alpha \tag{4}$$

$$\begin{aligned} h^\alpha{}_\mu h_\beta{}^\nu G^\mu{}_\nu = {}^3G^\alpha{}_\beta \; &+ \; (K^\alpha{}_\beta - h^\alpha{}_\beta K)_{,n} - KK^\alpha{}_\beta + \frac{1}{2}h^\alpha{}_\beta(K_{\mu\nu}K^{\mu\nu} + K^2) \\ &- \; 2(K^\alpha{}_\beta - h^\alpha{}_\beta K)n^\lambda \kappa_\lambda \end{aligned} \tag{5}$$

which generalize the corresponding equations in Riemannian geometry (cf. Ref.3) and which are to be equated to the corresponding components of the stress-energy tensor $T^\mu{}_\nu$. The intrinsic stress-energy tensor on Σ is defined by

$$S^\mu{}_\nu \equiv \lim_{\varepsilon \to 0} \int_{-\varepsilon}^{\varepsilon} T^\mu{}_\nu dn \tag{6}$$

and is proportional to the jump in the extrinsic curvature $\gamma^\mu{}_\nu \equiv [K^\mu{}_\nu]$, where $g_{\mu\nu}$, κ^μ and β are assumed to be continuous across Σ, but their normal derivatives discontinuous. Integrating Eqs.3-5 in the normal direction from $-\varepsilon$ to $+\varepsilon$ across Σ and then taking the limit as $\varepsilon \to 0$ yields the usual result:

$$h^\alpha{}_\mu h_\beta{}^\nu S^\mu{}_\nu = \gamma^\alpha{}_\beta - h^\alpha{}_\beta \gamma, \qquad S^n{}_\mu = 0. \tag{7}$$

The GMC formalism is here applied to Dirac's choice[5] for a conformally invariant action:

$$I_D = \int \{-\frac{1}{4}f_{\mu\nu}f^{\mu\nu} + \beta^2 R + 6\beta_{,\mu}\beta^{,\mu} - \lambda\beta^4\}\sqrt{-g}d^4x, \tag{8}$$

where $w(\beta) = -1$. The field equations that follow from Eq.8 are Maxwell's equations $\Box_\nu f^{\mu\nu} = 0$, Einstein's equations

$$G_{\mu\nu} = \frac{1}{2\beta^2}E_{\mu\nu} + \frac{2}{\beta}(\Box_\nu \Box_\mu \beta - g_{\mu\nu}\Box_\alpha \Box^\alpha \beta) - \frac{1}{\beta^2}(4\beta_{,\mu}\beta_{,\nu} - g_{\mu\nu}\beta_{,\alpha}\beta^{,\alpha}) + \frac{1}{2}\lambda g_{\mu\nu}\beta^2, \tag{9}$$

where $E_{\mu\nu}$ is the Maxwell tensor, and the scalar field equation which is identically the trace of Eq.9.

In the present model, a particle is represented by a spherically symmetric shell that carries a charge q. Application of the GMC formalism requires the determination of the interior and exterior line elements

$$ds^2_{I,E} = -e^{\nu(r)}_{I,E} dt^2_{I,E} + e^{\mu(r)}_{I,E} dr^2 + r^2(d\theta^2 + \sin^2\theta d\phi^2), \tag{10}$$

as well as the intrinsic stress-energy tensor $S^{\mu}{}_{\nu}$. In V^I, the assumptions that $\kappa_\mu = 0$ (which establishes length integrability in the spacetime region occupied by the particle) and $\beta = \beta_0$, a constant (which breaks the interior conformal invariance and fixes the scale of the particle), reduce Eq.9 to that of anti-de Sitter geometry (with pressure $\lambda\beta_0^2/2$) so that

$$e_I^{-\mu} = 1 + \frac{1}{6}\lambda\beta_0^2 r^2 = e_I^{\nu}. \tag{11}$$

In V^E, $\beta = \beta(r)$ and the static solution is found to be

$$e_E^{-\mu} = (1 + r\beta^{-1}d\beta/dr)^{-2}[1 - \frac{2m}{\beta r} + \frac{q^2}{4\beta^2 r^2} + \frac{1}{6}\lambda\beta^2 r^2] \tag{12}$$

$$e_E^{\nu} = (\ell_0\beta)^{-2}[1 - \frac{2m}{\beta r} + \frac{q^2}{4\beta^2 r^2} + \frac{1}{6}\lambda\beta^2 r^2], \tag{13}$$

where the dimensionless integration constants m and q have been chosen such that for $\beta = 1$, $\ell_0 = 1$, the corresponding Riemannian solution is recovered. The integration constant ℓ_0, which arises in a radial integration from the boundary of the spherical shell to the field point r, has a dimension of length. Hence, the interesting result is obtained that a length scale appears in the conformally invariant exterior geometry. From Eq.6 the only nonvanishing components of the intrinsic stress-energy tensor are calculated to be

$$h^{\alpha}{}_{\mu} h_{\beta}{}^{\nu} S^{\mu}{}_{\nu} = -2\alpha h^{\alpha}{}_{\beta}, \qquad \alpha \equiv [n^{\mu}(\ln\beta)_{,\mu}], \tag{14}$$

corresponding to a domain wall with surface energy density 2α and equal surface tension that opposes the interior pressure and Coulomb repulsion.

To determine the properties of the spherical shell one considers the intrinsic metric $ds^2 = -d\tau^2 + R^2(\tau)(d\theta^2 + \sin^2\theta d\phi^2)$, where the equation $r = R(\tau)$ describes the motion of the shell. The equation $[K^{\theta}{}_{\theta}] = \sigma$, which follows from Eqs.7 and 14 can be solved to yield the result

$$\begin{aligned} m = \frac{q^2}{8\beta R} &+ \frac{\alpha R^2}{\ell_0^2\beta}\Omega^4(1 + \dot{R}^2 + \frac{1}{6}\lambda\beta_0^2 R^2)^{\frac{1}{2}} + \frac{1}{12}\lambda R^3(\beta^3 - \frac{\beta_0^2}{\ell_0^2\beta}\Omega^4) \\ &- \frac{1}{2}\frac{\alpha^2 R^3}{\ell_0^2\beta}\Omega^4 + \frac{1}{2}\beta R[1 + \dot{R}^2\Omega^2 - \frac{\Omega^4}{\ell_0^2\beta^2}(1 + \dot{R}^2)], \end{aligned} \tag{15}$$

where the relationship $R(K^{\theta}{}_{\theta})^{I,E} = \mp(e_{E,I}^{-(\mu+\nu)}\dot{R}^2 + e_{E,I}^{-(2\mu+\nu)})^{\frac{1}{2}}$ (see Ref.6) and Eqs.11-13 have been used and $\Omega \equiv (1 + R\beta^{-1}d\beta/dr)$. Perhaps the most interesting contribution to the energy in Eq.15 comes from the last term that is purely a consequence of the Weyl geometry since, in the Riemannian limit ($\beta = 1$, $\ell_0 = 1$), this term vanishes. The remaining terms have Riemannian anologues that, in order of appearance in Eq.15 are: the electrostatic energy; a surface energy; a volume energy; and a surface-surface interaction energy.

In order for the above model to reasonably represent a particle, the spherical shell should be stable against elliptical deformations. To investigate this the Newtonian limit of Eq.15, which contains a Coulomb energy term, a surface energy term and volume energy term for net interior pressure, was analyzed as follows. First, for $q = 0$, a neutral spherical shell was shown to be stable under prolate and oblate deformations by Taylor expanding the energy about zero eccentricity. Then, a charged shell with no net interior pressure was similarly shown to be stable, contrary to a previous claim[7]. Hence, for sufficient tension to balance a net interior pressure and Coulomb repulsion, a charged shell will have a spherical equilibrium configuration.

3. Summary

The possibility of rehabilitating Weyl's unified theory of gravitation and electromagnetism by breaking the conformal symmetry in some manner rather than adopting Dirac's two metric approach was proposed. Application of the GMC formalism to spherical shells in Weyl space with the conformal invariance broken only in the interior anti-de Sitter space led to a model of stable neutral or charged particles in which the existence of a standard of length could be reconciled with Weyl's geometric interpretation of the exterior electromagnetic field.

4. Acknowledgements

This research was supported by the Natural Sciences and Engineering Research Council of Canada.

5. References

1. H. Weyl, *Sitzungsber. Preuss. Akad. Wiss.* (1918) 465.
2. For a more detailed summary see D. Gregorash and G. Papini, *Nuovo Cim.* **B63** (1981) 487, where the gauge-covariant derivative is denoted by a star.
3. C.W. Misner, K.S. Thorne and J.A. Wheeler, *Gravitation* (W.H. Freeman and Company, New York, 1973).
4. W. Pauli, *Theory of Relativity* (Pergamon Press, Oxford, 1967).
5. P.A.M. Dirac, *Proc. R. Soc. London Ser. A* **333** (1973) 403.
6. W.R. Wood and G. Papini, In preparation.
7. P. Hasenfratz and J. Kuti, *Phys. Rep.* **40** (1978) 75.

GRAVITATION ET NEUTRINOS

CLAUDE GAUTHIER
Département de mathématiques, Université de Moncton
Moncton (Nouveau–Brunswick), Canada E1A 3E9

SOMMAIRE

Nous proposons d'expliquer l'origine de l'inertie et de la gravitation par l'interaction des ondes associées aux particules fondamentales d'une nouvelle théorie unitaire. La notion de masse correspond alors à celle de section efficace. Des indices nous incitent à croire que ces particules pourraient être identifiées aux neutrinos.

1. De nos jours, la presque totalité des tentatives d'unification de la physique théorique visent à adapter le formalisme quantique à la gravitation. Ce problème peut toutefois être traité de façon non quantique. Dans cet esprit, nous avons déjà montré que l'utilisation d'une variété cadre représentant un hyperespace–temps avec un espace supérieur multiconnexe permettait d'interpréter plusieurs caractéristiques de la physique quantique en termes de ses propriétés géométriques[1,2]. L'objet principal de cette présentation est de préciser la place qu'occupe la gravitation dans l'ébauche de théorie unitaire construite à partir de cette variété et que nous appelons théorie des toriions.

2. L'hyperespace–temps est supposé de la forme: $V_{11} = M_4 \times (S^7(K)/\Gamma)$, où M_4 désigne l'espace–temps, $S^7(K)$ est la 7–sphère de faible courbure K et Γ est l'un de ses groupes discrets d'invariance ayant un ordre N très élevé. Appelons hyperéther la substance de V_{11}, c'est–à–dire ce qui sert de support aux qualités par lesquelles les phénomènes se manifestent dans M_4. Nous supposerons que l'hyperéther possède une rigidité intrinsèque Φ s'harmonisant à la structure topologique de V_{11}. Le champ Φ est ainsi défini sur V_{11} et prend ses valeurs dans l'espace d'une représentation de Γ. On supposera donc que $\Phi \equiv |\Phi| \exp(i\xi)$, possède une densité énergétique décrite par un potentiel du type:

$$V(\Phi,R) = (2 + \cos N\xi)\,(\Phi^4 - 2B^2 R^{-2} \Phi^2),$$

où l'angle $\xi \in [0,2\pi)$ est mesuré par rapport à une valeur initiale donnée, R représente la courbure scalaire de V_{11} et B est la rigidité minimale de l'hyperéther. Lorsque $R \neq 0$, ce potentiel a N minimums situés sur un cercle:

$$\Phi_k = BR^{-1} \exp(2k\pi i/N), \quad k = 0, ..., N-1. \tag{1}$$

La multiconnexité de l'espace supérieur implique l'existence d'une jauge discontinue appelée jauge supérieure. Pour la théorie unitaire anticipée, l'unicité de la structure de base associée à un M_4 minkowskien requiert de fixer un même $k = k_0$ pour tout $x \in M_4$, c'est–à–dire de fixer la jauge supérieure au–dessus de chaque point de M_4. Le choix d'une telle structure de base brise spontanément la symétrie de la jauge supérieure. Ce phénomène se produit lors de toute interaction dans M_4 et fait apparaître des entités géométriques intrinsèques à V_{11} que nous appelons toriis. Cependant, la latitude relative à la localisation de l'angle zéro pour l'ensemble (1) met en évidence une nouvelle jauge locale dans M_4 dont le groupe d'invariance est U(1): nous l'appelons jauge du zéro. La phase entre la structure de base autocohérente d'un observateur et celle, libre, d'une structure de base non encore observée donne lieu à une onde nouvelle appelée onde toriidale. Nous nommons toriion l'entité qui se déplace dans M_4 sous la forme d'une onde toriidale et s'y manifeste sous forme ponctuelle, à cause du torii présent au moment où celle–ci est impliquée dans une interaction.

3. Nous savons déjà qu'au déplacement d'une onde toriidale dans un M_4 minkowskien correspond la giration dans V d'une structure de base libre relativement à une structure de base autocohérente, ce qui détermine sa fréquence apparente. L'invariance des ondes toriidales par rapport aux déplacements dans un M_4 minkowskien équivaut donc à l'invariance de la jauge du zéro relativement aux rotations dans V. Selon le théorème de Noether, l'invariance précédente entraîne la conservation de l'énergie–impulsion associée aux ondes toriidales. Ce résultat se généralise directement au cas où M_4 est riemannien. L'impulsion des toriions devient alors une manifestation de l'invariance de la jauge du zéro par rapport au groupe U(1). Nous allons maintenant montrer qu'il en est de même pour l'inertie et la gravitation.

A cet effet, considérons un corps, c'est–à–dire une agrégation de toriions dans une région spatiale limitée de M_4. Supposons que M_4 soit rempli d'ondes toriidales isotropiquement distribuées et pour lesquelles les corps sont presque transparents, une onde n'entrant en interaction avec un corps que si elle rencontre l'un de ses toriions. Lors d'une telle interaction avec un corps, ce dernier subit l'impulsion du toriion associé à l'onde. La masse inertielle du corps s'interprète ici comme la section efficace totale des toriions qui le composent relativement aux interactions toriidales. Cette masse se manifestera donc seulement au moment où le corps sera impliqué dans une interaction toriidale.

Afin de préciser la relation entre les ondes toriidales et la gravitation considérons deux corps dans une région minkowskienne de M_4 et initialement au repos l'un par rapport à l'autre. Chaque corps joue pour l'autre le rôle d'un écran à une partie des ondes toriidales venant du milieu ambiant. Dans la direction déterminée par la droite spatiale passant par le centre de chacun des corps, ceux–ci recoivent en fait plus d'ondes venant de l'extérieur que de la région située entre les corps. Plus les corps seront près l'un de l'autre et plus ce déséquilibre sera grand. L'ensemble des ondes impliquées fait que ce phénomène se manifestera par une poussée tendant à rapprocher les corps l'un de l'autre. De plus, cette poussée sera a priori proportionnelle aux nombres de toriions formant chacun des corps et variera comme l'inverse du carré de la distance spatiale entre les corps. L'argument précédent se développe aisément à l'intérieur d'un modèle simplifié où les corps sont des sphères[3]. On retrouve alors la loi newtonienne comme première approximation de la force d'attraction entre les corps en définissant la constante gravitationnelle G par:

$$G = \frac{h}{\pi c} \int_0^\infty \nu\, n(\nu)\, d\nu,$$

où h et c sont respectivement la constante de Planck et la vitesse de la lumière, tandis que $n(\nu)$ est le flux des ondes toriidales de fréquence ν dans une direction orientée de l'espace tridimensionnel usuel.

Complétons l'identification de la force d'attraction engendrée par la symétrie giratoire de V à la gravitation en la reliant à la relativité générale dans M_4. Lorsqu'un corps se déplace dans M_4, l'onde toriidale de chacun de ses toriions demeure associée au minimum radial de V. La constance de la courbure scalaire de l'espace supérieur fait alors que le mouvement s'effectuera toujours suivant une trajectoire pour laquelle la courbure scalaire de M_4 correspond au minimum radial de V, c'est–à–dire selon le cercle passant par le fond de chacune de N cuvettes de $V(\Phi,R)$. Entre deux points donnés de M_4, sa trajectoire sera donc celle qui rendra minimale, relativement au potentiel V, l'intégrale de la courbure scalaire de V_{11}, étendue à V_{11} entre les instants de départ et d'arrivée. Cette prescription justifie l'application du principe variationnel de Hilbert à V_{11}, que l'on peut réduire à M_4 parce que la courbure scalaire de l'espace supérieur est constante. Notons enfin que si le corps participe à une interaction toriidale, l'impulsion impliquée se manifestera toujours dans la direction spatio–temporelle correspondant à la tangente au cercle déterminé par la valeur de R au point d'interaction. Ce phénomène est identique à l'inertie.

4. Chaque torii peut être vu comme un lieu de transition entre deux sections horizontales du fibré de groupe Γ sur la base M_4. Le caractère sinusoïdale du fond circulaire de V implique en outre que les toriions se manifestent avec une

chiralité. Ces deux caractéristiques permettent d'associer un spin aux toriions. Pour le voir, notons d'abord que la structure topologique d'un torii est $S^1 \times S^7$. Lorsqu'une onde toriidale rencontre un torii, sa chiralité l'oblige alors à effectuer le circuit correspondant à S^1. L'onde toriidale se manifestera donc avec un moment cinétique propre discrétisé que nous assimilerons au spin du toriion représenté par cette onde. Soit $h/2$ l'unité de discrétisation du spin. Le toriion peut alors être représenté par une onde qui se déplace à la vitesse de la lumière dans toute région minkowskienne de M_4 et se manifeste avec une chiralité et un spin valant $h/2$. Ces propriétés nous incitent à l'identifier au neutrino. L'inertie et la gravitation seraient ainsi des conséquences de l'interaction des neutrinos avec les corps.

Le rapprochement que nous venons d'esquisser entre les neutrinos et la gravitation s'étend à l'électromagnétisme. Il s'ensuit que l'onde électromagnétique s'interprète comme une onde toriidale chiro–hélicoïdale dont le torii constitue l'électron ou le positron. La charge électrique correspond alors à la section efficace d'un tel torii vis–à–vis les ondes électromagnétiques. En adoptant la section efficace de Thompson pour l'interaction des ondes électromagnétiques avec un électron ($\sim 10^{-24}$ cm ^{2}) et la section efficace de la matière nucléaire pour l'interaction avec les neutrinos cosmiques ($\sim 10^{-45}$ cm ^{2}), dans la description de l'effet gravitationnel entre électrons, on retrouve la valeur ($\sim 10^{42}$) du rapport de la force électrique sur la force gravitationnelle entre deux électrons.

1. C. Gauthier et P. Gravel; *Jauge discontinue et particules dans un hyperespace–temps multiconnexe*, Il Nuovo Cimento, sous presse.

2. C. Gauthier et P. Gravel; *Le rôle d'un hyperespace–temps multiconnexe en mécanique quantique*, R. de recherche: Math–9, U. de Moncton, 1990.

3. C. Gauthier; *Inertie et gravitation dans la théorie des toriions*, R. de recherche: Math–10, U. de Moncton, 1991.

LORENTZ CHERN-SIMONS TERM AND THE BREAKDOWN OF THE COSMIC NO-HAIR CONJECTURE

NEMANJA KALOPER
School of Physics and Astronomy, University of Minnesota
Minneapolis MN 55455, USA

ABSTRACT

The theory of gravity and the Kalb-Ramond axion field with the Lorentz Chern-Simons coupling is considered. It is found that the extra term in the energy-momentum tensor which is produced by the coupling leads to the breakdown of the cosmic no-hair conjecture. This is illustrated by the discussion of a diagonal Bianchi II model.

One of the cornerstones of modern cosmology is the inflationary paradigm[1], which developed from the attempts to explain the initial condition problems that defied reason for a long time. In short, it proposes that there existed a stage of very fast growth of the length scale in the early universe which diluted initial abundances of matter yielding a smooth flat geometry at the end. In the framework of Einstein's general relativity the capacity of the inflationary stage to resolve the initial value problems, such as high degree isotropy and homogeneity, is guaranteed by the cosmic no-hair conjecture[2]. It stresses that under physically realistic conditions initial inhomogeneities and anisotropies are dissipated in the course of evolution and the universe reaches asymptotically a locally de-Sitter geometry. Hence it is vital for the success of inflation.

Here the version of the no-hair conjecture obtained by Wald[3] will be reviewed, as it applies to Bianchi models. We will examine the Bianchi models in the framework of the superstring motivated modification of Einstein's theory, which involves the Kalb-Ramond axion field with the Lorentz Chern-Simons (LCS) coupling to gravity. This coupling is incorporated in the action through redefinition of the axion field strenght $H_{\mu\nu\lambda}$ to ensure the gravitational anomaly cancelation in the string field theory, as explained by Green and Schwarz[4]. We want to investigate implications of this term on inflation and specifically the no-hair conjecture. Our interest is supported by the departures from the no-hair conjecture at smaller scales provided by the LCS terms, where it produces extra hair on black holes[5].

The explicit calculations show that although for certain Bianchi cases our model obeys the no-hair conjecture for others the corrections can invalidate it. This is demonstrated by numerical solutions of the equations of motion for the diagonal Bianchi II model with only one anisotropy function, $\beta_2 = \beta_3 = -\beta_1/2$, featuring recollapse. It is also manifest in a class of particular solutions that experience the exponential growth of the scale factor, but retain hair, as the rate of change of anisotropy doesn't vanish.

Our model is defined by the action that is a part of the effective action for light modes in superstring theory and includes the graviton and the Kalb-Ramond axion[6]. We presume a source of supersymmetry breaking and add an effective cosmological constant Λ to it. Hence in 4D our action is

$$S = \int d^4x \sqrt{g} \left(\frac{1}{2\kappa^2} R - H_{\mu\nu\lambda} H^{\mu\nu\lambda} - \Lambda \right) \tag{1}$$

where $H_{\mu\nu\lambda}$ is the 3-form Kalb-Ramond field strength:

$$H = dB + \omega_L \tag{2}$$

and B is the Kalb-Ramond two form, with

$$\omega_L = Tr(\omega \wedge d\omega + \frac{2}{3} \omega \wedge \omega \wedge \omega) \tag{3}$$

the LCS form and ω the spin connexion.

We ignore explicit dilaton and gauge terms, as we are not interested in them here. The content of the no-hair conjecture has been mentioned in the introduction. It gives the sufficient conditions for an expanding universe to continue to expand forever, with the expansion rate eventually approaching a constant. A more precise statement, due to Wald[3] is as follows:

All Bianchi models except Bianchi IX dynamically described by Einstein's equations

$$R_{\mu\nu} - \frac{1}{2} g_{\mu\nu} R = T_{\mu\nu} - \Lambda g_{\mu\nu} \tag{4}$$

which are initially expanding, and where the energy-momentum tensor satisfies the strong and dominant energy conditions, SEC and DEC respectively,

$$(T_{\mu\nu} - \frac{1}{2} g_{\mu\nu} T) t^\mu t^\nu \geq 0 \tag{5}$$

$$T_{\mu\nu} t^\mu t^\nu \geq 0 \tag{6}$$

for all future directed timelike vectors t^μ, asymptotically approach isotropic expansion with $K = (3\Lambda)^{\frac{1}{2}}$ and appear to be matter free. Moreover, the conclusion holds true for the Bianchi IX model too if initially the cosmological constant is larger than the initial spatial curvature with β=0.

Apparently, the the LCS terms do not alter the guise of the problem as the equations of motion have the form of Eq. (4). But we note that the energy-momentum of the axion involves a term linear in the field strength[7], and therefore there is no guarantee that it will obey the energy conditions. On the contrary, it violates SEC and DEC. With explicit computation one can see that in certain Bianchi models the SEC and DEC hold due to the properties of the structure constants: for example, for general Bianchi I and V, the LCS correction in the energy-momentum vanishes. This occurs also in the diagonal Bianchi III and IV models. However, if it is non-zero, it can become large and negative, and hence lead to the breakdown of the cosmic no-hair conjecture[8].

The simplest Bianchi case which exhibits breakdown of the cosmic no-hair conjecture in our model is the diagonal Bianchi II model with only one anisotropy function, $\beta_2 = \beta_3 = -\beta_1/2$. The equations of motion for this model are given in ref. (7). The free parameters determining the solutions are Λ and the initial values of the anisotropy, shear and the axion field rate of change in time[7]. A class of exact solutions is easy to obtain. A guess suggested by the form of the equations of motion $R = \frac{1}{\eta}e^{2\beta}$, with $\eta = (e^{2\beta}/R)|_{t_0}$ the effective coupling constant. This ansätz eliminates R and β from the equations of motion and gives the identity $K = 6\dot{\beta}$. The consistency of the system requires $\dot{\beta} = \rho = const.$, and then ρ and η are, in the parametric form with a positive parameter $p > 2/9$:

$$\begin{aligned}
\eta^2 &= \sqrt{\frac{18p + 2}{15(3p + 2/15)^2 - 64/15}} \\
\rho^2 &= p\sqrt{\frac{18p + 2}{15(3p + 2/15)^2 - 64/15}} \\
\Lambda &= \frac{3(126p^2 - 15p - 2/9)}{2(15p + 2)}\eta^2
\end{aligned} \tag{7}$$

and they exist for all $\Lambda \geq 1.64144$. The axion equation reads $\dot{b} = \frac{\eta^3}{6} - \frac{3}{4}\rho^2\eta$. Our interest in these solutions is obvious, as they actually violate the no-hair conjecture! Although the cosmological constant is positive, and the universe is initially expanding, and the volume grows exponentially, according to $\sqrt{g} = \exp(6\rho t)$ it is not matter-free.

The violation of the no-hair conjecture we found above indicates that even more severe departures from it can emerge. However, complexity of the equations of motion forces us to resort to numerical methods to investigate the possibility further. The results are presented in fig. 1.–3.

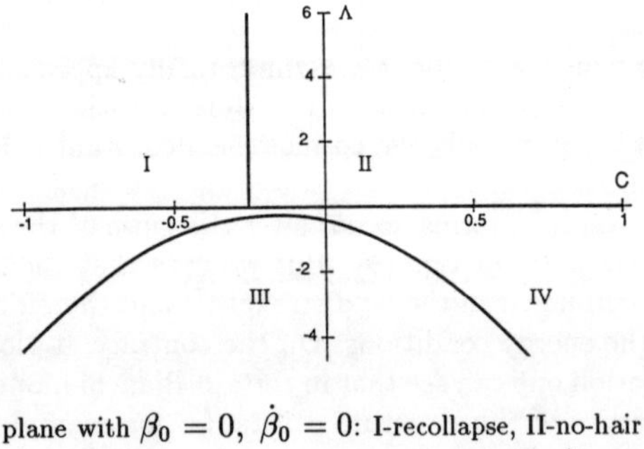

1. $\Lambda - C$ plane with $\beta_0 = 0$, $\dot{\beta}_0 = 0$: I-recollapse, II-no-hair conjecture, III-kinematically forbidden, IV-negative Λ.

Figures 1.–2. correspond to $\beta_0 = 0,\ \dot{\beta}_0 = 0$. We find the half-plane of positive Λ to be divided in two regions, one where the no-hair conjecture holds and the other where the universe recollapses (fig. 1.). A typical recollapsing solution is given in fig. 2. The time when initial singularity is reached is $\Delta t \simeq t_P$, where t_P is the Planck time. Qualitatively the same picture is obtained for nonvanishing but small $\beta_0,\ \dot{\beta}_0$.

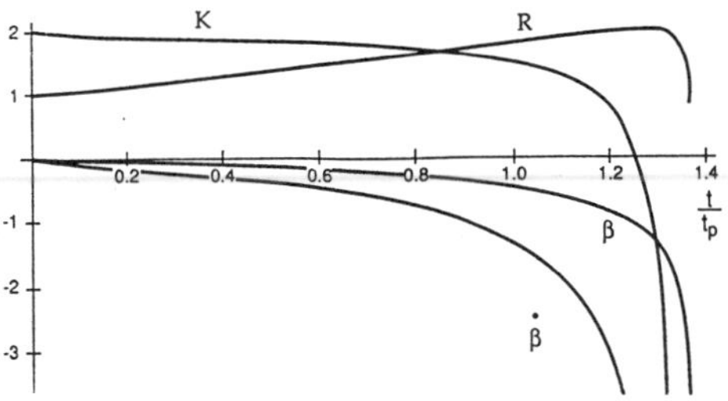

2. A typical recollapsing solution: $\Lambda = 1,\ C = -0.26$.

When initial values are large, the situation becomes more complicated. It turns out that all solutions violate the no-hair conjecture. Besides the solutions which feature recollapse, we find those which asymptotically tend to the exact solutions Eq. (7) instead of obeying the no-hair scenario (fig 3.). This is no surprise since this solution actually replaces the corresponding asymptotia $K = const.$, $\dot{\beta} = 0$ of the Einstein's theory. The numerical investigation of these solutions displays that they are stable in a large region of the parameter plane.

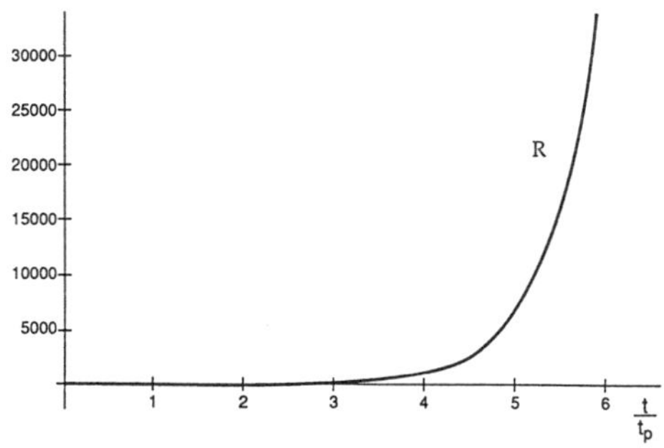

3.a A solution with axion hair: $\beta_0 = 0,\ \dot{\beta}_0 = 0,\ \Lambda = 10,\ C = 0$; the scale factor.

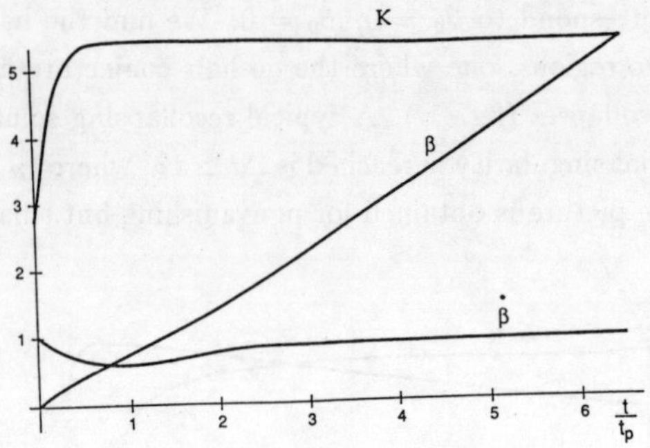

3.b The expansion rate and anisotropy.

In summary, the validity of the cosmic no-hair conjecture in the superstring motivated theory of gravity and Kalb-Ramond axion field is discussed. In the definition of the axion field strength the Lorentz Chern-Simons form has been included. It produces extra contributions to the equations of motion which violate the no-hair conjecture. The solutions featuring this are provided in a diagonal Bianchi II cosmology. They suggest that a similar situation occurs in other Bianchi models.

Acknowledgements

The author would like to thank K.A. Olive for suggesting this problem.

References

1 A.D. Linde, *Inflation and Quantum Cosmology*, Academic Press, Boston, Ma. 1990; K.A. Olive, *Phys. Rep.* **190** (1990) 302
2 G.W. Gibbons and S.W. Hawking, *Phys. Rev.* **D15** (1977) 2738; S.W. Hawking and I.G. Moss, *Phys. Lett.* **B110** (1982) 35; W.Boucher, G.W. Gibbons and G.T. Horowitz, *Phys. Rev.* **D30** (1984) 2447
3 R.M. Wald, *Phys. Rev.* **D28** (1983) 2118
4 M.B. Green and J.H. Schwarz, *Phys. Lett.* **B149** (1984) 117; A. Das, J. Maharana and S. Roy, ER13065-573/UR-1106
5 B.A. Campbell, M.J. Duncan, N. Kaloper and K.A. Olive, *Phys. Lett.* **B251** (1991) 34
6 B.A. Campbell, M.J. Duncan, N. Kaloper and K.A. Olive, *Nucl. Phys.* **B351** (1991) 778
7 N. Kaloper, UMN-TH-926/91, submitted to *Phys. Rev.* **D**
8 J.D. Barrow, *Phys. Lett.* **B180** (1986) 335; *Phys. Lett.* **B183** (1987) 285; *Phys. Lett.* **B187** (1987) 12

IMAGINARY WORMHOLES

JASON TWAMLEY AND DON N. PAGE

Theoretical Physics Institute, Department of Physics, University of Alberta Edmonton, Alberta T6G 2J1, Canada

Abstract

Using a Friedmann-Robinson-Walker minisuperspace model with a minimally coupled homogenous scalar field, we search for, and discover, wormhole-type solutions where (a) $V(\phi) = \frac{\lambda}{4}\phi^4$, (b) $V(\phi) = \frac{m^2}{2}\phi^2 + \frac{\lambda}{4}\phi^4$. For these models, stationary configurations satisfying the boundary conditions of asymptotic flatness necessitate that the scalar field be imaginary. In distinction from all previously found wormhole solutions, they do not possess a conserved charge. In the case of potential (a) all solutions found have negative actions whereas for potential (b), there exist solutions with negative, zero, and positive action. The existence of such solutions may seriously undermine current arguments concerning the resolution of the "Large Wormhole Problem" of Fischler and Susskind. Those wormholes with negative action would, presumably, not be included into the path-integral for the vacuum to vacuum quantum amplitude. The criterion used to choose those wormholes that should contribute to the path-integral differs from one previously conjectured by Halliwell and Hartle.

Introduction

Here we present new asymptotically flat Euclidean wormhole solutions. In distinction from all previously known wormhole solutions, the solutions presented here do *not* possess a conserved charge and may have positive, zero, or negative Euclidean action. These solutions are interesting for three reasons. Firstly, they dispell a common folklore in wormhole physics which states that a conserved charge is necessary for the existence of a wormhole solution with nonzero throat size. Secondly, they dispell a conjecture by Halliwell and Hartle stating that wormhole configurations with $Re(\sqrt{g}) > 0$ possess positive Euclidean actions and are thus suppressed in the Euclidean path-integral formulation of quantum cosmology [1]. Thirdly, the existence of these new solutions may seriously undermine current arguments attempting to resolve the "Large Wormhole Problem" of Fischler and Susskind [2].

To elaborate the above motivations, it has been conjectured that it is necessary to have a cyclic quantity appearing in the Lagrangian, the conservation of which holds the throat open. However, here, we present wormhole solutions without a conserved charge. In this case, the throat is 'kept open' by virtue of the nonlinear dynamics. It is well known that the Euclidean path-integral over real four-metrics and matters fields diverge. To surmount the conformal instability of the Euclidean action it has been proposed to rotate the contour of integration so as to include complex metrics and complex matter fields [3]. Halliwell and Hartle have proposed a number of criteria which the complex contour should obey in order to yield a physically appealing wavefunction [1]. The fifth and last criterion involves wormholes. In Ref [1] they conjecture that, in the same way that four-spheres with $Re(\sqrt{g}) > 0$ possess negative action, wormholes with $Re(\sqrt{g}) > 0$ possess positive action and will thus be suppressed by the Euclidean path-integral. Here, we explicitly demonstrate Euclidean wormhole solutions with $Re(\sqrt{g}) > 0$ possessing negative actions. Thus, although $\sqrt{g}$ appears to serve as good indicator concerning the appropriate inclusion of certain saddle points into the contour of integration for pure gravity, it must be augmented when one includes complex matter fields into the path-integral. The 'Large Wormhole Problem' arrises when one includes small wormhole configurations into the Euclidean path integral which not only connect the base four-geometry to itself but to other large four-geometries. It has been shown that the effective interactions generated by these small wormholes may cause the fundamental coupling constants to float unpredictably [4]. However, it was also quite generally shown that these interactions also forces wormholes to be dense in spacetime on all possible scales irrespective of their actions! [2]. This is the 'Large Wormhole Problem'. The most successful argument to date attempting to resolve this disastrous result is that proposed by Coleman and Lee [5]. They argue that, in the case where the wormholes possess conserved charges, the large wormholes with large charges will be destabilised through interactions with smaller wormholes, i.e., the small wormholes will bleed away the large charge of the large wormholes. However, for the wormholes described below possessing no conserved charges, this resolution may not be possible and the Large Wormhole Problem may appear in these theories when one allows wormhole configurations to connect separate large four-geometries. We now very briefly describe the models and end with a short discussion. For a more detailed discussion of the wormholes presented in this paper see Ref [12].

Model 1

For simplicity we use the $k = +1$ Euclidean FRW O(3)-invariant metric ansatz, appropriately scaled for later convenience, $ds_4^2 = \frac{2G}{3\pi}\left[N^2(t)dt^2 + a^2(t)d\Omega_3^2\right]$ where $d\Omega_3^2$ is the standard O(3)-invariant metric on the unit S^3, and $\hbar = c = 1$. The scalar field will be taken to be $\Phi = \sqrt{\frac{3}{4\pi G}}\,\phi(t)$, with a self-interaction potential $\tilde{V}(\Phi) = (\frac{9}{8G^2})V(\phi)$. The Euclidean Einstein-Hilbert action, with the York-Gibbons-Hawking boundary term [6] at the S^3 boundaries at $t_\pm$, is

$$I_E = \frac{1}{2}\int_{t_-}^{t_+} N dt\left(-a\dot{a}^2 + a^3\dot{\phi}^2 - a + 2a^3V\right) + \frac{1}{2}a^2(t_+) + \frac{1}{2}a^2(t_-)\ ,$$

where the overdot denotes $N^{-1}d/dt$. By varying the action with respect to the metric and matter field, one arrives at two second order ordinary differential equations for a and ϕ (ϕ real) and one first order constraint equation. We are looking for finite action wormhole solutions connecting two asymptotically flat regions, so $\dot{a} \to \pm 1$ as $t \to \pm\infty$, and $\phi \to \phi_\pm$ with $V(\phi_\pm) = 0$ and $dV(\phi_\pm)/d\phi = 0$.

It is known that no real wormhole solutions exist for pure gravity [7,8]. It is essentially necessary that the Ricci tensor (and thus the matter stress-tensor), have at least one negative eigenvalue [9]. This does not occur for a real scalar field with a non-negative potential $V(\phi)$. This restriction can be circumvented by going to imaginary $\phi = i\varphi$ (φ real). For a massless imaginary scalar field ($V = 0$), wormhole solutions have been found [10]. These have a conserved charge, the value of π_ϕ, which is the momentum conjugate to ϕ. Solutions exist for any value of π_ϕ and have a linear throat size proportional to $\pi_\phi^{1/2}$.

We are interested in discovering solutions with no conserved charge and so consider non-constant $V(\phi)$. Imposing the reasonable condition $V(\Phi) \in \mathbf{R}^+$, $\forall \Phi \in \mathbf{R}$, the simplest potential one might consider is $V = \frac{\lambda}{2p}\phi^{2p} = (-)^p \frac{\lambda}{2p}\varphi^{2p}$. However, Jungman and Wald [11] show that no finite action wormholes are possible for a real scalar field with $\phi dV/d\phi \geq 0$, and their argument also applies virtually unchanged for an imaginary scalar field with $\varphi dV/d\varphi \leq 0$. For real ϕ with V satisfying the above condition, no finite action, asymptotically flat wormholes are possible and one must consider imaginary scalar fields. For p odd and $\phi = i\varphi$ imaginary (φ real), wormholes are disallowed by Jungman and Wald's argument. We thus consider the potential $V = \frac{\lambda}{4}\phi^4 = \frac{\lambda}{4}\varphi^4$.

One can now eliminate time from the equations of motion and obtain a second order differential equation for φ in terms of $\alpha \equiv \log a$. From this equation, in the limit of large α, we get an asymptotically conserved quantity $E = \frac{1}{2}(\frac{dz}{d\alpha})^2 - \frac{1}{2}z^2 + \frac{1}{4}z^4$, where $z \equiv a\varphi$. The asymptotic behaviour for large α is that of the Giddings-Strominger solution and thus all classical solutions are asymptotically flat. However, for the action to be finite in one asymptotically flat region, the asymptotic constant E must vanish there. Eliminating the arbitrary constant associated with the zero of time, the classical equation of motion possess a two-parameter set of solutions. Imposing the conditions $E_+ \equiv E(t \to +\infty) = E_- \equiv E(t \to -\infty) = 0$ for a finite action wormhole, we obtain a discrete set of solutions.

One can now, in a perfectly standard way, obtain a large time asymptotic solution to the equations of motion in one of the asymptotically flat regions when $E = 0$, say $E_- = 0$. This will involve one arbitrary constant, c_- say, where asymptotically $\varphi \approx c_- a^{-2}$. Numerically integrating forward to large $+t$, the zeros of $E_+(c_-)$, indicating finite action wormholes, occurs for two positive values of c_-. [Note: because of the symmetry $\varphi \leftrightarrow -\varphi$, solutions with opposite values of φ and c_- also exist.] These values of c_- are (i) $c_- = c_1 \approx 0.486421\,\lambda^{-1} \approx 9.6015\,\tilde{\lambda}^{-1}$, and at (ii) $c_- = c_2 \approx 6.0055\,\lambda^{-1} = 118.5438\,\tilde{\lambda}^{-1}$. The behaviour of these solutions near the throat is shown in Figures 1 and 2.

Since these are classical solutions, the vanishing of the first variation of the action with respect to an infinitesimal constant or scaling transformation of the metric or the matter field, gives that the gravitational action vanishes, while the total action is negative. Numerically, the total action for type (i) wormhole is, $I_1 \approx -0.479314\,\lambda^{-1} \approx -9.46128\,\tilde{\lambda}^{-1}$, while for type (ii) wormhole is, $I_2 \approx -2.26035\,\lambda^{-1} \approx -44.6175\,\tilde{\lambda}^{-1}$. Since the asymptotic behaviour is that of a Giddings-Strominger wormhole, these wormholes possess an asymptotic charge, i.e., the Lorentzian momentum conjugate to Φ is asymptotically a constant. However this charge is not conserved though the throat. For type (i) wormhole: $\pi_\Phi(t = -\infty) = -\pi_\Phi(t = +\infty) = \pi_1$, whereas for type (ii) wormhole: $\pi_\Phi(t = -\infty) = \pi_\Phi(t = +\infty) = \pi_2$.

Model 2

Except for the two critically damped solutions (wormholes), the imaginary scalar field in Model 1, generically exhibits damped oscillations about $\varphi = 0$ and thus possesses an infinite action. By introducing a slight barrier at $\varphi = 0$ with $d^2V(i\varphi = 0)/d\varphi^2 < 0$ and arranging $V(i\varphi = 0) = 0$ (false vacuum) we introduce a small friction which will eventually stop the oscillations of φ about zero and cause the field to decay into a true vacuum ($V(i\varphi) < 0$) or (if one is careful) asymptotically approach the false vacuum at $\varphi = 0$. We start the scalar field at the false vacuum so that, after a number of zero crossings of φ, it finally ends at the false vacuum. We will then have a wormhole solution. It has finite action since the matter field falls off exponentially in proper time because $d^2V(i\varphi = 0)/d\varphi^2 < 0$. A suitable form for the potential that also satisfies the above mentioned reality and positivity conditions for real scalar fields is $V(i\varphi) = -\frac{m^2}{2}\varphi^2 + \frac{\lambda}{4}\varphi^4$ or $V(\phi) = \frac{m^2}{2}\phi^2 + \frac{\lambda}{4}\phi^4$, where $\phi = i\varphi$ (φ real). One can proceed as before and obtain a large time ($t \to -\infty$) asymptotically flat solution to the classical Euclidean equations of motion near $\varphi = 0$. This will involve one arbitrary constant, χ say, where asymptotically for large a, $\varphi(a) \sim \mu\,(\chi/x)^{3/2}\,e^{\chi - x}$ where $\mu = m/\sqrt{\lambda}$, and $x = \mu a$. One can either numerically or analytically integrate towards large positive t and evaluate a certain test function $F(\chi)$, the roots of which indicate the presence of asymptotic flatness on the other side of the throat. We, however, also choose to consider μ as a parameter and thus the equation $F(\chi, \mu) = 0$ will, in general possess, a discrete set of one parameter continuous families of solutions. A relatively complex computer program was written to map out the first six solution curves to the equation $F(\chi, \mu) = 0$ (see Figure 3) and to calculate pertinent properties of the resulting wormholes (e.g. minimal scale factor, action, etc.). It was found that no wormholes occur for $\mu > \mu_c \sim 0.402$. Only two types of wormhole were found (i) those with $\dot{\alpha} = \dot{\varphi} = 0$ at the throat, and (ii) those with $\dot{\alpha} = \varphi = 0$ at the throat. For a fixed $\mu < \mu_c$, there are $N_{MAX} + 1$ pairs of wormholes, where N_{MAX} is the maximum number of zero crossings of φ. These results were also obtained, to a good approximation, by analytical methods. The total Euclidean action my be positive, zero, or negative depending on the values of μ and χ of the wormhole. The solutions with zero action are indicated in Figure 3, while all solutions with the same N but with smaller χ have negative actions while those with larger χ have positive actions. For more information concerning the numerical and analytical derivations, properties and behaviours of these wormholes see Ref [12].

Discussion

We have explicitly demonstrated the existence of wormhole solutions possessing no conserved charge which may have negative action. As regards the use of a minisuperspace model one can show that the classical solutions found are solutions to the full field equations and thus the actions found will be the actions of solutions to the full Euclidean Einstein field equations. Thus, the results of the minisuperspace calculation does give some indication towards the behaviour of the full theory. Regarding the presence of an imaginary scalar field, if one follows the program of complexifying the contour of integration, one is inevitably led to complex matter fields appearing the the path integral and, in particular, saddle points with complex scalar fields. Imaginary scalar fields also arise when Lorentzian momentum boundary conditions are imposed within the Euclidean path integral formalism. Such boundary conditions are necessary to match a Lorentzian classical solution with a Euclidean classical solution, i.e. tunneling. Imaginary scalar fields arise naturally in the above models by virtue of the boundary conditions of asymptotic flatness. One is also interested in the stability properties of these new wormhole solutions. However, it is not clear at present, which types of perturbations in the metric and matter fields should the wormhole be stable under, i.e. real, imaginary, or complex?

Acknowledgments

We thank S. Davidson, J. Louko, and A. Lyons for useful discussions. This research was supported in part by NSF Grants PHY-8701860 and PHY-9006031 and by the National Sciences and Engineering Research Council of Canada.

References

1. J. J. Halliwell and J. B. Hartle, *Phys. Rev.* D**41** (1990) 1815.
2. W. Fischler and L. Susskind, *Phys. Lett.* **217**B (1989) 48.
3. J. J. Halliwell and J. Louko, *Phys. Rev.* **D42** (1990) 3997 and references therein.
4. S. Coleman, *Nucl. Phys.* B**307** (1988) 867; J Preskill, *Nucl. Phys.* B**323** (1989) 141.
5. S. Coleman and K. Lee, *Nucl. Phys.* B**341** (1990) 101.

6. G. W. Gibbons and S. W. Hawking, *Phys. Rev.* **D15** (1977) 2752.
7. G. W. Gibbons and C. Pope, *Commun. Math. Phys.* **66** (1979) 267.
8. R. Schoen and S. T. Yau, *Commun. Math. Phys.* **79** (1982) 281.
9. J. Cheeger and D. Grommol, *J. Diff. Geom.* **4** (1971) 119.
10. S. Giddings and A. Strominger, *Nucl. Phys.* **B306** (1988) 890.
11. G. Jungman and R. M. Wald, *Phys. Rev.* **D40** (1989) 2615.
12. J. Twamley and D. N. Page, "Wormholes without a conserved charge", in progress; J. Twamley, *"Complex Wormholes and the Contour of Integration in Quantum Cosmology"*, PhD thesis, University of Alberta, Canada.

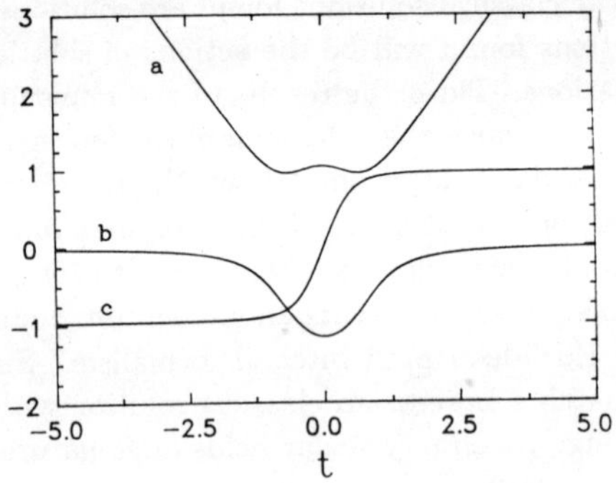

Figure 1: Behaviour of wormhole solution (i) in the throat region where the abscissa is t in the gauge $N = \lambda^{-1/2}$ with the origin of t at the point of symmetry of $a(t)$. This solution corresponds to $c_- \approx 0.486421\,\lambda^{-1}$. The separate graphs refer to (a) scale factor $\lambda^{1/2}a(t)$, (b) behaviour of φ (which does not scale with λ) and (c) $\lambda a^3\dot{\varphi}$.

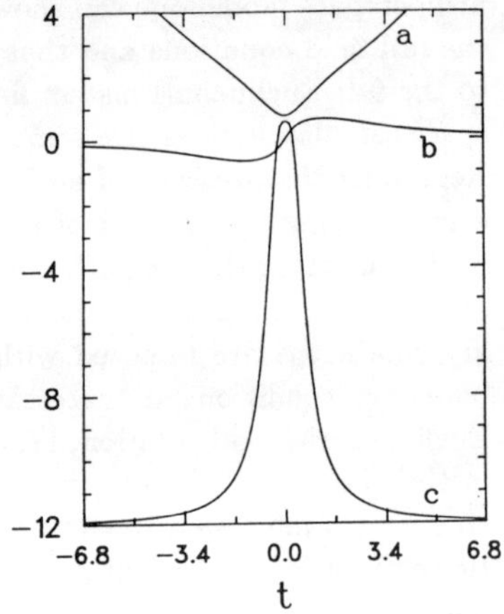

Figure 2: Behaviour of wormhole solution (ii) in the throat region. This solution corresponds to $c_- \approx 6.0055\,\lambda^{-1}$. The separate graphs refer to: (a) $\lambda^{1/2}a(t)$, (b) φ, and (c) $\lambda a^3\dot{\varphi}$. Note: we have compressed t by a factor of 5 for curve (c) so that the large t behaviour is apparent.

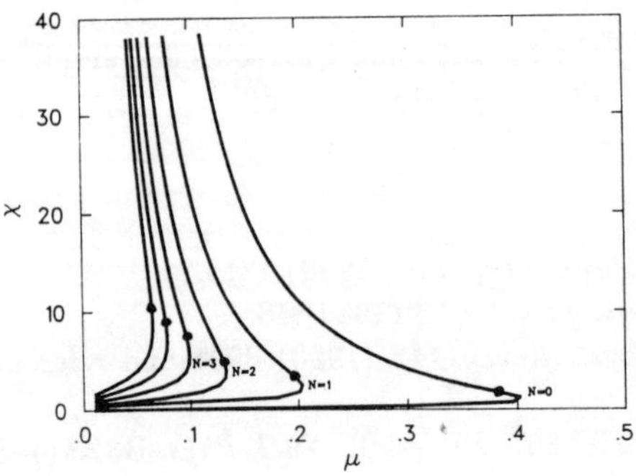

Figure 3: Numerical calculation of loci of $N = 0,..,5$ wormhole solutions in the (χ, μ) parameter space where N is the number of zero crossings of φ. Solutions with zero action are marked with black dots.

ON THE INEQUIVALENCE BETWEEN DIRAC AND REDUCED PHASE SPACE QUANTIZATION

G. Kunstatter
Physics Department
University of Winnipeg
Winnipeg, CANADA R3B 2E9

ABSTRACT

Simple geometrical techniques are used to show that the inequivalence between the Dirac and reduced phase space quantization of a certain class of gauge theories can be most easily understood as an ambiguity in the factor ordering of gauge invariant operators in the reduced theory. This operator ordering ambiguity arises naturally in Dirac quantization because the induced Hilbert space measure contains an extra contribution from the volume of the gauge orbit over each point in the physical configuration space.

There exist at least two distinct methods for handling the canonical quantization of constrained Hamiltonian systems[1,2]: reduced quantization and Dirac quantization. The reduced method first solves the constraints classically and then uses operator techniques to quantize only the physical degrees of freedom. Since the reduction to the physical phase space is done at the classical level, all information concerning the gauge structure is abandonned prior to quantization. The Dirac method, on the other hand, quantizes the full unconstrained theory first, and then imposes the constraints as operator constraints on the physical states which comprise the reduced Hilbert space. As shown in numerous recent examples[3-8], these two approaches lead to different quantum systems in general. It is fairly well known that the difference between the Dirac and reduced schemes has its origin in an operator ordering ambiguity that arises due to the extra geometrical structure present in gauge theories[4,9]. Kuchar[4], for example, has shown that for a certain class of theories one can define a Dirac quantization scheme that agrees with the reduced method by choosing an appropriate factor ordering for the operators prior to imposing constraints. In a gauge theory, however, one would like to understand all quantum effects in terms of gauge invariant quantities. It is therefore reassuring that, at least for a simple class of models, the Dirac vs. reduced ambiguity can in fact be understood in terms of an ambiguity that exists in the ordering of gauge invariant operators defined on the reduced Hilbert space[1]. Moreover, as shown below, this operator ordering ambiguity arises in the Dirac method because

[1] This has been shown for a particular quantum mechanical model in [8], but to the best of my knowledge, no general proof exists in the literature.

the corresponding Hilbert space measure retains a (gauge invariant) contribution from the volume of the gauge orbits. The present discussion summarizes the more detailed analysis in ref.[10], where further references can also be found.

Consider the general class of quantum mechanical models described by the classical Hamiltonian[4-6]:

$$H(\phi^a, P_b; \lambda^\alpha) = \frac{1}{2} g^{ab}(\phi) P_a P_b + \lambda^\alpha K^a_\alpha(\phi) P_a + V(\phi) \tag{1}$$

In the above, ϕ^a are assumed to be local coordinates for a manifold $\mathcal{C}$, with metric g_{ab}. The λ^α are Lagrange multipliers enforcing the constraints $C_\alpha \equiv K^a_\alpha P_a$ which are linear in momentum, and first class, provided the vectors $\vec{K}_\alpha \equiv K^a_\alpha \partial_a$ close under Lie derivation:

$$\mathcal{L}_{\vec{K}_\alpha} \vec{K}_\beta = f^\epsilon_{\alpha\beta} \vec{K}_\epsilon \tag{2}$$

We will assume that the structure functions $f^\epsilon_{\alpha\beta}$ are constant, in which case the constraints span a Lie algebra. In order for the constraints to be preserved in time, it is sufficient that the vectors $\vec{K}_\alpha$ be Killing vectors for the metric g_{ab}:

$$\mathcal{L}_{\vec{K}_\alpha} g_{ab} = 0, \tag{3}$$

and that the potential V be constant along the orbits of the Killing vectors. It should be noted that by utilizing DeWitt's condensed notation, in which single indices are used to denote both discrete indices and spatial coordinates, Yang-Mills theories can be put into precisely the same form as (1). However, in this case many of the relevant geometrical quantities described below are infinite and care must be taken when interpreting the formal results.

Given the metric g_{ab} and the Killing vectors $\vec{K}_\alpha$, one can define the orbit space metric:

$$g^\perp_{ab} = \left(\delta^c_a - K^c_\alpha \gamma^{\alpha\beta} K_{a\beta}\right) g_{cb} \tag{4}$$

which measures displacements orthogonal to the gauge orbits. In Eq.(4), $\gamma^{\alpha\beta}$ is the inverse of

$$\gamma_{\alpha\beta} \equiv \vec{K}_\alpha \cdot \vec{K}_\beta = K^a_\alpha g_{ab} K^b_\beta\ . \tag{5}$$

which is the metric for displacements along the gauge orbits.

Reduced phase space quantization proceeds by expressing the classical Hamiltonian in terms of physical (i.e. gauge invariant) variables. This can be done by performing a point canonical transformation to an adapted set of coordinates[11]. We choose the coordinates:

$$\phi^a \to \tilde{\phi}^{\tilde{a}} = (\xi^i, \chi^\alpha) \tag{6}$$

where $\xi^i(\phi)$, $i = 1, 2, ...\mathrm{dim}(\Sigma)$ are a complete set of gauge invariant functions and the functions $\chi^\alpha(\phi)$ are arbitrary, except for the condition that the Faddeev-Popov matrix[11]

$$Q^\alpha_\beta \equiv \{\chi^\alpha, C_\beta\} = \chi^\alpha_{,a} K^a_\beta \tag{7}$$

be invertible. This guarantees that the surfaces of constant χ^α cut each gauge orbit exactly once, at least locally.

A straightforward calculation reveals that in terms of these phase space coordinates the Hamiltonian is:

$$H = \frac{1}{2}h^{ij}\tilde{P}_i\tilde{P}_j + \tilde{V}(\xi) + \tilde{\lambda}^\alpha \tilde{P}_\alpha \tag{8}$$

where h^{ij} is the inverse of the induced metric h_{ij} defined by:

$$h_{ij} \equiv g^{\perp}_{ab}\frac{\partial\phi^a}{\partial\xi^i}\frac{\partial\phi^b}{\partial\xi^j} \tag{9}$$

In the above $\tilde{P}_i$ are canonically conjugate to ξ^i, while $\tilde{P}_\alpha$, the conjugates to χ^α, are proportional to the constraints: $\tilde{P}_\alpha = (Q^{-1})^\beta_\alpha C_\beta$. Thus, up to terms which vanish on the constraint surface, the Hamiltonian depends only on the gauge invariant functions ξ^i and their conjugate momenta. The reduced Hamiltonian is therefore defined to be:

$$H_{red} = \frac{1}{2}h^{ij}(\xi)\tilde{P}_i\tilde{P}_j + \tilde{V}(\xi) \tag{10}$$

Note that although it was not necessary to restrict to a particular surface in $\mathcal{C}$, such as $\chi^\alpha = 0$, this can in fact now be done trivially.

We now quantize (10) via the Schrodinger approach, by first defining the Hilbert space as the space of complex valued functions $\psi(\xi)$ on Σ, with inner product

$$\langle\psi_1|\psi_2\rangle_{red} = \int\left(\prod_i d\xi^i\right)\sqrt{h}\psi_1^*\psi_2 \tag{11}$$

and defining momentum operators $\hat{P}_i \equiv -i\hbar(\partial_i + \frac{1}{2}\partial_i \ln\sqrt{h})$ that are self- adjoint with respect to this inner product. In order to define the Hamiltonian, one must choose a factor ordering. The standard choice is the one that yields a kinetic term equal to the covariant Laplacian acting on the wave-functionals. It is:

$$\begin{aligned}\hat{H}_{red} &= \frac{1}{2}h^{-\frac{1}{4}}\hat{P}_i h^{\frac{1}{2}}h^{ij}\hat{P}_j h^{-\frac{1}{4}} + \tilde{V}(\xi) \\ &= -\frac{\hbar^2}{2\sqrt{h}}\partial_i\sqrt{h}h^{ij}\partial_j + \tilde{V}(\xi)\end{aligned} \tag{12}$$

This completes the formal reduced phase space quantization of this system.

The basic idea behind Dirac quantization is to quantize the Hamiltonian (1) prior to reduction and then impose the constraints as operator constraints on physical states. If we follow the procedure outlined in recent work[5,6] this requires the definition of a Hilbert space of functions on the full space $\mathcal{C}$ with a measure deduced from the kinetic term in the Hamiltonian, namely

$$\langle\psi_1|\psi_2\rangle_{dir} = \int\left(\prod_a d\phi^i\right)\sqrt{g}\psi_1^*\psi_2 \tag{13}$$

The corresponding self-adjoint momentum operator is then:

$$\hat{P}_a = -i\hbar\left(\partial_a + \frac{1}{2}\partial_a \ln\sqrt{g}\right) \tag{14}$$

The constraints can be made self-adjoint by choosing the following factor ordering[1]

$$\hat{C}_\alpha = \frac{1}{2}\left(\hat{K}^a_\alpha\hat{P}_a + \hat{P}_a\hat{K}^a_\alpha\right) = -i\hbar K^a_\alpha\partial_a \tag{15}$$

If the factor ordering ambiguity in the kinetic term is resolved by again requiring it to correspond to the Laplacian acting on the wave functions[5,6], then the Dirac Hamiltonian in the Schrodinger picture is:

$$\begin{aligned}\hat{H} &= \frac{1}{2}g^{-\frac{1}{4}}\hat{P}_a g^{\frac{1}{2}}g^{ab}\hat{P}_b g^{-\frac{1}{4}} + V(\hat{\phi}) + \lambda^\alpha\hat{C}_\alpha \\ &= -\frac{\hbar^2}{2\sqrt{g}}\partial_a\sqrt{g}g^{ab}\partial_b + V(\phi) + \lambda^\alpha K^a_\alpha\partial_\alpha \end{aligned} \tag{16}$$

The reduction to the physical Hilbert space is achieved by imposing the constraints C_α as operator constraints on physical states. In terms of the adapted coordinates (6) this implies that the physical wave functions are independent of χ^α. Up to an overall normalization factor, the measure (13) reduces on the physical Hilbert space to:

$$< \psi_1|\psi_2 >_{\rm phys} = \int\prod_i d\xi^i\sqrt{h\gamma}\psi_1^*(\xi)\psi_2(\xi) \tag{17}$$

The corresponding reduced Hamiltonian acting on physical states is:

$$\hat{H}_{dir} = -\frac{\hbar^2}{2\sqrt{h\gamma}}\partial_i\sqrt{h\gamma}h^{ij}\partial_j + \tilde{V}(\xi) \tag{18}$$

which differs both in appearance and in physical content from (12) above. As shown in ref.[5] the two Hamiltonians are not unitarily equivalent.

---

[1] I am grateful to B. Dolan for discussions on this point.

The key to understanding the relationship between (12) and (18) lies in the Hilbert space measure and its role in defining the Hamiltonian. Instead of constructing operators on the full configuration space, one can ask what are the appropriate operators on the reduced space that are self-adjoint with respect to the induced measure (17). The correct momentum operator is:

$$\hat{P}_i^{dir} = -i\hbar(\partial_i + \frac{1}{2}\partial_i \ln \sqrt{\gamma h}) \tag{19}$$

A natural, self-adjoint Hamiltonian that one can construct using this momentum operator and the metric (17) is:

$$\hat{H}_{dir} = \frac{1}{2}(h\gamma)^{-\frac{1}{4}}\hat{P}_i^{dir}(h\gamma)^{\frac{1}{2}}h^{ij}\hat{P}_j^{dir}(h\gamma)^{-\frac{1}{4}} + \tilde{V}(\xi) \tag{20}$$

A simple calculation shows that this is precisely the Hamiltonian in Eq.(18).

There is, however, another "natural" choice for the Hamiltonian factor ordering, namely:

$$\begin{aligned}\hat{H}'_{dir} &= \frac{1}{2}h^{-\frac{1}{4}}\hat{P}_i^{dir}h^{ij}h^{\frac{1}{2}}\hat{P}_j^{dir}h^{-\frac{1}{4}} + \tilde{V}(\xi)\\ &= -\frac{\hbar^2}{2}\frac{1}{\sqrt{h\gamma}}\partial_i\sqrt{h\gamma}h^{ij}\partial_j + \frac{\hbar^2}{2}\frac{D^iD_i(\gamma^{\frac{1}{4}})}{\gamma^{\frac{1}{4}}} + \tilde{V}(\xi)\end{aligned} \tag{21}$$

where D_i denotes covariant differentiation with respect to h_{ij}. As shown in [10], (21) is equivalent by a simple unitary transformation to the reduced Hamiltonian in Eq.(12). Thus, as claimed above, the difference between the reduced and Dirac quantizations (locally at least) can be traced to a factor ordering ambiguity in the Dirac method after reduction. This factor ordering ambiguity is a direct consequence of the possible contribution of the metric along the gauge orbits, $\sqrt{\gamma}$, to the Hilbert space measure. Unfortunately, the present analysis does not apply directly to systems with Hamiltonian constraints quadratic in the momenta, and hence can not be used for quantum gravity. One would nonetheless hope that similar qualitative results concerning the relationship between Dirac and reduced phase space quantization would still be valid in this case.

Acknowledgements I am grateful to A. Ashtekar, B. Dolan, M. Dubois-Violette, M. Henneaux, K. Kuchar, J. Madore, A. Rebhan and C.M. Viallet for many helpful discussions and the L.P.T.H.E.-Orsay for its kind hospitality. This research was supported in part by the Natural Sciences and Engineering Research Council of Canada.

REFERENCES

[1] P.A.M. Dirac, Can. J. Math. **2**, (1950) 129; *Lectures in Quantum Mechanics* (New York: Academic 1965).

[2] A.J. Hanson, T. Regge and C. Teitelboim, *Constrained Hamiltonian Systems* (Rome: Accademia Nazionale dei Lincei 1976).
[3] A. Ashtekar and G.T. Horowitz, Phys. Rev. **D26** (1982) 3342.
[4] K. Kuchar, Phys. Rev. **D34** (1986) 3031; Phys. Rev. **D34** (1986) 3044; Phys. Rev.**D35**, (1987) 596; Contemp. Math. **71**, (1988) 285.
[5] J. Romano and R. Tate, Class. Quant. Grav.**6** (1989) 1487.
[6] K. Schleich, Class. Quant. Grav. **7** (1990) 1529.
[7] R. Loll, "The Dirac Condition: a U(1) gauge model", Imperial College preprint /TP/88-89/15, 1989.
[8] R. Jackiw, in *Physics, Geometry and Topology*, ed. H.C. Lee (Plenum: New York 1990), p. 191.
[9] O.F. Dayi, Phys. Letts. **B228** (1989) 435.
[10] G. Kunstatter, LPTHE-Orsay preprint no. 91/17.
[11] L. Faddeev, Theor. Math. Phys. **1**, (1970) 1. See also C. Itzyskon and B.-J. Zuber, *Quantum Field Theory* (McGraw-Hill, New York, 1980).

WORMHOLES IN STRING THEORY AND LOW ENERGY QUANTUM GRAVITY

ALEX LYONS
Department of Physics, University of Alberta
Edmonton, Alberta, T6G 2J1, CANADA

ABSTRACT

We discuss the wormhole effective interactions in string theory, thought of as a sum over two-dimensional field theories on different world-sheets. The effective interactions are calculated in the 'dilute wormhole approximation' by considering the effect of a complete basis of wave functions on scattering amplitudes for a surface with boundary. The sum over wormholes is equivalent to having a world-sheet of trivial topology and summing over different space-time and matter field backgrounds. To leading order these consist of the massless fluctuations, since the tachyon cancels out when a sum is done over different spin structures going through the wormhole. In this way we recover quantized general relativity as an effective theory, from a sum over field theories on higher-genus Riemann surfaces.

1. Introduction

It has been widely suggested that manifolds with non-trivial topology might play a role in quantum gravity [1–3]. Most work on this has concentrated on wormholes, which are like handles connecting two large flat regions of space-time. They have been shown to have dramatic effects on the apparent coupling constants of low-energy physics. The ability of wormholes to shift coupling constants has led to a mechanism by which the cosmological constant can naturally be forced to zero [3]. However, such calculations are currently plagued with problems. The main problem is that the work done so far is based on a path integral for euclidean quantum gravity which does not converge, even naively, because the euclidean action is unbounded from below. This sickness (and others) has led many workers to doubt the validity of the 'wormhole calculus'. The idea of this paper is to show that one can calculate the effective interactions associated with wormholes in a much simpler theory, also based on a euclidean path integral over geometries, but which is much better defined than the path integral for euclidean quantum gravity. This is within the Polyakov approach to string theory [4].

One can regard string theory wormholes as providing a toy model for wormholes in full quantum gravity. They might reveal new ways of investigating some of the problems with the arguments in the four-dimensional case [5]. However, we shall pursue a different motivation here. We propose that the problems with euclidean quantum gravity, such as the unboundedness of the action, are so bad that it is likely that euclidean quantum gravity is just an effective theory which is a low-energy approximation to something more fundamental, such as superstrings. In this case it is important to see how euclidean quantum gravity emerges, in a low-energy limit, from superstrings. Wormholes in two dimensions can provide an explanation, the first to our knowledge, of how this comes about [6,7]. The metric and matter fields in four-dimensional space-time would emerge as a set of Coleman 'alpha parameters' for the two-dimensional wormholes. It is this idea which we shall have in mind in the following development.

The phenomenon of handles being replaced by bilocal effective interactions on a manifold with the handles removed should be present in any quantum theory defined on manifolds of non-trivial topology [6–8]. In this paper we describe how the wormhole effective interactions are obtained in the Polyakov approach to string theory. Further details of these calculations are contained in [7]. We find, using a suitable basis of 'wormhole wave functions', that the leading interactions correspond to tachyonic and massless fields in space-time. We choose the basis to consist of eigenfunctions of the Wheeler-DeWitt operator for the string. For superstrings, the tachyon interaction cancels out when one sums over different spin structures going through the wormhole [9]. The dominant effective interaction is then associated with massless space-time particles, and gravitons. We then describe how quantum gravity in space-time could emerge as an effective theory, from a 'wormhole resummation' [10]. The effective theory is a string theory (at tree level) interacting with a background of free massless quantum fields, including gravitons. Possible extensions of this work are then outlined. This paper is largely based on work done in collaboration with S. W. Hawking which will appear in published form elsewhere [7].

2 Effective interactions and wormhole wave functions

We need to have a general scheme for relating a correlation function

$$\langle X^{\mu_1}(z_1)\cdots X^{\mu_n}(z_n)\rangle \quad , \tag{2.1}$$

on a Riemann surface of arbitrary genus, to a correlation function on the complex plane, with suitable operators inserted at the places 'where the ends of the handles were'. For this purpose, a suitable representation of the surface is in terms of the Schottky parameters. For a surface of genus g, these can be roughly thought of as the radii of g pairs of circles, together with the locations of each of the centres of these circles. The circles are identified in pairs under the action of a finitely generated subgroup of $SL(2,\mathbf{C})$ generated by certain transformations T_R ($R = 1,\cdots,g$). In this representation, the limit in which the correlation functions (2.1) become approximated by correlation functions on the complex plane is given by the limit in which the radii of all the circles shrink to zero. We use an approach to finding the effective interactions which has some similarities to that presented in

[1,11,12]. This starts by cutting the Riemann surface across each of the wormholes. This produces a surface with each wormhole replaced by two boundaries. In the Schottky representation this is done by taking the fundamental region, exterior to the circles, and *not* making the $SL(2,\mathbf{C})$ identifications.

The n-point functions on the original surface can then be obtained from those on the cut surface by inserting delta functions (which ensure that the fields take the same values on each boundary) and then integrating over the boundary values [6,7]. Thus we have, formally,

$$\langle X^{\mu_1}(z_1)\cdots X^{\mu_n}(z_n)\rangle = \int \mathcal{D}X \prod_{R,z} \delta(X(z)-X(T_R z))\, X^{\mu_1}(z_1)\cdots X^{\mu_n}(z_n)\, e^{-I[X]}\,, \tag{2.2}$$

where $\prod_z$ is a continuous product running around the circle, R_-, which encloses z_{R-}. The Schottky generator T_R maps the circle R_- to the circle R_+. Now the field, X, on the circles $R_\pm$, is expanded in a complete set of modes as $X(z)=\sum_n X_n^{\pm} e^{in\theta}$, where $\theta = \arg(z-z_{R\pm})/(z_{R\mp}-z_{R\pm})$. The delta function becomes a product of delta functions in each of the X_n:

$$\prod_z \delta(X(z)-X(T_R z)) = \prod_n \delta(X_n^- - e^{in\phi} X_{-n}^+)\ , \tag{2.3}$$

where $\phi_R = \arg(q_R)$ is the twist angle between the wormhole ends.

The delta functions can then be expanded in a complete basis of wave functions on the boundaries. The effective interactions take the simplest form if we use a basis of eigenfunctions of the world-sheet hamiltonian, $L_0 + \tilde{L}_0$. This is the world-sheet analogue of the Wheeler-DeWitt operator. It is a sum over harmonic oscillator hamiltonians in each of the non-zero n modes, together with a free particle hamiltonian in the zero mode. This means that the basis wave functions can be interpreted as corresponding to wormholes containing different numbers of 'particles' in each non-zero n mode and carrying a certain amount of 'momentum' in the zero mode. This is similar to the situation encountered in [1,11,12], for the discussion of wormholes in space-time coupled to conformal matter. However, an important difference here is that the wave functions for wormholes need not satisfy the hamiltonian constraint, i.e. they need not be 'on-shell'. This contrasts with the approach of [1,13] in which only solutions of the Wheeler-DeWitt equation are allowed.

We further simplify the discussion by working only to leading order in the 'dilute wormhole approximation'. This is valid in the limit that the radii of the circles all tend to zero. In this case the problem reduces to one of computing correlation functions in the region exterior to a disk, with a wave function on the boundary. Correlation functions similar to this have been calculated in [14]. By explicitly comparing these correlation functions to those on the complex plane with a vertex operator insertion at the origin, we are able to find the correspondence between wave functions and effective interactions [7]. Each of the factors in the wave function turns out to correspond to an effective interaction according to the rule

$$e^{ik\cdot X_0} \leftrightarrow r^{k^2} : e^{ik\cdot X(0)} :$$
$$\psi_m(Y_n^{\mu(1)}) \leftrightarrow r^{nm} : (\Re e\, \partial^n X^\mu(0))^m :$$
$$\psi_m(Y_n^{\mu(2)}) \leftrightarrow r^{nm} : (\Im m\, \partial^n X^\mu(0))^m : \quad , \tag{2.4}$$

where $Y_n^{(1)} = \frac{1}{2}(X_n + X_{-n})$ and $Y_n^{(2)} = \frac{1}{2i}(X_n - X_{-n})$, for $n > 0$, and ψ_m denote the standard harmonic oscillator wave functions. The details of this calculation are in ref. [7]. The vertex operators we have found can be interpreted in space-time terms as corresponding to the emission of space-time particles. To leading order in the radius of the circle, r, the dominant effect comes from the tachyon vertex operator. This is because there is a factor of $r^{k^2+\sum_{n>0} nm}$ in the effective interaction, so the higher states are damped by extra powers of r. This power is the eigenvalue of $L_0 + \tilde{L}_0$ for the wormhole wave function. There is also an average over the orientation of the circle, which ensures that the only interactions which survive are those which have equal powers of ∂X and $\bar{\partial} X$. Thus, after the tachyon interaction term, the next most important interactions are those involving massless fields.

We also find in [7], by considering $SL(2,\mathbf{C})$-invariant amplitudes, that there is a pole at the on-shell value $k^2 + \sum_{n>0} nm = 2$ when an integral is performed over r. To find the correct measure in the r integration we use the Fadeev-Popov gauge-fixing procedure to fix the $SL(2,\mathbf{C})$-invariance of the integrand. This leads to a factor of r^{-3}, so the r integral takes the form

$$\int dr r^{k^2+\sum_{n>0} nm-3} \sim \Big(k^2 + \sum_{n>0} nm - 2\Big)^{-1} . \tag{2.5}$$

This is like a propagator for a space-time state with $(mass)^2 = \sum nm - 2$.

3. Low-energy quantum gravity from superstrings?

It is known [9] that unwanted tachyon states do not propagate in loops when one sums over spin structures on the world-sheet in such a way as to preserve full diffeomorphism invariance. We show explicitly in [7] how the sum over the two spin structures for spinors going *through* the wormhole ensures that the wormhole wave functions corresponding to space-time tachyons get projected out. In fact all states with $(-1)^F = 1$ become projected out, where F is the fermion number in each sector. This can be shown explicitly in the wave function formalism, with wormhole wave functions which depend on anticommuting quantities in the holomorphic representation [11]. This projection leaves the massless states as the dominant effective interaction.

Now consider applying the 'wormhole calculus' to the massless interactions. This means that, to each interaction, we associate an alpha parameter. The alpha parameters for the massless states form a whole 'alpha field' $\alpha_{\mu\nu}(k)$. The effective theory is then given by a path integral over α, and X, of the form

$$\int \mathcal{D}\alpha \mathcal{D}X \, e^{-\frac{1}{2}\int d^D k \alpha_{\mu\nu}(k)\alpha^{\mu\nu}(-k)/k^2} \, e^{-I[X,\alpha]} \ . \tag{3.1}$$

The $1/k^2$ comes from the propagator for the massless state to go through the wormhole. In (3.1), X is a mapping from a sphere into a curved space-time background, and $\alpha_{\mu\nu}(k)$ is the Fourier transform of the space-time graviton, anti-symmetric tensor, and dilaton fields. The action $I[X,\alpha]$ is the usual one for a string in a curved space-time background.

The quadratic weighting factor for α is the linearized form of the Einstein action for the space-time background (including the dilaton and anti-symmetric tensor as matter fields). The derivation of this effective theory from perturbation theory means that it is valid only if we take the space-time to be linearized about flat space-time. Nevertheless, it represents a way in which *quantum* gravity can arise as an effective theory, from a sum over conformal field theories on different Riemann surfaces. It is clearly important to see whether working beyond the dilute wormhole approximation can shed light on the effective theory of quantum gravity at the non-linear level. Although we have seen how the mass-shell condition appears, from the integral over r, it is not clear how the other L_n conditions appear. A possible resolution to this might involve including the world-sheet ghost fields. Work is in progress on these problems [15].

Acknowledgements

I would like to thank S. W. Hawking for useful discussions relating to this work. My thanks also go to NSERC and the Winspear Foundation for financial support.

References

[1] S. W. Hawking, *Phys. Rev.* **D37** (1988) 904.
[2] S. W. Hawking, *Nucl. Phys.* **B335** (1990) 155.
S. Coleman, *Nucl. Phys.* **B307** (1988) 867.
S. B. Giddings and A. Strominger, *Nucl. Phys.* **B306** (1988) 890.

S. B. Giddings and A. Strominger, *Nucl. Phys.* **B307** (1988) 854.

[3] S. Coleman, *Nucl. Phys.* **B310** (1988) 643.

[4] A. M. Polyakov, *Phys. Lett.* **B103** (1981) 207.

[5] J. Polchinski, *Nucl. Phys.* **B324** (1989) 123.

[6] S. W. Hawking, The effective action for wormholes (1990) DAMTP preprint, to appear in Nucl. Phys. B.

[7] A. Lyons and S. W. Hawking, Wormholes in string theory (1991) U. of Alberta preprint, submitted to Phys. Rev. D.

[8] A. Lyons, *Nucl. Phys.* **B333** (1990) 279.

[9] N. Seiberg and E. Witten, *Nucl. Phys.* **B276** (1986) 272.

[10] A. A. Tseytlin, *Phys. Lett.* **B251** (1990) 530.

[11] A. Lyons, *Nucl. Phys.* **B324** (1989) 253.

[12] H. F. Dowker, *Nucl. Phys.* **B331** (1990) 194.

[13] S. W. Hawking and D. N. Page, *Phys. Rev.* **D42** (1990) 2655.

[14] W. Fischler, I. Klebanov and L. Susskind, *Nucl. Phys* **B306** (1988) 271. A. Cohen, G. Moore, P. Nelson and J. Polchinski, *Nucl. Phys.* **B281** (1987) 127.

[15] A. Lyons, forthcoming (1991).

Is physics consistent with closed timelike curves?

John L. Friedman

Department of Physics, University of Wisconsin–Milwaukee,

Milwaukee, Wisconsin 53201

Local fluctuations of a Lorentz metric imply a microscopic sea of closed timelike curves (CTCs), unless one a priori demands global causality. Several recent investigations, suggesting that the laws of physics may be consistent with CTCs, are summarized here. In particular, the classical Cauchy problem is consistent for free fields on a class of spacetimes with CTCs; and a billiard-ball model of interacting particles also appears to have a well defined initial value problem. The class of spacetimes considered are asymptotically flat and have CTCs confined to a finite region. Free quantum fields on these spacetimes have a well-defined and unitary scattering matrix, but unitarity for interacting quantum fields is still in doubt.

I. Introduction

An underlying motivation for considering closed timelike curves (CTC's) is the possibility that at scales where quantum fluctuations of the metric are of order unity, the notion of spacetime — of manifold and Lorentz metric — retains its meaning. Local fluctuations of the metric would be independent of fluctuations several Planck-lengths away, and the resulting randomly oriented field of light cones would give rise to a sea of small CTC's. One apparently avoids microscopic CTC's only by adopting a theory in which causal structure is fundamental or one in which the metric is not meaningful or not Lorentzian at small scales.

Although CTCs arise from quantum fluctuations whether or not the spacetime topology is Euclidean, a theory that has an everywhere-defined Lorentz metric can allow topology change only if it allows CTCs. In particular, the evaporation of black holes is more readily described by a theory that allows topology change. When a black hole forms and evaporates, information is lost to external observers. If one does not know the initial state of the collapsing system, arbitrarily accurate measurements of the exterior spacetime cannot recover the state inside the horizon. If, however, the collapsing matter re-expands, forming a child universe, it would be possible to avoid loss of information in the full spacetime.

The usual picture of a trousers universe is a spacetime in which a single S^3 bifurcates to form two S^3's, and it allows no Lorentz metric. But when one removes three balls from CP^2, the resulting manifold admits a Lorentz metric in which each of the three spherical boundaries is spacelike, representing one initial and two final universes. The metric is everywhere smooth and nonsingular, but the spatial manifold has changed, and the metric necessarily has CTC's.

More generally, a spacetime with CTC's is the price of Lorentzian topology change. Any two 3-manifolds, S_1 and S_2, can be joined by a 4-manifold with a smooth Lorentz metric for which S_1 and S_2 are spacelike. Geroch[1] showed that if S_1 and S_2 are not homeomorphic or if M is not a product of the form $S_1 \times R$, then the metric must have CTC's. As stated in Geroch's paper, the spacetime must either have closed timelike curves or be time nonorientable, but by going to the covering space, one can easily prove that there must be CTC's in all but one case: the creation of a universe from nothing. (A simple extension of Geroch's theorem from closed to asymptotically flat spacetimes is given in Ref. 2).

Then if one allows spacetimes with CTC's in a sum-over-histories, topology change is implied by a Lorentzian path integral — there are nonzero amplitudes for continuous paths that start at any 3-manifold and end at any other.

CTC's have commonly been regarded as a pathologies of general relativity that do not arise in the real world, and which lead to an inconsistent theory - allowing almost no consistent initial data in the classical level and no consistent interpretation at the quantum level. This talk will serve as a brief introduction to recent work on whether the laws of physics might permit CTC's. A talk by Kip Thorne[3] and reference [4] are more detailed reviews, and this talk will spend somewhat more time on the minority of issues with which I have been directly involved.

II. Model spacetimes with CTC's

A 3-dimensional wormhole can be constructed by removing two balls from Euclidean space and identifying their spherical boundaries Σ_1 and Σ_2, as shown in Fig. 1. To describe a corresponding spacetime with a smooth, time-independent, traversable wormhole one can choose a metric of the form

$$ds^2 = -dt^2 + d\ell^2 + r^2(\ell)d\Omega^2;$$

for example,

$$\text{e.g.} \qquad ds^2 = -dt^2 + d\ell^2 + (\ell^2 + a^2)d\Omega^2.$$

Here ℓ is a radial coordinate that runs from $-\infty$ to ∞, and has the value zero halfway through the wormhole throat, and $2\pi r(\ell)$ is the circumference of a sphere surrounding the hole at coordinate ℓ.

Morris and Thorne[5] (see also Morris, Thorne and Yurstever[6], Friedman et. al.[4] considered spacetimes in which the two mouths of a wormhole move toward one another, as seen by the external spacetime. The relative boost alters the separation between identified points so that an initially spacelike separation becomes null and then timelike (Fig. 2). In an alternative spacetime of Frolov and Novikov,[7,8] one surrounds one wormhole mouth by a cloud of matter. With no matter, the timelike Killing vector has the same length everywhere, but the presence of the cloud means that the Killing vector field defined outside the two mouths meets at the identified spheres with a discontinuity in its length, due to the gravitational redshift of the cloud. Again identified points with initial spacelike separation become timelike separated, joined by CTC's.

There are several ways in which physics might prevent such spacetimes on a macroscopic scale. First, with matter satisfying the null energy condition, the wormholes would collapse. Second, the Cauchy horizon might be unstable, with the energy density of initial fields growing without bound before the region of CTC's is reached. Third, the Cauchy problem itself might not be well-defined: for generic initial data there might be no solutions to its field equations, or there might be more than one solution. Finally, there might be no consistent quantum field theory in a spacetime with CTC's, one that allows no decomposition into space$\times$time. The recent work has been primarily concerned with these questions.

Gott [9] recently proposed another class of models in which two parallel cosmic strings are moving by each other with relativistic speeds. Here, however, the CTCs are eternal, when they exist at all. There is no Cauchy horizon but work by Deser et. al.[10] and Carroll et. al.[11] show that one cannot construct spacetimes of this kind (flat everywhere except for the conical string singularities) unless the spacetime has negative mass. (The spacetimes are not asymptotically flat but have a natural definition of asymptotic 4-momentum).

It should again be emphasized that all one needs for closed timelike curves is a tipping of the light cones. Energy conditions that would prevent one from keeping wormhole mouths open or prevent Gott's strings need not prevent closed timelike curves. However, in order to have CTCs confined to a finite region, a spacetime must have a Cauchy horizon, and that turns out to imply that at least the weak energy condition *is* violated.[12]

III. Does Physics Make Sense on Spacetimes With CTC's?

A. Classical free fields.

It has commonly been thought that in spactimes with closed timelike curves, one cannot find consistent time evolutions of classical fields for generic initial data—that the Cauchy problem is not well defined. A simple example is the temporally compact spacetime obtained from the piece of Minkowski space between the $t = 0$ and $t = T$ planes by identifying the points $(t = 0, x)$ and $(t = T, x)$. A well-defined solution exists only if the Minkowski evolution is periodic in time with period T — there is no solution for generic data. For the wormhole spacetimes mentioned above, however, the set of closed timelike and null geodesics has measure zero. Morris, Thorne and Yurtsever *et al.*[6] noted that the evolution of free fields is well defined in the limit of geometrical optics; and this in turn makes it seem likely that a multiple scattering series converges to a solution for arbitrary initial data.[4,5,6]

Morris and I[13,14] have recently obtained a formal proof that the Cauchy problem is well-defined on a class of static wormhole spacetimes with CTC's. Because there are no complete spacelike hypersurfaces, one must pose data at past null infinity. Standard existence theorems for hyperbolic equations are based on energy inequalities or on the spectral theorem for self-adjoint operators, and they fail on spacetimes with

CTC's. It nevertheless turns out that one can build a solution as a convergent sum of nonorthogonal modes. The formal proof is technical, and a brief discussion of the heuristic argument is more illuminating.

An incoming wave from $\mathcal{I}^-$ may be thought of as initially scattering off of the wormhole mouths, producing at each mouth some purely outgoing scattered wave composed of a reflected part that comes from the initial scattering at that mouth and a transmitted part that comes from the initial scattering at the other mouth. These first-scattered waves in turn scatter again, producing new reflected and transmitted waves, and so on. In each scattering, in the geometrical optics approximation, the amplitude of the wave decreases by a factor $a/2d$ in case (a) or $a/[2d(1-v)]$ in case (b) in traversing the wormhole and then travelling from one mouth to the other. Again the series converges when $a/[2d(1-v)]$ is small compared to one. We have explicitly constructed this series for the static spacetime, (a), when the spacetime external to the wormhole throat is flat. A proof of existence can then be secured by a proof that this multiple-scattering series is convergent. For small mouths, one can prove convergence for wavelengths short enough that geometrical optics must be valid ($\lambda << a$); and one already had convergence in the geometrical optics limit. We therefore have little doubt that, at least for handles with small mouths, the multiple scattering series converges.[14] We thus claim an alternate, partly numerical, construction of the modes whose existence we have previously established.

The Cauchy problem makes sense for data posed at $\mathcal{I}^-$ or on a surface to the past of all CTCs. In a region with CTCs, however, there are no global spacelike hypersurfaces on which to specify data. One can at best freely specify local data on a finite spacelike surface or specify global but constrained data on an surface that is spacelike in a region exterior to the CTCs but includes timelike parts. Yurtsever[15] calls "benign" a spacetime on which data can be freely specified in some neighborhood of any point. It is plausible that asymptotically flat spacetimes are benign if their CTCs are confined to a compact region and their topology has the form R× (S - one point), where S is a closed 3-manifold.

B. Interacting Systems: Classical Billiard Balls

For a free scalar field, the grandfather paradox is easily resolved on the wormhole spacetimes. An initial field that propagates along a CTC returns to its starting point with a different value; but because the field is linear, a unique consistent solution is given by summing the result of such traversals, by a multiple scattering series of waves that have propagated through the wormhole an arbitrary number of times. For self-interacting fields, the grandfather paradox is much more interesting. Echeverria, Klinkhammer and Thorne [16] have examined a simple nonlinear problem in which a billiard ball traverses a time-tunnel, and on its return, it collides with its earlier self. The collision is violent enough to prevent the original ball from entering the wormhole, and it appears at first that one has no self-consistent solution. Remarkably, however, it appears that solutions always exist: The incident ball is struck a glancing blow on its way to the wormhole, deflecting it in such a way that when it emerges, it is aimed correctly to give its earlier self precisely that glancing blow! Solutions of this type are not unique — in fact many initial trajectories allow an infinite number of solutions with the ball traversing the wormhole an arbitrarily large number of times. Although they have no formal proof of existence, Echeverria et. al. searched for initial trajectories in which a solution fails to exist and they found none.

I.D. Novikov[17] suggested several years ago that one might be able to have CTCs without a grandfather paradox. His recent work[18] discusses a class of more complex classical interactions, with bombs whose debris return to prevent the initial explosion.

His examples all allow classical solutions, but it appears that by adding an extra degree of freedom to the billiard balls one *can* find a classical system for which some initial data has no classical evolution.[19] Suppose that one has a variable – call it spin – that can take two values – up and down, and that the interaction energy of the switch is small compared to the collision energy. One can choose a Hamiltonian so that when two down balls collide, they change to two up balls, but when up collides with down, the spins do not change. If one sends an up-ball through the wormhole, and if it collides with an up ball (its future self) before traversing the wormhole, its spin will change to down; this contradicts the fact that it emerged from the wormhole with spin up. If it collides with a down ball, it remains up, contradicting the fact that its future self was down. So there appears to be no self-consistent solution. If the billiard ball trajectories of Echeverria et. al. always allowed solutions with arbitrarily small glancing angle, one could get around the problem by choosing the glancing angle small enough that the collision energy was smaller than the spin-flip energy. Then the spin flip could disrupt the collision. But the glancing angles for a given initial trajectory appear to be bounded above some minimum value, and it therefore appears that one cannot find a consistent classical solution.

Although that the classical theory of interacting billiard balls is not well-defined, the evolution of a quantum mechanical wave function has a better chance for two reasons. First the Schrodinger eqn. and its relativistic counterpart are linear, it is plausible that a solution exists and is unique. Second, the classical models can be said to cheat by not making their matter out of particles. If one constructs the switch of the example above out of atoms, then in a sum-over-histories, one must allow any other configuration of atoms to traverse a closed timelike path and interact with the incoming system. This could make it inordinately difficult to show that any but simplest system is inconsistent. It is nevertheless possible to prove that free quantum fields can be consistently described, and to decide whether the scattering theory of simple interacting fields is perturbatively well-defined and unitary.

Before considering this question, let us return to the problem of maintaining traversable wormholes, and turning them into time-machines.

IV. Quantum Physics

A. Can generic properties of quantum fields forbid CTCs?

As mentioned in Sect. I, a spacetime must violate at least the null energy condition if it is to have CTCs that are not eternally present. This is not inconsistent with quantum field theory, but for the examples of spacetimes that have been studied, a more restrictive energy condition must be violated as well. That is, a spacetime with a wormhole that does not collapse of a wormhole, cannot everywhere satisfy the averaged null energy condition: [5,6,20] If k^a is tangent to a null geodesic with affine parameter λ, the averaged null energy condition has the form,

$$\int T_{ab}k^a k^b d\lambda < 0,$$

where T_{ab} is the energy-momentum tensor of the spacetime. The local null energy condition is violated by particle creation, by the Casimir effect, and by squeezed vacuum states[21], but the averaged null energy condition is more difficult to contravene.[22] It holds for the scalar field in 4-dimensional Minkowski space[23] and Wald and Yurtsever have recently constructed a proof for asymptotically flat 2-dimensional spacetimes,[24]; but they also show that it cannot in general hold for curved 4-dimensional spacetimes. The question of whether one can maintain traversable, macroscopic wormholes consequently remains open.

B. Quantum Instability of the Cauchy Horizon

The fact that the Cauchy problem is well defined for free classical fields means that the Cauchy horizon is classically stable to linear perturbations. In the quantum theory, however, Kim and Thorne[25,26,12] find that the renormalized stress tensor diverges at points joined by closed null geodesics. Their calculation uses the Hadamard form of the Green's function to compute $< T_{ab} >$ – by point splitting. In the geometrical optics approximation, the Green's function, $G^{(1)}(x', x)$, normally has two terms that diverge when the (squared) geodesic distance, $\sigma(x, x')$, goes to zero. In a normal neighborhood, σ is unique, and one regularizes $G^{(1)}$ by subtracting the divergent part. When there are CTC's, however, the remaining finite part of the Green's function contains a sum of large terms corresponding to null geodesics that are nearly closed, and which join x to x'. That is, $G^{(1)}$ has the form,

$$\begin{aligned} G^{(1)}(x', x) &=< \psi|\phi(x)\phi(x') + \phi(x')\phi(x)|\psi > \\ &= \sum_N \frac{\Delta_N^{1/2}}{4\pi^2}\left(\frac{1}{\sigma_N} + u_N \, ln|\sigma_N| + w_N\right). \end{aligned}$$

To regularize, one removes only the $N = 0$ term. Then $< T_{ab} >$, constucted by taking two derivatives of G, diverges as one approaches the Cauchy horizon. When the distance Δt to the Cauchy horizon is of the order of the Planck length, as measured by an observer moving along one of the wormhole mouths, T_{ab} is very small. For $\Delta t > \ell_{Planck}$, and the maximum fractional strain is less than 10^{-35}, when the wormhole separation is macroscopic (1m). This may mean that the semiclassical approximation cannot show more than a negligible disruption of the spacetime in its regime of validity. The smallness of the strain, however, is observer dependent, and Hawking[12] argues that the divergence is serious: An observer moving along a path joining the point x to the intersection of the Cauchy horizon with the wormhole mouth sees a fractional strain of order unity when she is within a Planck time of the Cauchy horizon. If no version of the null energy condition prevents one from maintaining spacetimes with macroscopic CTC's, the quantum instability of Kim and Thorne may prevent one from ever forming them.

C. Consistent quantum theory on a background spacetime with CTC's.

While we have encountered several possible ways in which one might rule out macroscopic CTC's, it is at least plausible that CTC's pervade a small-scale Lorentzian foam. A first question is whether one can generalize quantum field theory to accomodate a background spacetime with CTC's. An obvious candidate for such a generalization is a sum-over-histories framework.[27] If the CTC's are confined to a compact region, one could look at histories that end at a spacelike hypersurface S to the future of the region containing CTC's. Probabilities for different classes of histories in the CTC-region would have the same formal expression as probablilities for histories when there are no CTC's. That is, if $A(\Phi)$ is the amplitude for a field Φ with valuc $\Phi|_S$ on S, thcn thc probability for a class C of histories has the form[28]

$$P(C) = \frac{\sum_{\Phi|_S} \left|\sum_{\Phi\in C} A(\Phi)\right|^2}{\sum_{\Phi|_S} \left|\sum_{\Phi} A(\Phi)\right|^2}.$$

For a static time-tunnel spacetime, or one in which relative velocity of wormhole mouths is slow, one can consider a non-relativistic limit of the theory. Thorne and Klinkhammer[29] have looked at the quantum mechanics of a billiard ball on such a spacetime, using a WKB approximation to construct an explicit path-integral. They find that all classical histories short enough that wave packet stays together are equally likely, while paths long enough that the wave packet has spread (so that the WKB approximation is invalid) are expected to have almost zero probability. This is a reassuring picture: the lack of uniqueness present for the classical billiard balls has been replaced by a well defined probability for each of those classical paths, so quantum theory would seem to have overcome one of the classical difficulties.

However, one in general violates unitarity – probabilities can differ if one places the final hypersurfaces before the region of CTC's rather than after. At least for the case discussed earlier, where there appear to be no classical solutions, this leads to a loss of causality in regions that do not interact with the CTC's.[19] In some cases, Thorne and Klinkhammer show that one can avoid a loss of causality of this type by a prescription that discontinuously changes the surface S from a location to the past of the CTC-region to a location in its future. That is, to determine the probability of events before the Cauchy horizon, one uses a surface in the past of the CTC's, while for events after the Cauchy horizon one takes S in the future. However, one cannot in general avoid the problem in this way for spacetimes containing disjoint regions with CTC's.

Is the loss of unitary simply a difficulty with the WKB approximation or is a characteristic feature of interacting quantum fields? At first sight it appears that unitarity is preserved. For a free field theory, conservation of the Klein-Gordon inner product in the classical evolution implies a unitary evolution of the quantum field. For an interacting field, a path-integral derivation of the perturbation expansion (in effect, an LSZ reduction) for the S-matrix can be formally repeated for spacetimes with CTCs (this was suggested to us by J.B. Hartle). The resulting perturbation series expresses S-matrix elements solely in terms of in- and out- states and the Feynman propagator. One would think that the same unitarity relations that hold in curved spacetimes without CTCs would be valid here as well since the form of the propagator implies a unitary evolution for the free field. This appears however, *not* to be the case. When no CTCs are present, a series of identities that lead to diagrammatic unitary. The identities follow from the propagator's having the form

$$i\Delta_F(x,y) = \theta(x^o - y^o)D(x,y) + \theta(y^o - x^o)D(y,x),$$

where $D(x,y) =< |\phi(x)\phi(y)| >$, and x^o is a time-coordinate, a scalar whose constant surfaces are timelike. This form fails when there are CTCs, and the identities appear to fail as well.[19]

It may be that a nonunitary S-matrix can describe a consistent physics. An alternative, suggested by Deutsch[30], seeks to recover unitarity by allowing density matrices to represent states that propagate around CTCs. The quantum evolution is then underdetermined, and Deutsch must add a ad hoc requirement entropy be minimized in order to pick out a unique evolution of an arbitrary state.

Time spent in preparing this summary was supported in part by NSF Grant PHY 8603173.

[1]Geroch, R. P. , 1967. J. Math. Phys. **8**, 782-786.

[2]Friedman, J.L. & Higuchi, A. 1990. Nuc. Phys. B339, 491-515.

[3]Thorne, K.T. "Do the Laws of Physics Permit Closed Timelike Curves?", in Sixth Florida Workshop in Nonlinear Astronomy, New York Academy of Sciences, in press.

[4]Friedman, J. L. Morris, M. S. , Novikov, I. D. , Echeverria, F. , Klinkhammer, G. , Thorne, K. S. & Yurtsever, U. 1990. Phys. Rev. D **42**, 1915-1930.

[5]Morris, M. S. & Thorne, K. S. 1988. Am. J. Phys. **56**, 395-412.

[6]Morris, M. S., Thorne, K. S. &Yurtsever, U. ,1988. Phys. Rev. Lett. **61**, 1446-1449.

[7]Frolov, V. & Novikov, I. 1990. Phys. Rev. D, 42, 1057.

[8]Novikov, I. 1989, Zh. Eksp. Teor. Fiz. 95, 769 [Sov. Phys. JETP 68, 439].

[9]Gott, J. R. 1991. Phys. Rev. Lett., 66, 1126.

[10]Deser, S. Jackiw, R. & 't Hooft, G. 1991. Submitted to Phys. Rev. Lett.; 1984. Ann. Phys. 152, 220.

[11]Carroll, S. M. , Farhi, E. & Guth, A. H. 1991. Submitted to Phys. Rev. Lett.

[12]Hawking, S. 1991. "The Chronology Protection Conjecture", a transcribed seminar; paper in preparation.

[13]Friedman, J. L. & Morris, M. S. 1991. Phys. Rev. Lett.**66**, 401-404.

[14]Friedman, J. L. & Morris, M. S. 1991. Paper in preparation.

[15]Yurtsever, U. 1990. J. Math. Phys. 31, 3064.

[16]Echeverria, F. Klinkhammer, G. & Thorne, K. S. paper in preparation.

[17]Novikov, I. D. 1983. Evolution of the Universe. Cambridge University Press, Cambridge.

[18]Novikov, I. D. Nordita preprint 90/38 A.

[19]Friedman, J.L., Papastamatiou, N.J., & Simon, J., paper in preparation.

[20]Lee, C.-W. 1983. Gen. Rel. Grav. 15, 21.

[21]Braunstein & Morris, M., paper in preparation.

[22]Yurtsever, U. 1990. Classical and Quantum Gravity 7, L251.

[23]Klinkhammer, G. 1991. Phys. Rev. D 43, 2542.

[24]Wald, R. M. & Yurtsever, U. 1991. Phys. Rev. D 44, 403.

[25]Kim, S.-W. & Thorne, K. S. 1991. Phys. Rev. D, 43, 3929.

[26]Frolov, V. P. 1991. Phys. Rev. D43, 3878.

[27]Hartle, J. B. 1991. In Quantum Cosmology and Baby Universes, Proceedings of the Dec.27, 1989 - January 4, 1990 Winter School in Jerusalem.

[28]Sinha, S. & Sorkin, R. D. 1991. Foundations of Physics, in press.

[29]Klinkhammer, G. & Thorne, K. S. paper in preparation.

[30]Deutsch, D. 1991. "Quantum Mechanics Near Closed Timelike Lines", preprint.

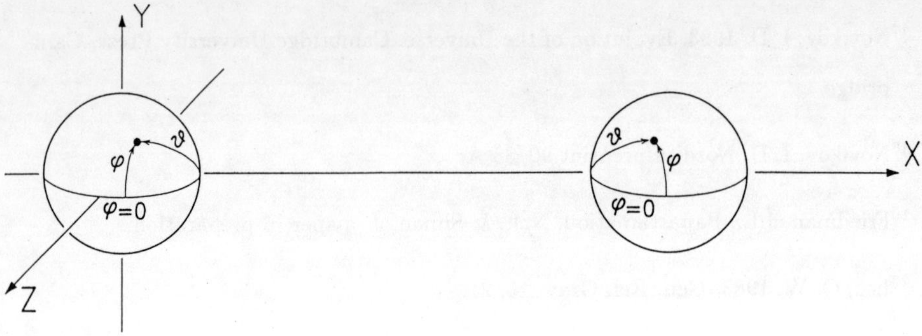

Fig. 1a

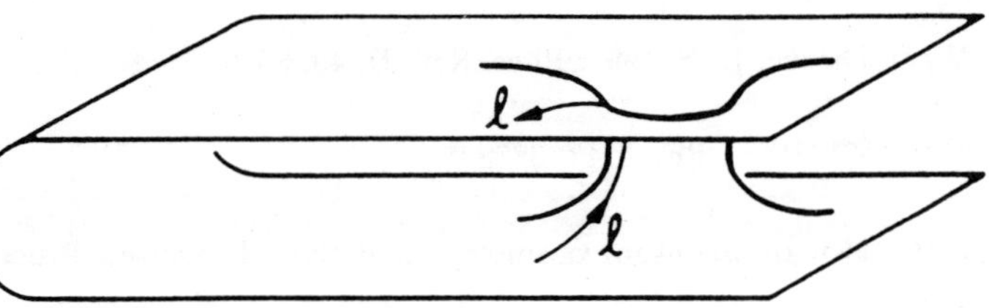

Fig. 1b

Figure 1. Two spatial diagrams showing how points on the two wormhole mouths are identified. Points with the same values of θ and ϕ are the same.

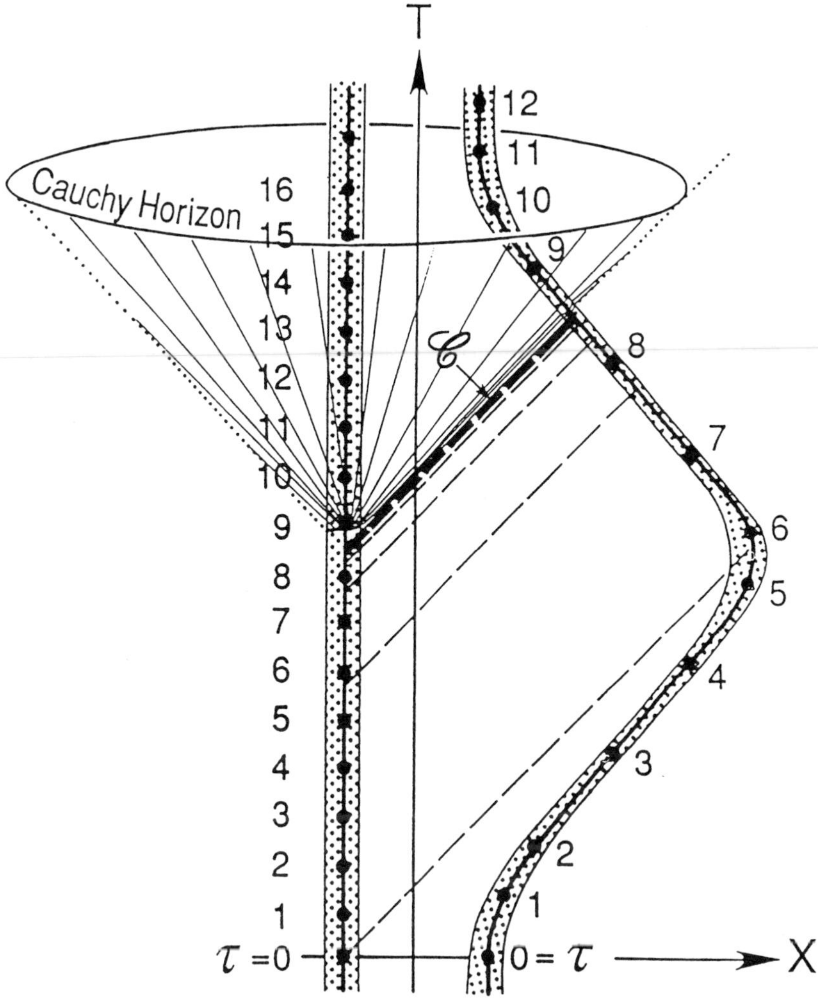

Figure 2. A spacetime depiction of the formation of CTC's arising from the relative motion of wormhole mouths. The stippled region is excised from the spacetime and its outer edges are joined together to make the wormhole throat. Numbered dots represent proper time read by a clock in the wormhole, and two dots at the same proper time represent the same spacetime point. Straight lines joining these points are closed curves; they are timelike in the future of the Cauchy horizon.

General laws of thermodynamics

Geoff Hayward
Department of Physics, University of British Columbia,
Vancouver, B.C. V6T 2A6, Canada

Abstract

I present four laws of thermodynamics which apply to general stationary horizon/matter systems. Each of the laws relates directly to properties of the Euclidean action and its variations.

A standard approach to gravitational statistical mechanics [1] posits that the grand canonical partition function for a stationary system is,

$$Z = \int \mathcal{D}\left[g_{\mu\nu}, \phi\right] \mathrm{e}^{-I_{\mathrm{E}}}, \tag{1}$$

where I_{E} is the Euclidean action appropriate to the system and the path integral extends over all paths compatible with boundary constraints on the metric and matter fields. [$G = c = k = \hbar = 1$.] From this, one obtains the Massieu function for the system [2],

$$J \equiv \ln Z. \tag{2}$$

So long as the partition function is sharply peaked and quantum gravitational effects are small [1, 3],

$$J = -I_{\mathrm{E}_0} + \text{ corrections} \simeq -I_{\mathrm{E}_0}, \tag{3}$$

where I_{E_0} is the classical, smooth action. Here, I concentrate on systems for which this 'stationary phase' approximation holds.

To define the Euclidean action for a stationary system, it is necessary to specify the sector of complexified spacetime over which the action integral is to be performed. The 'Euclidean' sector is obtained by defining a Euclidean time parameter $\tau \equiv it$ and identifying points (τ, x^i) with $(\tau + k, x^i)$, where t is the Lorentzian time coordinate associated with co–rotating observers and k is a constant determined by a constraint on the temperature at the boundary of the system [1, 4]. In fact, the metric induced on the 'Euclidean' sector is in general complex;

$$ds^2 = (N^2 - N_i N^i)\, d\tau^2 - 2i h_{ij} N^i\, dx^j\, d\tau + h_{ij}\, dx^i\, dx^j \tag{4}$$

where, since the metric is assumed to be stationary (i.e. independent of t), N, N^i, and h_{ij} are all real.

For definiteness, concentrate on a spacetime which has a connected boundary, B, and a unique black hole horizon, H, and in which the metric is minimally coupled to an electromagnetic gauge field, A_μ. Let $x^1 \equiv y$ be a 'radial' coordinate which parametrizes nested 3–surfaces extending from the boundary in to the horizon. Also, let p^{ij} and P^i be the momenta temporally conjugate to h_{ij} and A_i. The Euclidean action for the system is then [5, 4],

$$\begin{aligned} I_{\mathrm{E}} &= \int_{\mathcal{M}_E} \left\{ p^{ij} \dot{h}_{ij} + P^i \dot{A}_i - N\mathcal{H} - N^i \mathcal{H}_i + A_0 \partial_i P^i \right\} d^3x \, d\tau \\ &- \int_{\Sigma_{y_B}} \left(\frac{1}{8\pi} N\Gamma\sqrt{\sigma} - 2ip^1_i N^i + P^1 A_0 \right) d^3\tau \\ &- \lim_{y \to y_H} \int_{\Sigma_y} \left(\frac{1}{8\pi} \sqrt{h} N^{,1} - ipN^1 + 2ip^1_i N^i - P^1 A_0 \right) d^3x \end{aligned} \tag{5}$$

where $\mathcal{H}$ and $\mathcal{H}_i$ are the Hamiltonian and momenta densities, σ_{AB} is the induced 2–metric on $\Sigma_\tau \bigcap \Sigma_y$ and Γ_{AB} is the extrinsic curvature of this surface as embedded in Σ_τ. [Indices, A, B range from 2 to 3 and $\dot{} \equiv \frac{\partial}{\partial \tau}$.]

First Law:
Vary the action (5) to obtain,

$$\begin{aligned} \delta I_{\mathrm{E}} &= -\frac{1}{16\pi} \int_{\mathcal{M}_E} (G_{\mu\nu} - 8\pi T_{\mu\nu}) \, \delta g^{\mu\nu} \sqrt{g} \, d^3x \, d\tau + \int_{\mathcal{M}_E} \frac{\delta \mathcal{L}_m}{\delta A_\mu} \, \delta A_\mu \, d^3x \, d\tau \\ &+ \int_B \left(\tilde{p}^{ab} \, \delta\gamma_{ab} + \tilde{P}^a \, \delta A_a \right) d^2x \, d\tau + \int_H \left(\gamma_{ab} \, \delta\tilde{p}^{ab} - \tilde{P}^a \, \delta A_a \right) d^2x \, d\tau, \end{aligned} \tag{6}$$

where γ_{ab} is the induced metric on a surface of constant y and where $\tilde{p}^{ab}$ and $\tilde{P}^a$ are the momenta 'radially' conjugate to γ_{ab} and A_a. [Indices a, b assume values 0, 2, and 3.] Confining attention to configurations which satisfy the field equations and are regular at the horizon, the variation of the classical, smooth, Euclidean action, I_{E_0}, reduces to,

$$\begin{aligned} \delta I_{\mathrm{E}_0} &= \int_{\partial B} k \left\{ \tilde{p}^{ab} \, \delta\gamma_{ab} + \tilde{P}^a \, \delta A_a \right\} d^2x, \\ &= \int_{\partial B} k \left\{ \tilde{p}^{ab} \, \delta\gamma_{ab} - \tilde{P}^a \, \delta A_a \right\}_{(L)} d^2x, \end{aligned} \tag{7}$$

where $\tilde{p}^{ab}_{(L)}$ and $\tilde{P}^a_{(L)}$ are momenta conjugate to γ_{ab} and A_a on the Lorentzian sector and where ∂B is the two boundary of the system at a surface of constant τ. Making use of the stationary phase approximation, $J_{\mathrm{thermal}} = -I_{\mathrm{E}_0}$, the variational equation (7) can be identified with the first law of thermodynamics.

To see this, note that,

$$\begin{aligned}\left\{\tilde{p}^{ab}\,\delta\gamma_{ab}-\tilde{P}^a\,\delta A_a\right\}_{(L)} &= \Big\{-2N\tilde{p}^{00}\,\delta N+2\left[\tilde{p}^{00}N^B\sigma_{AB}+\tilde{p}^{0B}\sigma_{AB}\right]\delta N^A \\ &\quad -\tilde{P}^0\,\delta A_0-\tilde{P}^A\,\delta A_A+\left[\tilde{p}^{00}N^AN^B+2\tilde{p}^{0A}N^B+\tilde{p}^{AB}\right]\delta\sigma_{AB}\Big\}_{(L)} \\ &= \Bigg\{-\frac{\Gamma}{8\pi}\delta N+\frac{2}{\sqrt{\sigma}}p^{1j}h_{Aj}\,\delta N^A+\frac{P^1}{\sqrt{\sigma}}\,\delta A_0-\frac{\tilde{P}^A}{\sqrt{\sigma}}\,\delta A_A \\ &\quad +\left[\frac{2}{\sqrt{\sigma}}p^{1B}N^A-\frac{\Gamma}{16\pi N}N^AN^B+\frac{\tilde{p}^{AB}}{\sqrt{\sigma}}\right]\delta\sigma_{AB}\Bigg\}_{(L)}\sqrt{\sigma}. \end{aligned} \tag{8}$$

In this expression, note that

$$\mathcal{E}\equiv-\frac{1}{8\pi}\Gamma\Big|_{\partial B} \tag{9}$$

corresponds to the energy density of the system per unit boundary surface area [5]. Also, if n^i is the outward pointing unit normal, we have by virtue of Gauss' law that,

$$\mathcal{Q}\equiv-n_iP^i/\sqrt{h}=-P^1/\sqrt{\sigma} \tag{10}$$

is the charge per unit boundary spatial surface area. Further,

$$\hat{\mathcal{J}}_A\equiv-2n_ip^{ij}_{(L)}h_{Aj}/\sqrt{h}+\mathcal{Q}A_A=-2p^{1j}_{(L)}h_{Aj}/\sqrt{\sigma}+\mathcal{Q}A_A, \tag{11}$$

is the A component of the angular momentum per unit spatial boundary surface area. The inverse temperature as measured by locally non–rotating observers on the boundary of the system is $\beta(x^i)|_B=kN(x^i)|_B$, while $\hat{\omega}^A\equiv N^A/N\Big|_B$ is the proper angular velocity, and $\phi=\left(-N^{-1}A_0+\hat{\omega}^BA_B\right)\Big|_B$ is the proper electrostatic potential measured by these observers [5].

Thus,

$$\begin{aligned}\delta I_{\mathrm{E}_0} &= \int_{\partial B}\Big\{\left(\mathcal{E}-\hat{\mathcal{J}}_A\hat{\omega}^A+\mathcal{Q}\phi\right)\delta\beta-\beta\hat{\mathcal{J}}_A\,\delta\hat{\omega}^A \\ &\quad -\beta\lambda^{AB}\,\delta\sigma_{AB}+\beta\mathcal{Q}\,\delta\phi+\beta L^A\,\delta A_A\Big\}\sqrt{\sigma}\,d^2x. \end{aligned} \tag{12}$$

In the above,

$$\lambda^{AB}=-\hat{\omega}^Bj^A+\hat{\omega}^A\hat{\omega}^B\mathcal{E}/2+\tilde{p}^{AB}/(N\sqrt{\sigma}) \tag{13}$$

are surface pressure potentials and

$$L^A=-\mathcal{Q}\hat{\omega}^A-\tilde{P}^A/(N\sqrt{\sigma}) \tag{14}$$

are potentials thermally conjugate to the A_A.

To obtain a differential form of this variational equation consider an infinitesimal solid angle section of the system extending from x^A to $x^A+\delta x^A$. Let $I_{\mathrm{E}_0}\equiv\mathcal{I}_{\mathrm{E}_0}\sqrt{\sigma}|_{\partial B}\,d^2x$ be the classical smooth action for this solid angle section. Also note that since we have chosen boundary constraints appropriate to the grand canonical ensemble, the Massieu function for this solid angle section is

$$J_{\rm GC} \equiv \beta\mathcal{F}\sqrt{\sigma}\Big|_{\partial B} d^2x = -I_{\rm E_0}, \tag{15}$$

where $\mathcal{F}$ is the Gibbs free energy per unit spatial boundary surface area. Thus,

$$\delta\mathcal{F} = \beta^{-2}\mathcal{S}\,\delta\beta - \hat{\mathcal{J}}_A\,\delta\hat{\omega}^A - \left(\lambda^{AB} + \mathcal{F}\sigma^{AB}/2\right)\,\delta\sigma_{AB} + \mathcal{Q}\,\delta\phi + L^A\,\delta A_A, \tag{16}$$

where $\mathcal{S} \equiv \beta^2\left(\frac{\partial\mathcal{F}}{\partial\beta}\right)_{\hat{\omega}^A,\sigma_{AB},\phi,A_A}$ is given by

$$\mathcal{S}|_{\partial B} = \beta\left(\mathcal{E} - \hat{\mathcal{J}}_A\hat{\omega}^A + \mathcal{Q}\phi - \mathcal{F}\right)\Big|_{\partial B}. \tag{17}$$

Using (17), equation (16) may be reexpressed in the form;

$$\delta\mathcal{E} = \beta^{-1}\,\delta\mathcal{S} + \hat{\omega}^A\,\delta\hat{\mathcal{J}}_A - \phi\,\delta\mathcal{Q} - \left(\lambda^{AB} + \mathcal{F}\sigma^{AB}/2\right)\,\delta\sigma_{AB} + L^A\,\delta A_A, \tag{18}$$

where all quantities are as measured locally on some infinitesimal 2–patch of ∂B. Equation (18) is a differential form of the first law of thermodynamics.

Zeroth Law:
The classical zeroth law posits that when two isolated systems are brought into thermal contact they evolve until they share a common temperature. To derive a generalized form of this law, we require first a form of the action functional appropriate to an isolated system. Specifically, the action functional must have a well defined variational principle subject to boundary constraints on $\mathcal{E}$, $\hat{\mathcal{J}}$, $\mathcal{Q}$, σ_{AB}, and A_A. The appropriate action is [2, 4, 6],

$$I_{\rm E(MC)} = \left\{I_{\rm E(GC)} - \int_{\partial B}\beta\left(\mathcal{E} - \hat{\omega}\hat{\mathcal{J}} + \phi\mathcal{Q}\right)\sqrt{\sigma}d^2x\right\}\Big|_{\mathcal{H}=\mathcal{H}_i=0}. \tag{19}$$

Note that the *minimal* microcanonical action is

$$I_{\rm E(MC)_0} = -A_H/4. \tag{20}$$

Use the fact that the microcanonical Massieu function corresponds to the entropy of the system and equation (3) to find that the semi–classical entropy of the black hole system is given by $S_{\rm BH} = \frac{1}{4}A_H$. Generalize for a generic matter distribution to find that [7, 8, 4],

$$S_{\rm system} = \int s\,dV + A_H/4, \tag{21}$$

where s is the entropy density of the matter and the integral is over the spatial 3–volume of the system.

Now consider two initially isolated bodies at different temperatures brought into thermal contact (e.g. by replacing an adiabatic wall between them with a diathermal one). The total initial action for the composite system is

$$I_{\rm comp} = I_{\rm MC(1)} + I_{\rm MC(2)}, \tag{22}$$

Vary to obtain the usual variational terms plus an additional term:

$$\delta I_{\text{comp.}}|_{\Sigma_{y=\alpha}} = \int_{\Sigma_{y=\alpha}} \left(\beta|_{\alpha_+} - \beta|_{\alpha_-}\right) \delta\mathcal{E}\sqrt{\sigma}\, d^2x, \tag{23}$$

where $\Sigma_{y=\alpha}$ is the surface of contact between the two systems. If the action is to be extremized, all variational terms including (23) must vanish independently. So long as the adiabatic wall is present, $\delta\mathcal{E} = 0$ at $\Sigma_{y=\alpha}$ and, so, the variational term there is zero. When the diathermal wall is substituted, $\delta\mathcal{E} \neq 0$. In this case, we require;

$$\beta|_{\alpha_+} - \beta|_{\alpha_-} = 0. \tag{24}$$

Equation (24) may be interpreted as the generalized zeroth law of thermodynamics. Note that the general law is weaker than the classical law: it requires only that the equilibrium temperature be continuous across any diathermal surface. The extra flexibility allows for stable temperature gradients which arise due to gravitational redshifting.

Second Law:
The classical second law posits that it is not possible to vary the constraints of an isolated system in such a way as to decrease its entropy [9]. Since classical entropy is only defined for systems in equilibrium, this law applies only to comparisons of initial and final equilibrium states. Standard examples of the classical second law occur when one allows thermal contact between two initially isolated bodies, punctures a canister of gas inside a large chamber, or brings two inertial bodies into frictional contact. In each case, one relaxes an internal constraint on the composite system of interacting bodies and, in each case, this generally leads to an increase of entropy for the composite system. On the other hand, an ideal process which has no effect except to impose a constraint (for instance, the adiabatic separation of two equilibrium bodies which have been in thermal contact) results in no change in the composite entropy for the system.

A generalized second law of thermodynamics which applies to stationary horizon/matter systems derives directly from the principle of least action. Recall that the classical action for an isolated system is the negative of the entropy for the system in the stationary phase approximation. Also, note that in this approximation, an equilibrium system is in the state of least action accessible to it. Hence, if one is to change the state of a system initially in equilibrium, one must alter the constraints which determine the set of accessible states. One may either relax constraints or impose new ones.

Relaxing a constraint on the system increases the class of configurations accessible to it. The configuration which minimized the action in the initial class will not necessarily do so in the new larger class. Since the final equilibrium configuration will be that with least action accessible in the new class, $\Delta I_{E_0} \leq 0$ and $\Delta S \geq 0$.

On the other hand, imposing a constraint on the system, reduces the set of accessible states. The configuration which minimized the action in the initial class must also be minimal in any subset of that class. Hence, an ideal operation which merely imposes a constraint on a system does not alter its entropy.

We conclude that $\Delta S_{\text{total}} = \Delta S_{\text{matter}} + \Delta S_{\text{horizons}} \geq 0$ in the semi–classical approximation.

Third Law:
The classical third law posits that it is not possible to reduce a system to zero temperature within a finite amount of time. A general analogue of this law follows directly from the variational principle and the uncertainty principle.

Specifically, consider a system separated from a finite temperature heat bath by an adiabatic wall. Variations of the classical action at the boundary of the system are,

$$\delta I_{\mathrm{E}}\big|_{y=\alpha} = \int_{y=\alpha} \beta\big|_{-}^{+} \delta\mathcal{E}\, d^3x. \tag{25}$$

If the system were at zero temperature, then $\beta|_{\alpha_-} = \infty$. On the other hand, $\beta|_{\alpha_+}$ is finite. Hence, variations of the action diverge and nothing approximating an equilibrium state is possible unless $\delta\mathcal{E}|_B \equiv 0$. However, by virtue of the uncertainty principle,

$$\delta\mathcal{E}\delta t \geq \hbar. \tag{26}$$

Hence, it is not possible to measure a zero temperature state within a finite amount of time.

References

[1] G. W. Gibbons and S. W. Hawking, Physical Review D, **15**, 2752 (1977).

[2] J. D. Brown, G. L. Comer, E. A. Martinez, J. Melmed, B. F. Whiting and J. W. York, Jr., Classical and Quantum Gravity, **7**, 1433, (1990).

[3] J. W. York, Jr., Physical Review D, **33**, 2092 (1986).

[4] G. Hayward, "General laws of thermodynamics," University of British Columbia preprint, UBCTP–91–018.

[5] J. D. Brown, E. A. Martinez, and J. W. York, Jr., University of North Carolina preprint IFP–1000 (1990).

[6] G. Hayward, "Quantum gravity in the microcanonical ensemble," University of British Columbia preprint, UBCTP–91–019.

[7] E. A. Martinez and J. W. York, Jr., Physical Review D **40**, 2124 (1989).

[8] G. Hayward, Physical Review D **41**, 3248 (1990).

[9] A. B. Pippard, Elements of Classical Thermodynamics (Cambridge University Press, Cambridge, 1960).

New Demonstration of the Positivity of Mass

R.D. Sorkin*
Enrico Fermi Institute, 5640 S. Ellis Ave.
Chicago, IL 60637, USA
and
E. Woolgar**
Mail Stop 301-150, Jet Propulsion Laboratory
4800 Oak Grove Drive, Pasadena, CA 91109, USA

Abstract

A new demonstration of the positivity of mass in general relativity is given. It is based on spacetime causal structure and depends on conditions different from those assumed in other proofs of the positivity of mass. Herein we outline the proof for the case in which the spacetime metric approaches that of the Schwarzschild solution near infinity.

New Demonstration of the Positivity of Mass

Though proofs[1,2] have existed for more than a decade, the positive mass theorem remains somewhat enigmatic. For example, the Witten proof depends upon the existence of certain spinors on Cauchy surfaces, though it is not *a priori* obvious why this ought to be a crucial property of positive mass spacetimes. Herein we outline a proof of the positivity of mass based on causal structure. We show that in spacetimes whose causal structures have a particular nature near infinity, an inconsistency inevitibly develops, if a certain averaged energy condition is satisfied. Any spacetime whose metric asymptotes to a Schwarzschild metric will have this causal pathology if it has negative ADM mass. We claim the result generalizes well beyond this class of metrics. The idea for a proof of this sort arose in a discussion with Roger Penrose, who subsequently developed a sketch of a slightly different version of the argument[3]. We plan a comprehensive joint paper giving details of this work.

Consider the $t = 0$ slice of Minkowski space, where t is the usual time coordinate. In this slice, consider two parallel 2-planes, say $z = +1$ and $z = -1$. Let the $z = -1$ plane represent the wavefront of a family of null geodesics, which necessarily will have parallel wavevectors $k^\mu = (1,0,0,1)$. In Minkowski space, parallel rays meet at infinity, so we may think of this wavefront as formed by a family of null geodesics inward-bound from a single point $p \in \mathcal{I}^-$, *i.e.* a point p in past null infinity with $z = -\infty$ at p. Let us also interpret the $z = +1$ plane as the wavefront of a family of null geodesics bound for a point $q \in \mathcal{I}^+$. Then it is obvious that no non-spacelike curve in the spacetime can join p to q, for no null geodesic can travel fast enough to do this. The null geodesics from p have only reached the $z = -1$ plane by time $t = 0$, and they would have had to have reached the $z = +1$ plane to reach q. By the way, we note that the locus of points in $\mathcal{I}^+$ not reached from p by non-spacelike curves through spacetime is confined to (the lower part of) one null generator of $\mathcal{I}^+$, for points on any other generator are reached by null geodesics that meet the $t = 0$ surface in a 2-plane tilted with respect to the $z = +1$ plane, and this plane will always cross the $z = -1$ plane since these planes are not parallel.

Let us now consider the equivalent situation, but with the Schwarzschild metric. If the mass is positive, we ordinarily say there is a *time delay*[4] which retards null geodesics that pass near the origin, but one could equally well say that the rays that pass asymptotically far from the origin are advanced. The new forward-moving wavefront formed by null geodesics from p no longer resembles a flat plane – it warps forward as one moves away from the middle. Far away from the origin, the wavefront lying in the $t = 0$ surface is a 2-surface that obeys $z \sim \ln(b/M)$, where b is the impact parameter. Similarly, the null geodesics that converge to $q \in \mathcal{I}^+$ meet the $t = 0$ plane in a 2-surface that, for large b, obeys $z \sim -\ln(b/M)$. If one goes to large enough b, these surfaces always meet and cross each other, so one always can find a null geodesic (albeit one with large impact parameter) that leaves $p \in \mathcal{I}^-$ and meets the $t = 0$ surface at a point from which another null geodesic departs for $q \in \mathcal{I}^+$, so all of $\mathcal{I}^+$ is within the chronological future of each point of $\mathcal{I}^-$. For negative Schwarzschild mass, the wavefronts have precisely the opposite behaviour. They run logarithmically *away* from each other for large b, *i.e.* near spatial infinity. It is this behaviour near infinity which will be crucial to our proof.

Now let us consider a globally hyperbolic spacetime possessing a spatial and null infinity upon approach to which the metric becomes Schwarzschild. Fix a very large but still compact set C in this spacetime. No non-spacelike curve that meets C can meet $\mathcal{I}$ (null infinity) arbitrarily close to i^0, for suppose this were false. Then one could construct a sequence of non-spacelike curves with initial endpoints in C and final endpoints forming a sequence that converged to i^0. Then the limit of the sequence of initial endpoints would lie in $\mathcal{C}$ by its compactness and the limit of the sequence of final endpoints would be i^0. Since this is a globally hyperbolic spacetime, we are able to show (details will appear elsewhere) the existence of a limiting non-spacelike curve joining these limits of the sequences of endpoints. This is a non-spacelike curve joining a point of $\mathcal{M}$ to i^0, which violates the assumption that i^0 represents spatial infinity. Thus, spacetime is divided into two sets, $J^+(\mathcal{C}) \cup J^-(\mathcal{C})$ and its complement $\mathcal{N}$, the latter being a neighbourhood of i^0. If our initial compact set $\mathcal{C}$ is sufficiently large, the metric within $\mathcal{N}$ will approach its asymptotic form to sufficient accuracy for our needs; let us arrange that this is the case.

Now let us pick two points, $p \in \mathcal{I}^- \cap \mathcal{N}$ and $q \in \mathcal{I}^+ \cap \mathcal{N}$, so these are two points of null infinity that are within the neighbourhood $\mathcal{N}$ of i^0. In this spacetime, if p and q are to be joined by any non-spacelike curve through the spacetime, this curve must always remain in $\mathcal{N}$, which is a way of stating the requirement that the curve must have a large impact parameter. Otherwise, the curve would meet $J^+(\mathcal{C}) \cup J^-(\mathcal{C})$, whence it could not return to $\mathcal{N}$. But within $\mathcal{N}$, let us now assume that the metric is approximated by the (asymptotic) Schwarzschild metric to sufficient accuracy (if this accuracy proves insufficient, we start over again with a new and enlarged set $\mathcal{C}$). Then, as we saw two paragraphs previously, there exist non-spacelike curves in this spacetime joining all such p and q if $M > 0$ and *not* if $M < 0$. It is interesting that the logarithmic nature of the asymptotic behaviour of the wavefronts implies that again only (the bottom part of) one generator of $\mathcal{I}^+$ fails to meet $I^+(p)$ in the $M < 0$ case.

In the case $M < 0$, consider the boundary of the set $I^+(p)$, the chronological future of p. Clearly q as chosen above lies outside $I^+(p)$. Let it lie on a null geodesic generator ξ of $\mathcal{I}^+$ (as noted above, this is the *unique* generator of $\mathcal{I}^+$ containing points not in $I^+(p)$). Certainly other points farther up ξ lie within $I^+(p)$, so we may move q up ξ until moving it any higher would place it in $I^+(p)$ (contrast this with the $M > 0$ case, for which every point of $\mathcal{I}^+$ lies in $I^+(p)$). Of course, q will in general have to be moved up $\mathcal{I}^+$ and out of the neighbourhood $\mathcal{N}$ of i^0 to achieve this, but no matter. We may now consider a sequence of points $q_n \in \xi \cap I^+(p)$ such that this sequence converges to q. There is a neighbourhood of infinity (now including the q_n) on which the metric is Schwarzschild to sufficient accuracy. The null geodesics from p that remain in this neighbourhood will form a wavefront that recedes logarithmically with increasing impact parameter from the wavefronts formed by the null geodesics that remain in this neighbourhood and have their final endpoints in the set $\{q_n\}$. Therefore, since these wavefronts must overlap (since $q_n \in I^+(p)\forall n$), they do so within some compact region $\mathcal{K}$ of spacetime. Consider a sequence of non-spacelike curves γ_n joining p to each of the q_n. We will show elsewhere that this sequence possesses a limit curve which is a null geodesic joining p to q and that this limit curve meets the compact set $\mathcal{K}$ (which follows from the observation that the γ_n must all meet this set). This limit curve therefore does not lie on the boundary of spacetime (save at its endpoints of course). But this is a very interesting limit curve, for it extends to infinite values of its affine parameter (*i.e.* it joins $\mathcal{I}^-$ to $\mathcal{I}^+$ via a geodesic through spacetime) and yet it remains always on the boundary of the future set of p. It therefore cannot have developed conjugate points, for this would imply it had focussed, whence it would be compelled to leave the boundary of $I^+(p)$ and enter the interior of this set. Herein lies the problem, for conjugate points are inevitible whenever the conditions of the following theorem hold.

Theorem: Let $\mathcal{M}$ be an asymptotically flat spacetime that possesses a subset $\mathcal{D} = I^+(\mathcal{I}^-) \cap I^-(\mathcal{I}^+)$ (the *Domain of Outer Communications*) that satisfies the following conditions:

- Along every null geodesic in $\mathcal{D}$ with tangent $l^a = \frac{d}{d\lambda}$, λ an affine parameter, we have $\int_{-\infty}^{+\infty} d\lambda \; T_{ab} l^a l^b \geq 0$,
- Along every null geodesic in $\mathcal{D}$ for which $\int_{-\infty}^{+\infty} d\lambda \; T_{ab} l^a l^b = 0$, there is at least one point where $l^e l_{[a} C_{b]ef[c} l_{d]} l^f \neq 0$, and
- $\mathcal{D}$ admits a Cauchy surface.

Then $\mathcal{M}$ cannot possess a negative mass.

Proof: Note first that the conditions above are sufficient to guarantee that every null geodesic will develop conjugate points separated by finite values of the affine parameter[5]. Let $\bar{\mathcal{M}} = \mathcal{M} \cup \mathcal{I}^+ \cup \mathcal{I}^- \cup \{i^0\}$. If there is a neighbourhood of infinity on which the metric can be given as Schwarzschild to sufficient accuracy, then the proof follows from the arguments given above, which have it that one can always find a null geodesic of infinite affine length that does not develop conjugate points whenever the metric has a negative mass parameter. The result extends immediately to all spacetimes where the asymptotic metric produces the same behaviour in the wavefronts formed by null geodesics, namely that the wavefront formed by null geodesics from some $p \in \mathcal{I}^-$ recedes from that formed by the null geodesics that have final endpoint at some $q \in \mathcal{I}^+$

as one approaches spatial infinity. We will argue elsewhere that this behaviour is present in a large class of asymptotically flat metrics. □

Detailed arguments proving existence of various limits of sequences and giving details of other assertions will be published elsewhere. It remains to demonstrate the appropriate behaviour of the wavefronts of null geodesics in the largest possible class of metrics. In metrics that are not asymptotic to the Schwarzschild metric, it is still an open question whether it truly is the sign of the ADM mass that governs this behaviour. However, an argument of Ashtekar and Penrose[6] suggests that this is the case.

Acknowledgments

We would like to acknowledge support from the NSF (USA) uder grants PHY 9005790 and PHY 8918388. E.W. would like to thank NSERC (Canada) for a NATO Fellowship which supported the earlier phase of the work, the NRC (USA) for a Resident Research Associateship at the JPL during the later phase of the work, and Syracuse University Dept. of Physics for hospitality.

References

(1) Schoen, R., and Yau, S.-T., *Phys. Rev. Lett.* 43, 1457 (1979); *Commun. Math. Phys.* 79, 47 (1981); 231 (1981).

(2) Witten, E., *Commun. Math. Phys.* 80, 381 (1981); Parker, T., and Taubes, C.H., *Commun. Math. Phys.* 84, 223 (1982).

(3) Penrose, R., *Twistor Newsletter* 30, 1 (1990).

(4) Shapiro, I.I., *Phys. Rev. Lett.* 13, 789 (1964).

(5) Tipler, F.J., *J. Diff. Eq.* 30, 165 (1978); *Phys. Rev.* D17, 2521 (1978); Roman, T.A., *Phys. Rev.* D33, 3526 (1986); D37, 546 (1988); Borde, A., *Class. Quantum Gravit.* 4, 343 (1987).

(6) Ashtekar, A., and Penrose, R., preprint for *Twistor Newsletter*, (1990).

* Permanent Address: Dept. of Physics, Syracuse University, Syracuse, NY 13244-1130, USA.

** Talk given by E.W. at Fourth CCGRRA meeting, Winnipeg, Canada.

TIME AND INTERPRETATIONS OF QUANTUM GRAVITY

KAREL V. KUCHAŘ †
Blackett Laboratory
Imperial College of Science and Technology
South Kensington, London SW7 2BZ, UK

ABSTRACT

In canonical quantization of gravity, the state functional does not seem to depend on time. This hampers the physical interpretation of quantum gravity. I critically examine ten major attempts to circumvent this problem and discuss their shortcomings.

Mein sind die Jahre nicht, die mir die Zeit genommen;
Mein sind die Jahre nicht, die etwa möchten kommen;
Der Augenblick ist mein, und nehm ich den in acht,
So ist der mein, der Jahr und Ewigkeit gemacht.

Betrachtung der Zeit
Andreas Gryphius 1616-1664

1. Constraints and Dynamics

In the midst of the Thirty Years War, Andreas Gryphius contemplated the uncertainties of his time in an epigram about Time itself:[1]

Those years are not my years that were devoured by time,
Those years are not my years that fate may deem be mine;
This instant, though, is mine, and if I hold it dear,
Then He is mine who made the endless train of years.

---

† On sabbatical from Department of Physics, University of Utah, Salt Lake City, Utah 84112, USA.

I can hardly think of a better summary of the aims of canonical geometrodynamics.

Of course, to predict the future and reconstruct the past from the present data has been the main preoccupation of physicists long before Einstein's theory of gravitation was cast into canonical form. I do not want to read Gryphius' words in this trivial way. Canonical gravity is more than a complicated implementation of the Laplace program: When predicting the future and reconstructing the past, it relies on dynamical laws which are derived from the laws ruling the data at a single instant. While in every dynamical theory the future follows from the present, in relativity the laws according to which this happens also follow from the laws of the present.

To understand the relationship between the dynamical laws and the laws of an instant requires some explanation. The dynamical laws were introduced into physics by Newton; the first law of an instant was discovered by Gauss. In electrodynamics, the electric field cannot be arbitrarily prescribed, but it is subject to the Gauss law: In vacuo, the divergence of **E** must vanish. This is not a dynamical law, but a limitation on the instantaneous distribution of the electromagnetic data. The dynamical Maxwell equations preserve the Gauss law, but they do not follow from it.

In canonical geometrodynamics, the instantaneous state of a vacuum Einstein spacetime is described by the intrinsic geometry $g_{ab}(x)$ and the extrinsic curvature $p^{ab}(x)$ of a spacelike hypersurface $\Sigma \to \mathcal{M}$. These instantaneous data are not arbitrary; at any point $x \in \Sigma$, four combinations of the data,

$$H_a(x) = -2D_b p^b_a(x) \tag{1.1}$$

and

$$\begin{aligned} H(x) &= G_{abcd}(x;g)\, p^{ab}(x)\, p^{cd}(x) - |g|^{1/2} R(x;g] \,, \\ G_{abcd} &= \frac{1}{2} |g|^{-1/2} (g_{ac}g_{bd} + g_{ad}g_{bc} - g_{ab}g_{cd}) \,, \quad |g| := \det(g_{ab}) \end{aligned} \tag{1.2}$$

called the supermomentum and the super-Hamiltonian, are constrained to vanish:

$$H_a(x; g,p] = 0 \tag{1.3}$$

and

$$H(x; g,p] = 0 \,. \tag{1.4}$$

(The notation H(x; g,p] emphasizes that H is a function of x but a functional of g and p.) These are the laws of an instant in canonical gravity. The divergence law (1.1), (1.3) is analogous to the Gauss law; the super-Hamiltonian constraint (1.2), (1.4) has no counterpart in electrodynamics. It is this constraint that ultimately yields the dynamics of geometry.

Any reference to the hypersurface $\Sigma \to \mathcal{M}$ which carries the geometric data $g_{ab}(x)$, $p^{ab}(x)$ is conspicuously absent in the constraints (1.1) - (1.4). The hypersurface $\Sigma \to \mathcal{M}$ represents an instant of time; the fact that it drops out of the constraints (1.1) - (1.4) underlies the problem of time in quantum gravity.

One can surmise that the geometric data $g_{ab}(x)$, $p^{ab}(x)$ themselves specify the hypersurface which carries them (Baierlein, Sharp and Wheeler[2]). One suspects there is a canonical transformation[3]

$$g_{ab}(x),\ p^{ab}(x) \to X^A(x),\ P_A(x),\ \phi^r(x),\ p_r(x) \tag{1.5}$$

which separates the four *embedding variables* $X^A(x)$, $A = 0,1,2,3$ that specify the hypersurface from the two *true gravitational degrees of freedom* $\phi^r(x)$, $r = 1,2$. One could then solve the constraints (1.1) - (1.4) with respect to the *embedding momenta* $P_A(x)$ and, after doing so, replace them by an equivalent set of constraints

$$H_A(x) := P_A(x) + h_A(x;\ X,\phi,\pi] = 0\ . \tag{1.6}$$

The expressions $h_A(x)$ represent the energy density and the energy flux carried by the gravitational variables ϕ and π through the hypersurface $X^A(x)$.

This sets the stage for deriving the dynamical laws from the constraints (1.1) - (1.4), or (1.6), i.e., from the laws of an instant. Among several alternative routes, the one which is closest to quantum gravity relies on the Hamilton-Jacobi theory. One postulates that the momenta $p^{ab}(x)$ (or $P_A(x)$, $\pi_r(x)$) are generated by a Jacobi principal functional $S[g]$ (or $S[X,\phi]$):

$$\begin{gathered} p^{ab}(x) = \delta S[g] \,/\, \delta g_{ab}(x)\ , \quad \textit{or} \\ P_A(x) = \delta S[X,\phi] \,/\, \delta X^A(x), \quad \pi_r(x) = \delta S[X,\phi] \,/\, \delta \phi^r(x). \end{gathered} \tag{1.7}$$

By substituting the expressions (1.7) into the constraints (1.1) – (1.4) or (1.6),

$$-\,2D_b \frac{\delta S[g]}{\delta g_{ab}(x)} = 0\ , \tag{1.8}$$

$$H\left(x;\, g,\, \frac{\delta S[g]}{\delta g}\right] = 0 \quad , \tag{1.9}$$

or

$$\frac{\delta S[X,\phi]}{\delta X^A(x)} + h_A\left(x;\, X,\, \frac{\delta S[X,\phi]}{\delta \phi}\right] = 0 \quad , \tag{1.10}$$

one obtains variational differential equations for the Jacobi principal functional S[g] or S[X,ϕ]. These are the Hamilton-Jacobi equations in geometrodynamics.[4,5,3] A particular solution of Eqs. (1.8) - (1.9) (or Eq. (1.10)) generates a family of solutions of the Einstein equations.[6] A complete solution generates a canonical transformation which yields all the solutions of the Einstein equations.[7,5] In this way, the dynamical laws according to which geometry evolves follow from the constraints, i.e., from the laws of an instant.

The original constraints (1.1) - (1.4) lead to Eqs. (1.8) and (1.9). Equation (1.8) ensures that the Jacobi principal functional remains unchanged under Diff Σ, i.e., that it does not depend on the metric $g_{ab}(x)$, but only on the geometry **g**(x). Equation (1.9) is then the proper Hamilton-Jacobi equation for S[**g**]. In it, the dynamical degrees of freedom are not separated from the time (from the embedding variables) in which they evolve. This happens only in Eq. (1.10) which has the form of a first-order functional differential equation in $X^A(x)$.

• **Global Problem of Time**. Even at the classical level, the extraction of dynamics from the instantaneous constraints poses some problems. First, the separation of time (of the embedding variables) from the dynamical degrees of freedom may be globally impossible. Such a situation occurs already in particle dynamics. In parametrized Newtonian dynamics,[8] the constraints are automatically given in the form (1.6) with respect to the absolute time T:

$$H := P_T + h(T;\, Q,P) = 0 \ . \tag{1.11}$$

The linearity of the constraint (1.11) in P_T implies that the absolute time T always increases along any dynamical trajectory and hence every trajectory intersects a hypersurface of constant T once and only once. However, some other simple finite-dimensional constrained dynamical systems may admit no time function with this property.[9-12] This indicates that the super-Hamiltonian constraint of such systems cannot be globally replaced by an equivalent constraint (1.11). The analysis of homogeneous cosmological models reveals that geometrodynamics meets similar problems.[13] So far, no one seems to have analyzed in detail how serious and generic such problems are in full geometrodynamics (1.1) - (1.4). It is, however, quite likely that there does not exist any canonical tranformation (1.5) such that the original constraints (1.3) - (1.4) are fully equivalent to the new constraints (1.6). If so, the Hamilton-Jacobi equation (1.10) cannot generate all Einstein spacetimes. I shall call this purely classical problem *the global problem of time.*

• **The Sandwich Problem.** The Hamilton-Jacobi equation (1.9) has a related difficulty. The choice of the *metric representation* S[g] presupposes that the evolution of the gravitational field is described by a canonical transformation whose generating function (the Hamilton principal function[14,15]) is a functional of the initial and final metric: $S[g_{FIN} \mid g_{IN}]$. It is questionable to what extent the assignment of the *intrinsic metric* on the inital and final hypersurfaces uniquely determines a spacetime geometry in the sandwich between them. This constitutes the so called *sandwich problem.*[2,16-25] In general, one would like to know what choice of canonical variables, if any, as arguments of the Hamilton principal function, yields a unique spacetime geometry in the sandwich. This question leads back to investigating canonical transformations like those given by Eq. (1.5).

▪ **Independence of the Dynamic Evolution on the Foliation.** The last question one needs to address is whether the dynamics generated by the Jacobi function (1.8) - (1.9), or (1.10), preserves the constraints (1.3) - (1.4), or (1.6), which lead to this dynamics. It is well known[26] this is so if the constraints are first class, i.e., if the Poisson brackets between any two of them vanish on the constraint hypersurface. The constraints (1.3) and (1.4) are first class,

$$\{H(x),\, H(x')\} \approx 0 \quad , \textit{etc.}, \tag{1.12}$$

by virtue of the so called Dirac algebra.[27] One can show then by a general argument that if the constraints (1.3) - (1.4) can be brought into the new form (1.6) by a canonical transformation (1.5), the Poisson brackets of the new constraints (1.6) must actually vanish strongly rather than weakly:[28]

$$\{H_A(x),\, H_B(x')\} = 0 \ . \tag{1.13}$$

Equations (1.12) or (1.13) ensure that the evolution of the Jacobi principal functional by the functional Hamilton-Jacobi equation (1.8) - (1.9), or (1.10), from an initial hypersurface IN in a final hypersurface FIN does not depend on the foliation connecting IN with FIN. The functional nature of the time variable thus presents no problem in the classical theory.

2. Problems of Time in Quantum Gravity

The quantization of a system whose dynamics is completely described by first-class constraints can be accomplished by the Dirac constraint quantization.[26] First, the fundamental variables like $g_{ab}(x)$, $p^{ab}(x)$ or $X^A(x)$, $\phi^r(x)$, $P_A(x)$, $\pi_r(x)$ are turned into operators $\hat{g}_{ab}(x)$, $\hat{p}^{ab}(x)$ or $\hat{X}^A(x)$, $\hat{\phi}^r(x)$, $\hat{P}_A(x)$, $\hat{\pi}_r(x)$ which satisfy the fundamental commutation relations. By substituting these operators into Eqs. (1.1) - (1.2), or (1.6), one turns the constraint functionals into operators

$$\hat{H}_a(x) = -2\,\hat{D}_a \hat{p}^b_a(x) \tag{2.1}$$

$$\hat{H}(x) = \text{"}\,\hat{p}^{ab}(x) G_{abcd}(x;\hat{g})\,\hat{p}^{cd}(x) - |\hat{g}|^{1/2} R(x;\hat{g}]\,\text{"} \tag{2.2}$$

or

$$\hat{H}_A(x) = \hat{P}_A(x) + h_A(x;\hat{X},\hat{\phi},\hat{\pi}]\ . \tag{2.3}$$

The quotation marks in Eq. (2.2) are a reminder that the factor ordering is ambiguous. The constraints are then imposed as limitations

$$H_a(x;\hat{g},\hat{p}]\ \Psi[g] = 0\ , \tag{2.4}$$

$$H(x;\hat{g},\hat{p}]\ \Psi[g] = 0\ , \tag{2.5}$$

or

$$H_A(x;\hat{X},\hat{P},\hat{\phi},\hat{\pi}]\ \Psi[X,\phi] = 0 \tag{2.6}$$

on the physical states $\Psi \in \mathcal{F}_0$ of the system.

Equation (2.4) ensures that the physical states depend only on the three-geometry $\mathbf{g}(x)$, not on the metric $g_{ab}(x)$ which represents $\mathbf{g}(x)$: $\Psi[g] = \Psi[\mathbf{g}]$.[29-30] Equation (2.5) is a second-order functional differential equation for $\Psi[\mathbf{g}]$ called the *Wheeler-DeWitt equation.*[18,31-32] On the other hand, the constraint (2.6) has the form of a functional Schrödinger equation[33]

$$i\,\frac{\delta \Psi[X,\phi]}{\delta X^A(x)} = h_A(x;X,\hat{\phi},\hat{\pi}]\ \Psi[X,\phi] \tag{2.7}$$

in the hypertime variables $X^A(x)$.

In my schematic discussion of the Dirac constraint quantization, I swept under the rug a heap of technical problems. These are carefully spelled out, e.g., by Isham.[34] I must, however, pinpoint those technical and conceptual problems which constitute the problem of time in quantum gravity. Broadly speaking, there are three places in which things can go wrong in the transition from the classical to the quantum theory:

• **The Problem of Functional Evolution.** The involution conditions (1.12) or (1.13) do not need to be valid for the commutators of the constraint operators. Thus, instead of Eq. (1.13) one can get

$$[\hat{H}_A(x), \hat{H}_B(x')] \neq 0 \tag{2.8}$$

and instead of Eq. (1.12)

$$[\hat{H}(x), \hat{H}(x')] \Psi[g] \neq 0 \text{ , } \textit{etc., for a } \Psi \in \mathscr{F}_0 \text{ .} \tag{2.9}$$

If so, the functional equations (2.6), or (2.4) - (2.5), are inconsistent or, at best, admit only special solutions. Equations (2.8) and (2.9) mean that the evolution of state from an initial hypersurface IN to a final hypersurface FIN can depend on the foliation which connects FIN with IN. In other words, when one starts with the same initial state Ψ_{IN} on the initial hypersurface, and develops it to the final hypersurface along two different routes, one obtains two different states $\Psi_{1\,FIN} \neq \Psi_{2\,FIN}$ on the final hypersurface. Such a situation certainly violates what one would expect of a relativistic theory. I shall call this problem *the problem of functional evolution.*

• **The Multiple Choice Problem.** The Schrödinger equation (2.7) based on one choice of the time variable $X^A(x)$ may give a different quantum theory than the Schrödinger equation based on another choice of the time variable. I shall call this problem *the multiple choice problem.*

The multiple choice problem is quite distinct from the problem of functional evolution; the latter arises even if one sticks to a single choice of time. It is also distinct from the global problem of time in classical geometrodynamics. The multiple choice problem is one of an embarrassment of riches: out of many inequivalent options, one does not know which one to select. The global problem of time is an embarrassment of poverty: one really does not have any choice at all.

• **The Hilbert Space Problem.** The multiple choice problem stems from an attempt to separate time from the dynamical degrees of freedom. What happens when one sticks to the Wheeler-DeWitt equation in the metric representation? While the Schrödinger equation automatically brings in a conserved inner product (conserved in the selected time variable) and the possibility to construct, at a given time $X^A(x)$, meaningful observables $F[X,\hat{\phi},\hat{\pi}]$ from the dynamical degrees of freedom, nothing similar is granted by the Wheeler-DeWitt equation. As a second-order functional differential equation, the Wheeler-DeWitt equation presents familiar problems when one tries to turn the space of its solutions into a Hilbert space. One thus trades the multiple choice problem for a *Hilbert space problem.*

The problem of functional evolution, the multiple choice problem, and the Hilbert space problem are the three major classes of problems which quantum geometrodynamics encounters because classical geometrodynamics does not seem to possess a natural time variable, while standard quantum theory relies quite heavily on a preferred time. Under these three headings, more specific problems will start emerging as I shall examine the proposals for solving or circumventing these major problems of time.

3. The Role of Models

Quantum gravity is an ill-defined and little understood system. It is no wonder that most people who tackled its time problem tried to elucidate it on simpler models. Each of these models shares with Einstein's theory some relevant feature which gives rise to a time problem. However, none of them shares with Einstein's theory all the relevant features. Therefore, even if one knows how to solve the time problem of the model, the time problem in the full Einstein theory remains unsolved.

I shall arrange eleven widely explored models in a hierarchical order, starting with those which closely resemble Einstein's theory, and then moving further and further away from it. The first four models are field theoretical systems, the remaining seven are finite-dimensional.

▪ **Midisuperspace Models.** These provide a canonical description of Einstein spacetimes with a given group of isometries. Symmetry removes infinitely many degrees of freedom of the gravitational field, but there are still infinitely many degrees of freedom left.

A prototype of a midisuperspace model is the cylindrical gravitational wave (Kuchař[35]). Cylindrical symmetry restricts the spatial line element to the form

$$d\sigma^2 = e^{\gamma-\psi}\, dr^2 + R^2\, e^{-\psi}\, d\varphi^2 + e^{\psi}\, dz^2 \tag{3.1}$$

in which it is characterized by three functions ψ,γ and $R \geq 0$ of a radial coordinate r. Similarly, the extrinsic curvature of an arbitrary hypersurface which respects the cylindrical symmetry (i.e, which contains its two commuting Killing vectors) is completely characterized by the three conjugate momenta π_ψ, p_γ, and p_R. When expressed in terms of these canonical variables, the super-Hamiltonian (1.2) assumes the form

$$H(r) = e^{\frac{1}{2}(\psi-\gamma)}\left(- p_\gamma p_R + \frac{1}{2} R^{-1}\pi_\psi^{\ 2} + \frac{1}{2} R\psi'^2 + 2R'' - \gamma' R'\right) . \quad (3.2)$$

Only the radial component H_1 of the supermomentum is different from zero,

$$H_1(r) = - 2p_\gamma' + \gamma' p_\gamma + R' p_R + \psi'\pi_\psi . \quad (3.3)$$

In spite of these simplifications, the midisuperspace constraints (3.2) and (3.3) are still fairly complicated functionals of the canonical variables ψ, γ, R and π_ψ, p_γ, p_R.

Other widely studied midisuperspace models are Gowdy cosmologies (Gowdy,[36] Misner,[37], Berger,[38-40] and Husain[41]).

▪ **Strings (Membranes).** Canonical analysis of these systems[42-44] leads to a system of constraints whose structure closely resembles that of Einstein's theory of gravitation. In particular, one obtains the standard DiffΣ (supermomentum) constraint and a quadratic super-Hamiltonian constraint, which close according to the Dirac algebra.

▪ **Parametrized Fields.** A field theory on a given Riemannian background can be made covariant under Diff$\mathcal{M}$ by adjoining to the field variables the embedding variables specifying spacelike hypersurfaces in $\mathcal{M}$.[26,45-47] Canonical analysis of a parametrized field theory leads directly to a system of constraints which has the form (1.6), where ϕ^r are the field variables, π_r their conjugate momenta, and X^A the embedding variables. The projection of these constraints tangential and normal to the hypersurface leads to the supermomentum $H_a(x; X,P,\phi,\pi]$ and super-Hamiltonian $H(x; X,P,\phi,\pi]$, which again satisfy the Dirac algebra.

▪ **Inhomogeneous Perturbations of Homogeneous Cosmologies.** This class of models stands at the border between infinitely-dimensional and finite-dimensional systems. One takes a homogeneous cosmological model (see the next paragraph) and introduces infinitely many inhomogeneous modes by allowing its linear perturbations.[48]

▪ **Minisuperspace Models (Homogeneous Cosmologies).** As in the midisuperspace models, infinitely many degrees of freedom are frozen by symmetries. The reduction is so drastic that only a finite number of degrees of freedom is left. The models represent homogeneous (but in general anisotropic) cosmologies, in particular the various Bianchi-type vacuum spacetimes.[32,49-52] The requirement of homogeneity limits the allowed hypersurfaces to the leaves of a privileged foliation, which is labeled by a single time variable. A good example of a minisuperspace model is the Bianchi type IX *mixmaster universe*.[49,53-55] Its homogeneity group is SO(3). One can parametrize the hypersurfaces of homogeneity by the Euler angle coordinates and characterize the spatial metric uniquely by three real parameters, Ω and $\beta_{\pm}$. The first parameter, Ω, is related to the volume of Σ:

$$\Omega = \ln \int_{\Sigma} d^3x \; |g(x)|^{1/2} \; . \tag{3.4}$$

The remaining two parameters, $\beta_{\pm}$, describe the anisotropy of Σ.

The extrinsic curvature of a hypersurface of homogeneity is similarly determined by the three momenta, p_Ω and $p_{\pm}$, canonically conjugate to Ω and $\beta_{\pm}$. Due to the symmetry of the model, the supermomentum constraints are again identically satisfied, while the super-Hamiltonian constraints reduce to a single constraint

$$H := - p_\Omega^{\;2} + p_+^{\;2} + p_-^{\;2} + e^{-4\Omega} \left(V(\beta_+, \beta_-) - 1\right) = 0 \tag{3.5}$$

for the minisuperspace variables Ω, $\beta_{\pm}$; p_Ω, $p_{\pm}$. The potential $V(\beta_+, \beta_-)$ is a combination of exponential terms; it vanishes at the origin $\beta_+ = 0 = \beta_-$, and it is positive outside the origin.

▪ **2+1 Gravity.** A three-dimensional Einstein spacetime is necessarily flat.[56] Thus, 2+1 gravity lacks field degrees of freedom. However, there is still a finite number of variables characterizing the topology of Σ.[57-58] The problem of time in the 2+1 gravity was discussed by Moncrief,[59] Hosoya and Nakao,[60] and Carlip.[61,62]

▪ **A Relativistic Particle.** A relativistic particle of rest mass M moving in a Riemannian spacetime $(\mathcal{M}, G)$ is a dynamical system described by a single super-Hamiltonian

$$H := \frac{1}{2M} G^{AB}(Q) P_A P_B + \frac{1}{2} M V(Q) = 0 \; . \tag{3.6}$$

(I have allowed for the possiblity that the mass of the particle is influenced by a potential field $V(Q)$, as in Nordström's theory of gravitation[63-65]). By comparing Eq. (3.6) with Eq. (1.2), one sees that geometrodynamics is analogous to a relativistic particle, with the intrinsic geometry $g_{ab}(x)$ playing the role of the spacetime postion Q^A, DeWitt's supermetric $G_{abcd}(x; g)$ the role of the background metric $G^{AB}(Q)$, and the scalar curvature $-|g|^{1/2}R(x; g]$ the role of the potential $V(Q)$. By enlarging the spacetime by gauge degrees of freedom and cutting these down by additional gauge constraints, one can also mimick the supermomentum constraints (1.1).[66]

▪ **A Parametrized Newtonian Particle.** A Newtonian particle can be parametrized by labeling its trajectory by an arbitrary parameter and adjoining the Newtonian time to the position Q^a. Its motion is generated by the constraint (1.11). The constraint (1.11) can be rewritten in a geometric form in terms of the metric and affine structures of the Newtonian spacetime.[67,68] It resembles then the constraint (3.6) for a relativistic particle.

▪ **The BB Model.** A system of Newtonian particles $K=1, \ldots, N$ interacting by central forces can be parametrized by labeling the dynamical trajectories by an arbitrary parameter and adjoining the Newtonian time to the position variables $Q^a_{(K)}$. One can then constrain the system by the requirements that its total energy H, total momentum $\mathbf{P}$ and total angular momentum $\mathbf{J}$ vanish:

$$H = \Sigma_{K=1}^{N} \frac{1}{2M_{(K)}} \delta^{ab} P_{(K)a} P_{(K)b} + \Sigma_{K\neq L} \frac{1}{2} W(|\mathbf{Q}_{(K)} - \mathbf{Q}_{(L)}|) , \tag{3.7}$$

$$\mathbf{P} = \Sigma_{K=1}^{N} \mathbf{P}_{(K)} , \quad \mathbf{J} = \Sigma_{K=1}^{N} \mathbf{Q}_{(K)} \times \mathbf{P}_{(K)} . \tag{3.8}$$

This model was introduced by Barbour and Bertotti[69-73] to implement Mach's ideas. By comparing Eq. (3.7) with (1.2) and (3.8) with (1.1), one sees the analogy between the BB model and canonical geometrodynamics.

▪ **Two Particles in a Stationary State.** The first particle moves in a curved space with the metric $G^{AB}(Q)$ and an external potential $V(Q)$. The second particle moves in a curved space with the metric $h^{kl}(Q;q)$ and the potential $v(Q;q)$. An interaction of the two particles is brought in by a dependence of the metric h^{kl} and the potential v on the coordinates of the first particle. One constrains the motion of the system by requiring that it takes place with a given energy E:

$$H := \frac{1}{2M} G^{AB}(Q)P_A P_B + \frac{1}{2} MV(Q)$$
$$+ \frac{1}{2m} h^{kl}(Q;q)p_k p_l + \frac{1}{2}mv(Q;q) - E = 0 . \tag{3.9}$$

The model can be specialized in a number of ways. The particles can move in the same space, the space may be flat, or the particles need not interact with each other. The metrics may be positive definite or indefinite. This class of models plays a useful role when discussing semiclassical gravity and the role of internal clocks.

▪ **Two Harmonic (Antiharmonic) Oscillators.** One can build a super-Hamiltonian

$$H = \left(\frac{1}{2M} P^2 \pm \frac{1}{2}M\Omega^2Q^2\right) \pm \left(\frac{1}{2m} p^2 \pm \frac{1}{2}m\omega^2q^2\right) \tag{3.10}$$

from the Hamiltonians for two harmonic (antiharmonic) oscillators, and impose the constraint $H = 0$. This is a special case of the previous model. The super-Hamiltonian (3.10) was used by Hájíček to analyze the global problem of time.[11,12] It is also useful to illustrate "group quantization".[34]

One can clarify various aspects of the time problem on the listed models. Unfortunately, it is much too easy to forget those simplifying features which distinguish them from canonical gravity, and succumb to an illusion that time problems in quantum geometrodynamics can be solved in exactly the same way as the particular time problem of the model. To dispel such an illusion, let me discuss several typical differences between the models and Einstein's theory of gravitation.

• **Auxiliary Structures.** A model may possess symmetries (in superspace) or other geometric structures which geometrodynamics lacks. This happens for cylindrical waves, bosonic strings, 2+1 gravity, the parametrized nonrelativistic particle, the oscillator system; for parametrized fields, the relativistic particle, and strings and membranes in stationary spacetimes; and for special minisuperspace models. When generalizing to quantum gravity, one should be careful that the proposed solution to the time problem does not depend on such auxiliary structures.

A particular time problem may be absent in a model:

• **No Multiple Choice Problem.** The constraint system for a parameterized nonrelativistic particle or a parameterized relativistic field is directly given in the form (1.6). There is thus no multiple choice problem. For other systems (like cylindrical waves or bosonic strings) there exists a natural choice (1.5) of the embedding variables which yields the constraints (1.6).

• **No Functional Evolution Problem.** There is no problem of functional evolution for finite-dimensional models. It is thus easy to forget that such a problem exists in any field theory. In particular, it is dangerous to draw conclusions from minisuperspace models to full quantum geometrodynamics. Minisuperspace spacetimes possess a privileged foliation by leaves of homogeneity which does not exist in a generic spacetime.

• **Positivity Restrictions.** Two parallels which are often drawn, one between geometrodynamics and a relativistic particle, and another one between geometrodynamics and the BB model, lead to diametrically opposite insights about the nature of time in geometrodynamics. In the first parallel, the intrinsic metric is compared with the position Q^A of a relativistic particle in *spacetime*. In the second parallel, it is compared with the position Q^a of a non-relativistic particle in *space*. The first parallel leads one to believe that the intrinsic metric contains a purely *kinematical* time variable (like Q^A contains Q^0). The second parallel leads one to believe that time in geometrodynamics must be measured by observing a selected dynamical degree of freedom, say, Q^2, of a *realistic clock*.

It is important to realize that each of these parallels may be misleading. In all three systems (the relativistic particle, the BB model, and geometrodynamics) the super-Hamiltonian H is the sum

$$H = H_{KIN} + H_{POT} \tag{3.11}$$

of two terms, the kinetic term H_{KIN}, and the potential term H_{POT}. The kinetic term H_{KIN} is quadratic in the momenta, while the potential term H_{POT} depends only on the configuration variables. However, the individual terms have quite different *positivity properties* in the three systems I am considering. For a relativistic particle, the kinetic term is indefinite while the potential term is positive definite: $H_{POT} \geq 0$. In the BB model, the kinetic term is positive definite, $H_{KIN} \geq 0$, while the potential term is indefinite. In geometrodynamics, both the kinetic and the potential terms are indefinite.

The positivity of H_{POT} in the dynamics of a relativistic particle allows one to cast the super-Hamiltonian (3.6) into a form analogous to Eq. (1.6), where h is the (square-root) Hamiltonian of a relativistic particle. The expression under the square root is positive definite and the corresponding operator can be defined by spectral analysis. This highlights the kinematical nature of the time variable. In the BB model, the positivity of the kinetic term H_{KIN} leads to theorems (Pauli,[74] Hartle,[75] Wald and Unruh[76]) which limit the accuracy with which a dynamical degree of freedom Q^2 of a realistic clock can serve as an ideal kinematical clock. The indefiniteness of H_{KIN} in geometrodynamics prevents one from converting the Wheeler-DeWitt equation (2.5) into the Schrödinger equation (2.6) by square-rooting. The indefiniteness of H_{KIN} casts doubt on the relevance of the theorems proved by Hartle and by Wald and Unruh: geometrodynamics may still possess a kinematical time variable.

One should try to avoid a common hubris which consists of solving a time problem for a model way down in the hierachy, and jumping to the conclusion that the time problems of quantum gravity are removed by the same treatment.

4. The Problem of Functional Evolution

While people's response to the other issues of time tends to be intimately intertwined with their favorite interpretation of quantum gravity, the problem of functional evolution stands sufficiently apart to be discussed before I turn to the specific interpretative schemes. The actual form, (2.8) or (2.9), which the problem takes depends, of course, on whether one works within an internal time framework, or commits oneself to the Wheeler-DeWitt equation.

Much more than other time problems, the problem of functional evolution critically depends on our ability to construct gravity as a consistent quantum *field* theory, in particular, on giving sense to the infinite collection of constraint operators (2.1) - (2.3), and to their commutation relations (2.8) - (2.9). There is actually very little one can say about this problem in full quantum gravity before one succeeds in constructing the full quantum gravity. The best one can do is to elucidate what the problem means on sufficiently simple (and hence almost trivial) infinitely-dimensional models.

In this spirit, I have formulated and solved the problem of functional evolution for the parametrized massless scalar field propagating on a (1+1)-dimensional flat cylindrical Minkowskian background.[77-79] There is a canonical transformation which casts the super-Hamiltonian and supermomentum constraints in the midi-phase-space of the cylindrical gravitational waves into the constraints for a parametrized scalar field on a fictitious (1+1)- dimensional Minkowskian background.[35] This links the problem of functional evolution in the parametrized field theory to the one in quantum gravity. The classical steps needed to approach the problem of functional evolution for cylindrical gravitational waves were taken by Torre.[80] Similarly, there is a canonical transformation which casts the super-Hamiltonian and supermomentum constraints of a bosonic string moving in a d-dimensional target space into those of a parametrized theory of d − 2 independent massless scalar fields propagating on a 2-dimensional Minkowskian background. This allows one to formulate and solve the problem of functional evolution for a bosonic string.[44,81]

▪ **Functional Evolution for a Parametrized Scalar Field.** I shall introduce the problem of functional evolution on the example of a parametrized scalar field. Let $X^A = (T,X)$, $T\epsilon(-\infty,\infty)$ and $Z\epsilon(-\pi, \pi)$ be the Minkowskian coordinates on a Minkowskian cylinder whose radius was chosen as a unit of length, and $X^A(x)$ is any spacelike hypersurface (in this case, a curve) on the cylinder. The constraints of the parametrized massless scalar field theory then take the form (1.6), where the many-fingered Hamiltonian

$$h_A(x; X,\phi,\pi] = T_{AB}(x; X,\phi,\pi]\ |g(x; X]|^{\frac{1}{2}}\ n^B(x; X] \tag{4.1}$$

is obtained by projecting the energy-momentum tensor T^{AB} of the field into the (densitized) normal $|g|^{\frac{1}{2}}n^B$ to the embedding $X^A(x)$ and expressing the result in terms of the canonical variables $\phi(x)$, $\pi(x)$. These constraints have the vanishing Poisson brackets (1.13).

To quantize the theory, one must factor order the Hamiltonian (4.1). An obvious way of doing this is to adopt the normal ordering $:h_A(x; X,\hat{\phi},\hat{\pi}]:$ with respect to the (Heisenberg) creation and annihilation operators corresponding to the initial data $\hat{\phi}_o(x)$, $\hat{\pi}_o(x)$ on a flat slice (maximal circle) of the cylinder. The problem of functional evolution arises because the commutators of the energy-momentum tensor operators $:\hat{T}_{AB}(X):$ acquire Schwinger terms[82] which, due to the construction (1.6) and (4.1) of the constraints, creep into the commutator algebra of the constraints as an anomaly,

$$\frac{1}{i}\ [\ :\hat{H}_A(x):\ ,\ :\hat{H}_B(x'):\ ] = -\ F_{AB}(x,x'; X]\ . \tag{4.2}$$

The detailed form of F is not important for the further discussion. What is important is that F depends only on the embedding $X^A(x)$, not on the field variables $\hat{\phi}(x)$ and $\hat{\pi}(x)$.

The anomaly (4.2) makes the constraints (2.6) inconsistent: by applying Eq. (4.2) to a state Ψ which satisfies the constraints (2.6), one learns that the state functional must vanish. If the Dirac constraint quantization fails for such a simple system as a parametrized field on a flat background, how could one ever dare to apply it to quantum gravity?

The resolution of the problem requires an exploration of the structure of the anomaly as a functional of the embedding variables. The Jacobi identity for the commutator (4.2) amounts to the statement that the anomaly is a closed 2-form in the space of embeddings. Indeed, the anomaly 2-form is exact, i.e., it is a functional exterior derivative of a potential $A_A(x; X]$. This potential has two parts: the standard Casimir energy-momentum flux which is due to the finite spatial size of the Minkowskian cylinder, and the interesting piece, which depends only on the extrinsic curvature of the embedding $X^A(x)$, i.e., on the bending of the hypersurface.

One can redefine the constraint operators $:\hat{H}_A:$ by adding to them the anomaly potential:

$$\hat{H}_A(x) := :\hat{H}_A(x): + A_A(x; X]. \tag{4.3}$$

This is tantamount to a hypersurface-dependent reordering of the constraints. The anomaly potential exactly cancels the anomaly, so that the new constraint operators (4.3) commute with each other, Eq. (2.8). The Dirac constraint quantization based on the reordered constraints (4.3) solves the problem of functional evolution.

▪ **Possible Generalizations.** The same scheme helps to resolve the functional evolution problem in the Dirac constraint quantization of a bosonic string[44,81] and, I guess, in the quantum geometrodynamics of cylindrical gravitational waves. It should be possible to generalize these results to parametrized free field theories on a flat four-dimensional background, but this was actually never done. (One should somehow take care of a single infinite constant which in four dimensions multiplies the Schwinger terms.) It would then be worthwhile to spell out the difficulties one can expect to arise for parametrized free fields on a curved background, or for parametrized interacting fields on a Minkowskian background. In the first case, it is not clear how to factor order the Hamiltonian, in the second case, it is not clear if the right-hand side of Eq. (4.2) does not also depend on the dynamical variables. By analyzing such difficulties, one would gauge at least some of the problems of quantum gravity approached via the many-fingered time Schrödinger equation (2.7).

• **Functional Evolution Problem for the Wheeler-DeWitt Equation.** In the Wheeler-DeWitt approach to quantum gravity, the problem of functional evolution is encoded in Eq. (2.9). Its resolution is equivalent to solving the notorious factor-ordering problem: To factor order the constraints (2.1) – (2.2) so that the commutator (2.9) of two super-Hamiltonians does not engender more constraints.

To resolve the problem, one should find a factor ordering of the super-Hamiltonian and supermomentum constraints such that the commutator of the super-Hamiltonians yields an expression in which the supermomentum acts on the state function first, followed by the quantum version of the structure functions of the Dirac algebra. It was noticed by Anderson[83,84] that this task cannot be achieved if one insists on representing the super-Hamiltonian and supermomentum by self-adjoint operators on $\mathscr{F}$. A solution to the factor-ordering problem was offered by Schwinger[85] and criticized by Dirac.[86] The best, and certainly the shortest, exposition of Schwinger's solution may be found in a footnote of the paper[32] by DeWitt; this was later rediscovered by Komar.[87,88] DeWitt himself made a rather sweeping proposal on how to remove the problem by letting any two field operators taken at the same point formally commute.[32] Ashtekar has expressed the Hamiltonian theory of gravity in a new set of fundamental canonical variables and proposed a simple factor ordering of the constraints which (disregarding the regularization difficulties) satisfy the consistency requirement.[89] Unfortunately, all these results are purely formal: Tsamis and Woodard[90], and Friedman and Jack[91,92] have persuasively argued that by formal manipulations of the commutator one can obtain whatever result one wants. The problem of functional evolution in the Wheeler-DeWitt equation can be tackled only when the field-theoretical complications are honestly faced and the discussion is raised above the purely formal level.

The factor-ordering problem of the Wheeler-DeWitt equation can be divested of its field theoretical aspects by simulating it on finite-dimensional models, mainly on the relativistic particle with extra gauge degrees of freedom. The resolution of this model problem was given by Hájíček and Kuchař.[66,93]

▪ **Fixing the Foliation.** The problem of functional evolution is so formidable that one may be tempted to give up and accept a draconian solution: to fix the foliation. Instead of measuring the gravitational degrees of freedom $\phi^r(x)$ and $\pi_r(x)$ on an arbitrary hypersurface, one can ask what happens when one attempts to measure them only on the leaves of a constant internal time $T(x; g,p]$ in the frame $Z^a(x; q,p]$:

$$T(x; g,p] = t = const, \quad Z^a(x; g,p] = x^a . \tag{4.4}$$

For the purpose of evolving the state functional $\Psi[X,\phi]$ along the foliation (4.4), one can replace it by the reduced functional

$$\Psi(t; \phi] := \Psi[\, T(x) = t,\ Z^a(x) = x^a,\ \phi(x)\,] \tag{4.5}$$

which is evolved by an ordinary Schrödinger equation

$$i\partial_t \Psi(t; \phi] = h(t; \hat{\phi},\hat{\pi}]\ \Psi(t; \phi] \tag{4.6}$$

with the true Hamiltonian

$$\hat{h} = \int d^3Z \, h_0[\, T = t,\; Z^a = x^a;\; \hat{\phi},\; \hat{\pi} \,] \, . \tag{4.7}$$

It is easy to see that when the many-fingered time Schrödinger equation (2.7) is consistent, and $\Psi[X,\phi]$ is its solution, the restriction (4.5) of that solution satisfies the ordinary Schrödinger equation (4.6). However, Eq. (4.6) - (4.7) for $\Psi(t;\, \phi]$ can make sense even if the functional Schrödinger equation (2.7) is inconsistent, Eq. (2.8). There is no problem of functional evolution when there is only one foliation along which the evolution can take place.

The foliation-fixing equations (4.4) can be considered as additional constraints which, when adjoined to the original constraints (1.6), make the total system of constraints second class. When the second class constraints are eliminated at the classical level and one quantizes only the reduced system, one recovers the Schrödinger equation (4.6) - (4.7). This is how the foliation fixing is ordinarily described.[94]

- **Foliation Fixing Versus General Relativity.** The foliation fixing prevents one from asking what would happen if one attempted to measure the gravitational degrees of freedom on an arbitrary hypersurface. Such a solution to the problem of functional evolution amounts to conceding that one can quantize gravity only by giving up general relativity: to say that quantum gravity makes sense only when one fixes the foliation is essentially the same thing as saying that quantum gravity makes sense only in one coordinate system.

5. Interpretations of Quantum Gravity

Quite a number of different proposals were made over the years on how to interpret quantum gravity. There are basically three ways in which they tried to cope with the problem of time:

I. **Internal Time Framework.** Time is hidden among the canonical variables and it must be identified prior to quantization. The basic equation on which the interpretation is based is a Schrödinger equation, not the Wheeler-DeWitt equation. This class of interpretations is prone to the multiple choice problem.

II. **Wheeler-DeWitt Framework.** The constraints are imposed in the metric representation and they yield the Wheeler-DeWitt equation. One strives for a dynamical interpretation of the solutions of the Wheeler-DeWitt equation which would be insensitive to the identification of time from among the metric variables. This class of interpretations is prone to the Hilbert space problem.

III. **Quantum Gravity Without Time.** This class of interpretations is again mostly (though not necessarily) based on the Wheeler-DeWitt equation. It stresses the idea that one does not need time to interpret quantum gravity or, indeed, quantum mechanics in general. Time may emerge in particular situations, but even if it does not, quantum states still allow a probabilistic interpretation.

I find this classification of interpretations useful, even if the boundaries of the three classes are not sharply defined. I shall discuss ten representative attempts at interpreting quantum gravity under these headings.

I. **Internal Time Framework.**

- **Internal Schrödinger Interpretation.** Time (in the form of the embedding variables) is separated from the dynamical degrees of freedom by a canonical transformation in the geometric phase space prior to quantization. The constraints are resolved for the embedding momenta and in this form imposed on the physical states. The probabilistic interpretation is based on the resulting Schrödinger equation.

- **Matter Clocks and Reference Fluids.** The standard of time is provided by a matter system coupled to geometry rather than by the geometry itself. The introduction of matter facilitates the handling of the constraints which yield the Schrödinger equation.

- **Unspecified Cosmological Constant.** The cosmological constant λ is considered as a (conserved) dynamical variable. The super-Hamiltonian constraint yields the Schrödinger equation in the "cosmological time" conjugate to λ.

II. **Wheeler-DeWitt Framework.**

- **Klein-Gordon Interpretation.** The Wheeler-DeWitt equation is considered as an infinitely-dimensional analogue of the Klein-Gordon equation for a relativistic particle. The statistical interpretation of the theory is based on the Klein-Gordon norm, which is expected to be positive on a suitably restricted subspace of solutions.

- **Semiclassical Interpretation.** It is contended that the Wheeler-DeWitt equation has a probabilistic interpretation only if the Universe is in a semiclassical state. In this case, the Wheeler-DeWitt equation is approximated by the

Schrödinger equation.

▪ **Third Quantization**. The problems presented by the indefinite inner product of the Klein-Gordon interpretation are faced by suggesting that the solutions of the Wheeler-DeWitt equation are to be turned into operators. This is analogous to subjecting the relativistic particle whose state is described by the Klein-Gordon equation to second quantization.

III. **Quantum Theory Without Time.**

▪ **Naive Schrödinger Interpretation.** It is suggested that the square $|\Psi[\mathbf{g}]|^2$ of a solution $\Psi[\mathbf{g}]$ of the Wheeler-DeWitt equation is the probability density for finding a spacelike hypersurface which carries the geometry **g**.

▪ **Conditional Probability Interpretation.** This is an elaboration of the naive Schrödinger interpretation. Again, the space of solutions of the *Wheeler-DeWitt* equation is equipped with the *Schrödinger* norm. For any two projection operators $\hat{A}$ and $\hat{B}$, one gives the conditional probability that the answer to $\hat{A}$ is yes if the answer to $\hat{B}$ is yes. In special cases, $\hat{B}$ is supposed to define an instant of time at which $\hat{A}$ is measured.

▪ **Sum-Over-Histories Interpretation.** Path integrals are claimed to make sense even outside the Hilbert space framework. They are supposed to provide an interpretation of quantum systems which do not necessarily carry with them the notion of time.

▪ **Frozen Time and Evolving Constants of the Motion**. It is maintained that observables in quantum gravity are those dynamical variables which commute with all the constraints, i.e., which are the constants of the motion. An attempt is made to explain how such observables can describe an evolving Universe.

I shall devote the rest of my study to a critical examination of these ten interpretations. I have reviewed the issue of time in quantum gravity on two previous occasions, at the second Oxford Symposium on Quantum Gravity[33] and at the Osgood Hill Meeting.[95] My present analysis is parallel to that given by Isham at the 1991 Schladming Winter School.[96] Alternative reviews of some of these interpretations were given by Unruh and Wald[76] and by Barbour and Smolin.[97]

6. Internal Schrödinger Interpretation

I shall now discuss those interpretations of quantum gravity which insist that time is to be identified from among the geometric canonical variables and that the classical constraints must be cast into a very special form before quantization. Let me sketch the sequence of steps by which this is done:

I. The embedding variables $X^A(x)$ are separated from the true gravitational degrees of freedom $\phi^r(x)$ by a canonical transformation (1.5).

II. The constraints (1.1) - (1.4) are solved with respect to the embedding momenta and then replaced by a classically equivalent set of constraints (1.6).

III. Those new constraints are turned into operators (2.3) and imposed as limitations (2.6) on the physical states.

The outcome of this procedure is a functional Schrödinger equation (2.7).[33,35,78,79,98] Indeed, this is the justification of all the trouble. One knows pretty well how to interpret a state function(al) which satisfies a Schrödinger equation, while it is not at all clear how to interpret a state function(al) which satisfies the original form (2.4) - (2.5) of the quantum constraints.

Provided that the functional evolution problem is solved, the functional Schrödinger equation is internally consistent and one can find its solutions $\Psi[X,\phi]$. At least formally, the functional integral

$$\langle\Psi|\Psi\rangle := \int D\phi \, |\Psi[X,\phi]|^2 \tag{6.1}$$

does not depend on the embedding $X^A(x)$, i.e., on the internal many-fingered time.[33] The inner product (6.1) turns the space $\mathscr{S}_0$ of solutions $\Psi[X,\phi]$ of the Schrödinger equation (2.7) into a Hilbert space. The Hilbert space structure provides the customary statistical interpretation of the quantized system. In particular,

$$D\phi \; |\Psi[X,\phi]|^2 \tag{6.2}$$

is interpreted as the probability to find the true gravitational degrees of freedom $\phi^r(x)$ lying in the cell $D\phi$ on a hypersurface $X^A = X^A(x)$ specified by the internal embedding variables. More generally, the inner product (6.1) allows one to construct meaningful *quantum observables*. Any functional

$$\hat{F} = F[X,\hat{\phi},\hat{\pi}] \tag{6.3}$$

of the true gravitational variables $\hat{\phi}(x),\hat{\pi}(x)$ and of the embedding $X^A(x)$ which is self-adjoint under the inner product (6.1) is an observable. An observable does not need to be a *constant of the motion*, i.e., the operator $\hat{F}$ does not need to commute with the constraints (2.3) on the space of solutions:

$$[\hat{F}, \hat{H}_A(x)]\ \Psi \neq 0 \quad for \quad \Psi \in \mathscr{F}_0 . \tag{6.4}$$

Nevertheless, the familiar rules of quantum theory allow one to ask and answer questions about the spectrum of $\hat{F}$ and the probability that, in the state Ψ, the observable $\hat{F}$ on the hypersurface $X^A(x)$ assumes a value F allowed by its spectrum.

The basic question in this program is how to choose the internal time variable. Three broad classes of options have been explored:

- **Intrinsic Time** $T(x; g]$. One assumes that the time variable is constructed entirely out of the intrinsic metric g_{ab} carried by the hypersurface.

- **Extrinsic Time** $T(x; g,p]$. To identify the hypersurface, one needs to know not only its intrinsic metric, but also its extrinsic curvature, i.e., how the hypersurface is bent in the encompassing Einstein spacetime.

- **Matter Time**. Time is to be constructed, not from the geometric data, but from matter fields coupled to gravity. I discuss this possibility in the next section. For the time being, I will confine my attention to vacuum gravity.

Very little is known about how to choose an internal time and make sense of the resulting Schrödinger equation in full quantum gravity. Indeed, the only general scheme which was seriously contemplated was to foliate spacetime either by maximal hypersurfaces or by hypersurfaces of constant mean extrinsic curvature, and call these the hypersurfaces of constant time. The remaining proposals were made on models. I shall mention these first, and then return to the mean extrinsic curvature scheme.

- **Homogeneous Cosmologies.** A single real variable, rather than a function, is needed to label the leaves of the privileged foliation on which both the intrinsic metric and the extrinsic curvature are homogeneous. The early work on homogeneous models revealed that a useful internal time variable labeling the leaves is the logarithm (3.4) of their total volume.[53,54] The choice (3.4) is a primary example of an intrinsic time. Other choices, including the mean extrinsic curvature of the leaves, were also explored.[13,96]

To bring the super-Hamiltonian constraint of the model (like Eq. (3.5) for the mixmaster universe) to the Schrödinger form, one must solve a quadratic equation for p_Ω.

▪ **Linearized Gravity**. The advantages of an extrinsic time as a tool for casting geometrodynamics into the Schrödinger form first emerged in the linearized theory of gravitation.[99-101] The state functionals of the linearized theory are defined on embeddings which only slightly differ from the hyperplanes of the flat background. When one uses the Minkowskian coordinates $X^K = (T, Z^k)$ of the background, one can assume that $T_{,a}(x)$ and $Z^k{}_{,a}(x)$ are small quantities. The intrinsic geometry of a hypersurface $T = T(x)$ in the Minkowski spacetime depends on $T(x)$ only through the terms of the second order. On the other hand, the extrinsic curvature $K_{ab}(x)$ is influenced by a small deformation from the hyperplane already in the terms linear in the deformation:

$$K_{ab}(x) = T_{,ab}(x) . \tag{6.5}$$

It is thus easier to determine $T(x)$ by looking at the extrinsic curvature of the hypersurface rather than at its intrinsic geometry.[101] A deformation $Z^k(x)$ of the Cartesian coordinates on the hyperplane, however, affects the intrinsic metric in the first order,

$$g_{ab}(x) = \delta_{ab} + Z_{(a,b)}(x). \tag{6.6}$$

This enables one to reconstruct $Z^k(x)$ from the metric.

In linearized gravity, the true degrees of freedom are superimposed on the coordinate-induced effects (6.5) and (6.6). To separate them from the embedding variables, one uses the decomposition of symmetric tensors $f_{ab}(x)$ into transverse traceless $f^{TT}{}_{ab}$, transverse $f^{T}{}_{ab}$ and longitudinal $f^{L}{}_{ab}$ parts with respect to the flat background:

$$\begin{gathered} f_{ab} = f^{TT}{}_{ab} + f^{T}{}_{ab} + f^{L}{}_{ab} , \\ f^{TT}{}_{ab,b} = 0 = f^{TT}{}_{ab} , \quad f^{T}{}_{ab,b} = 0 , \end{gathered} \tag{6.7}$$

with

$$f^L{}_{ab} = f_{(a,b)}\ ,\quad f_a := \Delta^{-1}(f_{ab,b} - \frac{1}{2}\,\Delta^{-1}f_{bc,bca})\ ;$$
$$f^T{}_{ab} = \frac{1}{2}(f^T\delta_{ab} - \Delta^{-1}f^T{}_{,ab}),\quad f^T := f_{aa} - \Delta^{-1}f_{ab,ab}\ ; \tag{6.8}$$
$$f^{TT}{}_{ab} = f_{ab} - f^T{}_{ab} - f^L{}_{ab}\ .$$

(The indices are raised and lowered by the background metric δ_{ab}, and Δ^{-1} is the inverse Laplacian.)

The decomposition (6.7) - (6.8) can be applied to the linearized canonical variables $h_{ab} = g_{ab} - \delta_{ab}$ and p_{ab}. Equations (6.5) and (6.6) which are valid in the Minkowski spacetime can be written as

$$T(x) = -\frac{1}{2}\Delta^{-1}p^T(x)\ ,\quad Z^k(x) = h^k(x)\ . \tag{6.9}$$

One adopts them as definitions of the embedding variables in linearized gravity. The first equation (6.9) represents a choice of extrinsic time.

The conjugate momenta to the canonical coordinates (6.9) are

$$P_T = \Delta h^T(x),\ \ P_k(x) = -\,2\,(\Delta p_k(x) + p_{i,ik}(x)). \tag{6.10}$$

The canonical transformation (1.5) which separates the embedding variables from the true dynamical degrees of freedom takes the form

$$h_{ab}(x),\ p^{ab}(x)\ \rightarrow\ \begin{matrix} T(x),\ Z^k(x),\ h^{TT}{}_{ab}(x)\ ; \\ P_T(x),\ P_k(x),\ p^{TT\,ab}(x)\ . \end{matrix} \tag{6.11}$$

The meaning of the quantities p^T and h^k as parameters specifying the choice of spacetime coordinates was discovered by Arnowitt, Deser, and Misner (ADM).[99,100]

To exclude these surplus variables, ADM fixed the foliation and the reference frame by the coordinate conditions

$$T(x) = t = \mathit{const},\quad Z^k(x) = x^k. \tag{6.12}$$

In the many-fingered time formalism, one prefers to keep the variables (6.9) and (6.10) in the theory.

In linearized gravity, the constraints are to be written up to the quadratic terms in the perturbations of g_{ab} and p^{ab}. When expressed in the new canonical variables (6.11), they are to this order linear in the embedding momenta. Their imposition on the states $\Psi[T,Z,h^{TT}]$ leads then to a functional Schrödinger equation.[101]

Arnowitt, Deser and Misner proposed to use the extrinsic time (6.9) even for large perturbations. Unfortunately, the flat background used in the decomposition (6.7) - (6.8) loses then its geometric and physical significance, and the constraints written in the new variables become unmanageable.

▪ **Cylindrical Gravitational Waves.**[35,102] Cylindrical symmetry allows an infinite number of gravitational degrees of freedom and, by permitting ∞^1 many deformations of a hypersurface, it does not fix a single foliation. There are, however, natural spacetime coordinates T,R (the Einstein-Rosen time and radial coordinates) in which the vacuum Einstein equations for the cylindrical wave assume a simple form. The Einstein-Rosen radial coordinate is simply the function R(r) characterizing the intrinsic metric (3.1). The Einstein-Rosen time can be reconstructed from the function $p_\gamma(r)$ characterizing the extrinsic curvature:

$$T(r) = T(\infty) - \int_\infty^r dr\ p_\gamma(r)\ . \tag{6.13}$$

It is a good example of extrinsic time. One can complete Eq. (6.13) into a canonical transformation

$$\gamma,\ R,\ \psi;\ p_\gamma,\ p_R,\ \pi_\psi\ \rightarrow\ T,\ R,\ \psi;\ P_T,\ P_R,\ \pi_\psi \tag{6.14}$$

by putting

$$\begin{aligned} P_T &= -\gamma' + (\ln(R'^2 - p_\gamma^2))' \quad \textit{and} \\ P_R &= p_R + (\ln(R' - p_\gamma)/(R' + p_\gamma))'. \end{aligned} \tag{6.15}$$

The midisuperspace restrictions (3.2) - (3.3) of the super-Hamiltonian and supermomentum constraints are linear in the $2\infty^1$ momenta P_T and P_R and they are thus easily resolved:

$$\begin{aligned} P_T + (R'^2 - T'^2)^{-1}[\tfrac{1}{2}(R^{-1}\pi_\psi{}^2 + R\psi'^2)R' - \psi'\pi_\psi T'] &= 0\ , \\ P_R + (R'^2 - T'^2)^{-1}[-\tfrac{1}{2}(R^{-1}\pi_\psi{}^2 + R\psi'^2)T' + \psi'\pi_\psi R'] &= 0\ . \end{aligned} \tag{6.16}$$

The resulting constraints have the same form as those for a massless scalar field

ψ propagating on a flat Minkowskian background which is parametrized by the introduction of arbitrary curved spacetime hypersurfaces T(r), R(r). This procedure illustrates how an appropriate choice of the embedding variables can simplify the originally unwieldy constraints.

The canonical quantization based on the constraints (6.16) leads to a $2\infty^1$-fingered time Schrödinger equation. For the adopted foliation (6.12), this equation reduces to an ordinary Schrödinger equation for a single massless scalar field in Minkowski space.

▪ **The Mean Extrinsic Curvature as Time.** The simplifications achieved by a judicious choice of extrinsic time in some midisuperspace models and in linearized gravity encourages the search for extrinsic time in full geometrodynamics. An obvious choice seems to be

$$T(x) := \frac{2}{3}\,|g(x)|^{-1/2}p(x), \quad P(x) := -\,|g(x)|^{1/2}\ ; \tag{6.17}$$

indeed, T(x) and P(x) are a set of $2\infty^3$ canonically conjugate variables. Dirac proposed that gravity in asymptotically flat spacetimes should be quantized on maximal hypersurfaces $T(x) = 0$.[94] The choice (6.17) was elaborated into an effective and elegant scheme of solving the classical constraint equations (1.3) - (1.4) by York and his collaborators.[103-105]

Equations (6.17) can be completed into a canonical transformation by introducing the conformal metric σ_{ab} and its conjugate momentum π^{ab}:

$$\sigma_{ab} := |g|^{-1/3}g_{ab}\,, \quad \pi^{ab} := |g|^{1/3}(p^{ab} - \frac{1}{3}pg^{ab})\ . \tag{6.18}$$

Each of these quantities has only $5\infty^3$ independent components. To simplify the description I shall not attempt to split them further into the $3\infty^3$ frame variables $Z^k(x)$ and $2\infty^3$ gravitational degrees of freedom, in contrast to what is done in York's theory.

In terms of the new variables T and P, the super-Hamiltonian constraint (1.2), (1.4) becomes a quasi-linear elliptic equation for $\phi := P^{1/6}$:

$$\Delta_\sigma\phi - \frac{1}{8}R[\sigma]\phi + \frac{1}{8}\pi_{ab}\pi^{ab}\phi^{-7} - \frac{3}{64}T^2\phi^5 = 0\ . \tag{6.19}$$

Assume that one can find the solution $\phi(x; T,\sigma,\pi]$ of this equation and hence the corresponding Hamiltonian density $h = -\phi^6$. One can then replace the super-

Hamiltonian constraint (1.4) by an equivalent constraint

$$\Pi(x) := P(x) + h(x; T,\sigma,\pi] = 0 \tag{6.20}$$

whose quantum version is a functional Schrödinger equation[33]

$$i \frac{\delta \Psi[T,\sigma]}{\delta T(x)} = h(x; T,\hat{\sigma},\hat{\pi}] \ \Psi[T,\sigma] \ . \tag{6.21}$$

(The task of separating the frame variables $Z^k(x)$ from the gravitational degrees of freedom and solving simultaneously the supermomentum constraints for the conjugate momenta is much more involved.[103-105] The equations decouple only if one fixes the foliation: T = t = *const.*)

Having summarized the situations in which the identification of an internal time variable proved useful in quantum canonical gravity, I must turn to the problematic features of such a procedure. The two main problems, the global time problem and the multiple choice problem, representing the opposites of the embarrassment of poverty and the embarrassment of riches, were already identified in Sections 1 and 2. Besides them, there are more specific difficulties which remain to be pointed out.

• **Global Time Problem.** It may happen that the constraint system (1.6) cannot be made globally equivalent to the constraint system (1.3) - (1.4) by any choice of internal time. This must happen if there does not exist a global choice of the time function on the phase space such that each classical trajectory intersects every hypersurface of constant internal time once and only once. (This is the analogue of the Gribov problem[106] for parametrized dynamical systems). These problems were studied for simple model systems - a couple of harmonic (antiharmonic) oscillators in a stationary state (Section 3)[10-12] and in homogeneous cosmologies[13] - by Hájíček. The analysis of the harmonic oscillator systems reveals that many such models do not allow a global time function, and that such models get more and more frequent as the number of the oscillators increases. It is also known that some homogeneous cosmological models - e.g., the Friedmann universe driven by a massive scalar field - do not possess a global time function.[13,107] A systematic analysis of the global time problem for homogeneous cosmological models was not done, but it seems likely that the lack of a global time function is quite generic within this class. Further, even if such a function exists for a given model, it is fairly obvious that some popular choices of the time function are not globally valid. Thus, e.g., the volume time (3.4) is not globally permitted in oscillating cosmological models, because the universe attains a given value of $\Omega < \Omega_{MAX}$ at least twice, once when it is expanding, and once when it is recontracting. (This was often adduced as a reason why an extrinsic time is to be preferred to an intrinsic time). The global problems of the maximal and K = *const* foliations have been studied in the literature.[108-114] Still, very little is known about the global problem of time in full classical geometrodynamics. These topics definitely deserve more attention.

- **Spectral Analysis Problem.** The solution of the constraints with respect to the embedding momenta may not exist for all values of the dynamical variables and, even if it does, it may be too complicated. One can notice this already on minisuperspace models,[115,96] like the mixmaster universe: the constraint (3.5) contains a square root and, because the potential $V(\beta_+, \beta_-) - 1$ is negative in the region close to the origin of the β_+, β_- plane, the expression under the square root becomes negative for some values of the variables β_+, β_-, p_+, p_-. In quantum theory, one is called to replace the canonical variables $\beta_\pm$, $p_\pm$ by operators and give a meaning to the square root. This can be done in two steps, by first defining the operator under the square root, and then defining the square root itself by *spectral analysis*. There are two difficulties with this procedure. First, if the operator under the square root is not a positive definite operator, spectral analysis yields a Hamiltonian which is not self-adjoint. The Schrödinger equation (2.7) then does not yield unitary evolution. Second, even if the operator under the square root were a positive definite operator, spectral analysis is an involved nonlocal procedure which requires solving an eigenvalue problem for the operator in question. It may be difficult to write such a solution in an explicit form.

The spectral analysis problem becomes even more severe in full geometrodynamics because finding the many-fingered Hamiltonian $h_A(x; X,\phi,\pi]$ often requires solving nonlinear differential equations and because the definition of the operators involves renormalization problems. To give an example, if one tries to convert the York algorithm into a Schrödinger equation, one faces the problem of defining the operator version of the classical expression (6.20), whose dependence on the dynamical variables σ_{ab} and π^{ab} is known only implicitly, through the statement that the conformal factor ϕ must satisfy the Lichnerowicz-York equation (6.19).

The study of midisuperspace models and covariant field systems like string models strongly indicates that, if there is an internal time which converts the old constraints into the Schrödinger form, such a time is a nonlocal functional of the geometric variables. It is also likely that the many-fingered Hamiltonian $h_A(x; X,\phi,\pi]$ corresponding to such a choice of time is defined only implicitly, by solving PDE's or by an even more complicated procedure, and that the resulting expression is a rather horrendous functional of the new dynamical variables. This makes the whole program of interpreting quantum gravity through an identification of internal time rather remote and unattractive.

• **Multiple Choice Problem.** If there is not a geometrically natural choice of internal time, all choices of time are equally good or equally bad. One faces then the problem of comparing quantum theories which follow from different choices of an internal time variable.

The multiple choice problem can be neatly analyzed on the model (3.6) of a relativistic particle moving in a Riemannian spacetime.[95] Any time function T(Q) whose level surfaces are spacelike and foliate $\mathcal{M}$ has the property that each dynamical trajectory of the particle (which is timelike) intersects each leaf of the foliation T(Q) = *const* (which is spacelike) once and only once. There is no global problem of time for a relativistic particle, but there is a multiple choice problem: There are infinitely many permissible time functions T(Q). The time functions are constructed from the configuration variables Q ; they are analogous to different choices of intrinsic time in canonical geometrodynamics.

Construct the reference frame of worldlines orthogonal to the leaves of the foliation T(Q) = *const* and label the worldlines by the comoving coordinates q^a. The normal separation of the leaves of the T(Q) = *const* foliation is then characterized by the lapse function N(Q). Perform the point transformation

$$Q^A,\ P_A \rightarrow T,\ P_T\ ;\ q^a,\ p_a \tag{6.22}$$

on the phase space of the particle and solve the constraint (3.6) with respect to the momentum P_T :

$$\begin{aligned} H_T &:= P_T + h_T(T,q,p) = 0\ , \\ h_T &:= \left(N^2(T,q)(h^{ab}(T,q)p_a p_b + M^2 V(T,q))\right)^{1/2}\ ; \end{aligned} \tag{6.23}$$

here, h_{ab} is the intrinsic metric on the leaves of the T = *const* foliation. For future-oriented timelike trajectories, the square-rooted constraint (6.23) is equivalent to the quadratic constraint (3.6). So are any two square-rooted constraints adapted to two different foliations T_1 and T_2. The dynamical evolutions generated by all such constraints are mutually consistent. This is ensured by the closure of the Poisson brackets

$$\begin{aligned} \{H_{T_1}\ ,\ H_{T_2}\} &= \text{"C"}\ H_{T_1}H_{T_2} + \text{"C"}\ H\ , \\ \{H_T\ ,\ H\} &= \text{"C"}\ H_T + \text{"C"}\ H\ . \end{aligned} \tag{6.24}$$

The structure functions "C" are fairly hideous, but they are all regular on the constraint surface. There is no problem of time in classical relativistic mechanics.

One can turn the dynamical variables q^a, p_a into operators and define the Hamiltonians h_T by spectral analysis. One can then impose the constraints (6.23) as restrictions on the state functions $\Psi(Q)$:

$$\hat{H}_{T_1} \Psi = 0 \Rightarrow \Psi \epsilon \mathscr{F}_1 , \tag{6.25}$$

$$\hat{H}_{T_2} \Psi = 0 \Rightarrow \Psi \epsilon \mathscr{F}_2 . \tag{6.26}$$

One obtains thereby the Schrödinger equations based on the foliations $T_1(Q) = const$ and $T_2(Q) = const$. I have denoted the spaces of solutions of these Schrödinger equations by $\mathscr{F}_1$ and $\mathscr{F}_2$, respectively. One can also impose the original constraint (3.6) on the states Ψ and obtain thereby the Klein-Gordon equation

$$\hat{H}\Psi = 0 \Rightarrow \Psi \epsilon \mathscr{F}_\circ \tag{6.27}$$

whose solution space is called $\mathscr{F}_\circ$.

For different choices of the *intrinsic time* $T(Q)$, the spaces $\mathscr{F}_1$, $\mathscr{F}_2$, and $\mathscr{F}_\circ$ can be compared with each other, because each $\Psi(Q)$ can be considered to be a scalar function on $\mathcal{M}$. The multiple choice problem arises because these spaces do not coincide. Technically, one cannot find a factor ordering of the constraints which, based purely on the geometric structures used when writing down each individual constraint, would make the redundant constraint system (6.25) - (6.26) consistent, i.e., which would take the classical closure relations (6.24) into commutator relations in which the operators $\hat{H}_{T_1}$, $\hat{H}_{T_2}$, and $\hat{H}$ on its right-hand side would act on Ψ before the "C"'s and annihilate it. If the foliations T_1 and T_2 are such that they share the same initial and the same final hypersurface, the initial state Ψ_{IN} evolved by the Schrödinger equation (6.25) yields in general a different final state Ψ_{FIN} from that obtained by evolving Ψ_{IN} by the Schrödinger equation (6.26). The quantum mechanics (6.25) can in principle *be compared* with the quantum mechanics (6.26) and they are in general found to be mutually *incompatible*. When the two choices of the time function are based on an *extrinsic time*, it is even difficult to see how one can *compare* the ensuing quantum mechanical schemes. The multiple choice problem for the intrinsic time functions can be resolved only for a particle which moves in a stationary spacetime with a stationary potential V; the solution depends on the existence of a privileged background structure, namely, a timelike Killing vector field.

The multiple choice problem in canonical geometrodynamics is at least as serious as the multiple choice problem explained here for the relativistic particle.

• **Spacetime Problem.** I pointed out that time in general relativity is a function variable rather than the single parameter of Newtonian mechanics. In classical geometrodynamics, one is asked to reconstruct the time coordinate T of an event X in an Einstein spacetime $(\mathcal{M}, \gamma)$ from the canonical data $g_{ab}(x)$, $p^{ab}(x)$ on an embedding X(x) passing through X. The time should emerge from the canonical formalism as a spacetime scalar which does not depend on the embedding from which it was reconstructed. Thus, if two embeddings intersect at the same event X and if the canonical data on these two embeddings are related by Einstein's equations, the internal time T(x; g,p] obtained from such two sets of the canonical data should be the same.[95] This imposes the requirement

$$\{T(x), \int_\Sigma d^3x'\ H(x')N(x')\} \approx 0 \quad for\ \ \forall N(x'): N(x) = 0 \tag{6.28}$$

on the possible candidates T(x; g,p] for an internal time function.[116,117] The same limitation, of course, should also be imposed on the internal spatial coordinates Z^k(x; g,p]. To find internal embedding variables which satisfy Eq. (6.28) constitutes the *spacetime problem.*

The requirement (6.28) is indeed quite strong. It is fairly obvious that it excludes any intrinsic time T(x; g] which, like R(x; g], is a local functional of the three-geometry. It also excludes some obvious choices of extrinsic time. Thus, e.g., the time function (6.17) is blatantly not a spacetime scalar: When one bends a hypersurface around a given event in a vacuum Einstein spacetime, the extrinsic curvature $\frac{1}{2}|g|^{-\frac{1}{2}}p$ changes even if the event remains the same. The time function (6.17) helps one to define a unique foliation T(x) = t = *const* of a vacuum Einstein spacetime by hypersurfaces of constant mean extrinsic curvature, but the value of T(x) on a hypersurface which is transverse to that foliation gives the mean extrinsic curvature on that hypersurface, not the mean extrinsic curvature on the leaf of the foliation passing through the same event. The canonical coordinate (6.17) thus cannot be turned into a coordinate in spacetime. For this reason, the time variable (6.17) is unsuitable as a many-fingered time variable in the functional Schrödinger equation (2.7).

It should in principle be possible to take the data g_{ab}, p^{ab} on an embedding and construct from them a time function $T_\circ$(x; g,p] whose value at x would equal (4/3 of) the mean extrinsic curvature of the hypersurface of constant mean extrinsic curvature passing through the same event. Such a time function would be a spacetime scalar. However, one can expect $T_\circ$(x; g,p] to be a hideously nonlocal functional of the hypersurface data g_{ab}, p^{ab}. Essentially, the construction of $T_\circ$(x; g,p] calls for building the whole spacetime by the evolution of the initial data g_{ab}, p^{ab}, finding in this spacetime the leaves of constant mean extrinsic curvature, and assigning this constant mean curvature to the points at which the leaves intersect the initial hypersurface.

There are, of course, local functionals F(x; g,p] of the canonical data which are spacetime scalars. As an example, one can take the square $R_{\alpha\beta\gamma\delta}(X)R^{\alpha\beta\gamma\delta}(X)$ of the Riemann curvature tensor of a vacuum Einstein spacetime, and reexpress it in terms of the canonical data on a spacelike hypersurface. However, such an F(x) can be a candidate for the time function T(x) only if $\{F(x), F(x')\} = 0$ and if the hypersurfaces F(x) = *const* are spacelike. Even then, to use F(x) as a time function one must split it from the rest of the canonical variables by a canonical transformation, which is by no means a trivial task. So far, I do not know a single concrete example of a decomposition of the fundamental canonical variables based on a scalar time function T(x).

7. Matter Clocks and Reference Fluids

The intrinsic geometry and extrinsic curvature of a spacelike hypersurface enter into the constraints in a complicated way. Nothing in the structure of the constraints tells us how to distinguish the true dynamical degrees of freedom from the quantities which determine the hypersurface. This makes the task of constructing an internal time from the geometric variables so difficult and ambiguous.

The main role of embedding variables is to provide an orientation in spacetime, i.e., to identify spacetime events. The founding fathers of general relativity invented a conceptional device which does exactly that: the notion of a *reference fluid*. The particles of the fluid identify space points, and clocks carried by them identify instants of time. These fix the reference frame and the time foliation. In that frame and on that foliation, the metric itself, rather than the geometry, becomes measurable.

The concept of reference fluid goes back to Einstein,[118] who coined for it the charming term "molusc", and to Hilbert who, in the second of his famous communications on foundations of physics,[119] formalized the idea that the coordinate system should be realized by a realistic fluid carrying clocks which keep a causal time: He imposed a set of inequalities ensuring that the worldlines of the reference frame be timelike and the leaves of the time foliation be spacelike.

The reference fluid is traditionally considered as a tenuous material system whose back reaction on the geometry can be neglected. There is just enough matter to tell us where we are but not enough of it to disturb the geometry. Instead of deriving the motion of the fluid from its action, one encodes it in the coordinate conditions. These are statements about the metric which hold in the coordinate system of the fluid and are violated in any other system. Unfortunately, such a standpoint makes it difficult to view the reference fluid as a physical object which, in quantum gravity, could assume the role of an apparatus for identifying spacetime events.

To turn the reference fluid into a physical system, one can follow one of two alternative routes. The first one is to picture the fluid as a realistic material medium and devise a Lagrangian which describes its properties. By adding this Lagrangian to the Hilbert Lagrangian, one couples the fluid to gravity. The second route imposes the coordinate conditions *before* variation by adjoining them to the action with Lagrange multipliers. The additional terms in the action are then parametrized and interpreted as a matter source.

▪ **Phenomenological Fluids in Quantum Gravity.** The first route of using phenomenological fluids in quantum gravity was pioneered by DeWitt. In the covariant approach, DeWitt coupled the gravitational fluid to an elastic medium carrying mechanical clocks.[120] From these objects he constructed idealized apparatuses which, in the spirit of the Bohr-Rosenfeld analysis of measurability in quantum electrodynamics,[121] were able to detect appropriate projections of the quantized Riemann curvature tensor. A little later, he used the same device for interpreting the canonical minisuperspace quantization of the Friedmann universe:[32] he introduced a cloud of clocks into the model and studied their correlations with the radius of the universe. DeWitt's approach was recently rediscovered by Rovelli.[122] Rovelli used a square-root Hamiltonian for the clocks. He argued that when the clock's momentum is large in comparison with its mass, the Dirac constraint quantization approximately leads to a Schrödinger equation.

▪ **Reference Fluids Associated With Coordinate Conditions.** The association of a material medium with coordinate conditions stems from the problem of representing spacetime diffeomorphisms in canonical gravity.[123-126] Isham and Kuchař[127-129] approached this problem by breaking the invariance of general relativity by Gaussian coordinate conditions and restoring it again by parametrization. This procedure leads to the modification of the constraints by terms which are linear in the momenta conjugate to the Gaussian coordinates. Hartle[130] discussed the Schrödinger equation obtained by imposing the new constraints as restrictions on the physical states. In the end, he dismissed the new terms as devoid of physical reality. Halliwell and Hartle[131] related the new form of the constraints to the sum-over-histories approach to quantum gravity. Kuchař and Torre[28] showed that the modified constraints follow from an action principle when one enforces the Gaussian coordinate conditions with Lagrange multipliers, and that the new terms can be identified with energy-momentum densities of the corresponding reference fluid. This method is applicable to any coordinate conditions. The harmonic coordinate conditions were treated by Kuchař and Torre[132] and Kuchař and Stone.[133] The K = *const* slicing condition was related to a reference fluid by Kuchař.[134] The same approach can be used for introducing the conformal, harmonic, and light-cone gauges in the canonical theory of a bosonic string[81] and in the canonical treatment of (induced) two-dimensional gravity.[135] I shall describe the Schrödinger interpretation of quantum gravity based on the matter clock associated with a Gaussian reference fluid.

▪ **Gaussian Reference Fluid.** In the normal Gaussian coordinates $X^K = (T,Z^k)$ four components of the spacetime metric γ^{KL} are fixed by the conditions

$$\gamma^{00} + 1 = 0, \quad \gamma^{0k} = 0. \tag{7.1}$$

These follow from the action

$$S^F[\gamma_{KL}, M, M_k] = \int_{\mathcal{M}} d^4X\,|\det(\gamma_{KL})|^{1/2}\,\Big(-\frac{1}{2}M(\gamma^{00}+1) + M_k\gamma^{0k}\Big) \tag{7.2}$$

by varying the Lagrange multipliers $M_K = (M,M_k)$. By expressing the privileged (Gaussian) coordinates X^K as functions of arbitrary (label) coordinates x^α, $X^K = X^K(x^\alpha)$, one arrives at the parametrized action

$$S^F[\gamma_{\alpha\beta}, M_K, X^K] = \int_{\mathcal{M}} d^4x\,|\det(\gamma_{\kappa\lambda})|^{1/2}$$
$$\Big(-\frac{1}{2}M(\gamma^{\alpha\beta}T_{,\alpha}T_{,\beta} + 1) + M_k\gamma^{\alpha\beta}T_{,\alpha}Z^k{}_{,\beta}\Big)\,. \tag{7.3}$$

This is added to the Hilbert action $S^G[\gamma_{\alpha\beta}]$ of vacuum gravity. The total action

$$S[\gamma_{\alpha\beta}, M_K, X^K] = S^G[\gamma_{\alpha\beta}] + S^F[\gamma_{\alpha\beta}, M_K, X^K) \tag{7.4}$$

describes the Gaussian reference fluid coupled to gravity.

The variation of the total action (7.4) with respect to $\gamma_{\alpha\beta}$ leads to the Einstein law of gravitation

$$G^{\alpha\beta} = \frac{1}{2}\,T^{\alpha\beta} := |\gamma|^{-1/2}\delta S^F/\delta\gamma_{\alpha\beta}\,, \tag{7.5}$$

in which there appears a source, $T^{\alpha\beta}$, the energy-momentum tensor of the reference fluid. The variation of the multipliers M_K yields the parametrized Gaussian coordinate conditions. The variation of the function variables $X^K(x^\alpha)$ gives the Euler hydrodynamic equations of the reference fluid. The variables

$X^K(x^\alpha)$ thus play the role of the velocity potentials of the fluid.[136,137] Because the action (7.4) is invariant under arbitrary transformations of x^α, the Euler hydrodynamic equations follow from the Einstein law and the Gaussian coordinate conditions. They amount to the conservation law $\nabla_\beta T^{\alpha\beta} = 0$ of energy and momentum of the reference fluid.

By introducing the mutually orthogonal fields

$$U^\alpha := -\gamma^{\alpha\beta}T_{,\beta} \quad \textit{and} \quad M_\alpha := M_k Z^k{}_{,\alpha}\,, \tag{7.6}$$

the energy-momentum tensor (7.5) takes the form

$$T^{\alpha\beta} = MU^\alpha U^\beta + M^{(\alpha}U^{\beta)} \tag{7.7}$$

of the Eckart energy-momentum tensor of a heat-conducting fluid.[138,139] The four velocity U^α of the fluid is the four-velocity of the Gaussian reference frame, the multiplier M has the meaning of the *proper mass density*, and the spacelike vector M^α is the *heat flow*. If the heat flow vanishes, the reference fluid turns into an incoherent dust.

▪ **Energy Conditions.** The use of Gaussian conditions as auxiliary conditions in the Hilbert action principle is a purely formal device. It is quite surprising that the energy-momentum tensor which emerges from this procedure seems to describe a simple physical system, namely, heat conducting dust. If we had such a material at our disposal and scattered it throughout space, we could identify the Gaussian coordinates with its physical state variables. Figuratively speaking, we could anchor the Gaussian coordinates to the physical worldlines of the dust. However, unlike an abstract Gaussian system, the dust can be real only if its energy-momentum tensor satisfies appropriate energy conditions.[140]

For the energy-momentum tensor (7.7), the weak, the strong, and the dominant energy conditions are all equivalent, and they are satisfied if the multipliers M_K satisfy the inequality

$$M - 2|\mathbf{M}| \geq 0\,, \quad |\mathbf{M}|^2 := \gamma^{\alpha\beta}M_\alpha M_\beta\,. \tag{7.8}$$

Unfortunately, even if Eq. (7.8) holds at one instant, it can be violated later by the dynamical evolution. Only if the heat flow vanishes, are the energy conditions preserved in time. This foreshadows the problems encountered in interpreting the Schrödinger equation obtained by quantizing geometry whose source is the Gaussian reference fluid.

▪ **Canonical Description of the Gaussian Reference Fluid.** As in all theories with nonderivative gravitational coupling, the canonical description of gravity interacting with the reference fluid is obtained by adding to the gravitational super-Hamiltonian (1.2) (which we now call $H^G(x)$) the energy density $H^F(x)$ of the fluid, and to the gravitational supermomentum (1.1) (which we now call $H^G{}_a(x)$) the momentum density $H^F{}_a(x)$ of the fluid, and by imposing the constraints

$$H := H^G + H^F = 0,\ \ H_a := H^G{}_a + H^F{}_a = 0\ . \tag{7.9}$$

The standard ADM analysis yields

$$H^F{}_a(x) = X^K{}_{,a}(x)P_K(x)\ \ \textit{and}\ \ H^F(x) = n^K(x; X]\ P_K(x)\ , \tag{7.10}$$

with

$$n^K = \left((1 + g^{ab}T_{,a}T_{,b})^{½},\ (1 + g^{ab}T_{,a}T_{,b})^{-½}g^{cd}T_{,c}Z^k{}_{,d}\right)\ . \tag{7.11}$$

The densities (7.10) are functionals of the embedding variables $X^K(x)$, the embedding momenta $P_K(x)$, and the intrinsic metric $g_{ab}(x)$. The Lagrange multipliers were eliminated in the process.

The spectacular feature of Eq. (7.10) is that not only the momentum density $H^F{}_a(x)$, but also the energy density $H^F(x)$ is a linear homogeneous function of the embedding momenta $P_K(x)$. It is thus straightforward to solve the constraints (7.9) with respect to $P_K(x)$. The only thing one needs to do is find the cobasis $(-n_K,\ X^a_K)$ to the basis $(n^K,\ X^K{}_{,a})$. Geometrically, the coefficients (7.11) are the components of the unit normal to the embedding $X^K(x)$ in the Gaussian coordinate system X^K, and $X^K{}_{,a}$ are the tangent vectors to the embedding. The cobasis is formed by the dual covectors.

This brings the constraints into the form (1.6) on the extended phase space X^K, P_K, g_{ab}, p^{ab}, with

$$\begin{aligned} h_K(x; X,g,p] &= -\ n_K(x; X,g]\ H^G(x; g,p] \\ &\quad + X^a_K(x; X,g]\ H^G{}_a(x; g,p]\ . \end{aligned} \tag{7.12}$$

■ **Probabilistic Interpretation**. By imposing the constraints as operator restrictions on the state functional $\Psi[X,g]$, I get the many-fingered time Schrödinger equation. This suggests that the expression

$$|\Psi[T(x),\ Z^k(x),\ g_{ab}(x)]|^2 \tag{7.13}$$

constructed from a solution of this equation can be interpreted as the probability density that, on the embedding $T(x)$, $Z^k(x)$, the metric $g_{ab}(x)$ which is measured in the system of coordinates x^a connected to the Gaussian frame coordinates Z^k by the transformation $Z^k = Z^k(x^a)$, be found in the cell Dg_{ab} centered about $g_{ab}(x)$.

The advantages of the Gaussian matter time over an internal geometric time are rather obvious:

▪ Because the energy-momentum densities (7.10) are linear in the embedding momenta P_K, the many-fingered-time Hamiltonian (7.12) is a local functional of the variables X, g, and p which is quadratic in $p^{ab}(x)$. There is no spectral analysis problem.

▪ Because the variables $X^K(x^\alpha) = (T(x^\alpha), Z^k(x^\alpha))$ in the action (7.3) are spacetime scalars, the spacetime problem is automatically solved. Indeed, the embedding variables $X^K(x)$ satisfy Eq. (6.28).

▪ The introduction of a material reference frame and time foliation enables one to measure the metric rather than the mere (conformal) geometry.

An internal time based on the Gaussian reference fluid and the interpretation of quantum gravity which it offers come rather close to implementing the internal time program. However, there are two major objections against accepting such an interpretation of quantum gravity:

• **The Problem With Energy Conditions.**[141] The clock variable $T(x)$ and the frame variables $Z^k(x)$ are physically realizable only when the reference fluid satisfies the appropriate energy conditions. When the energy conditions are violated, the variables $T(x)$, $Z^k(x)$ cannot be identified by observing the physical properties of a real system and the interpretation of the expression (7.13) loses a sound epistemological foundation. It is thus necessary to consider the status of the energy conditions (7.8) in quantum gravity.

In the canonical formalism, the weak energy condition simply requires that $H^F(x) \geq 0$. The super-Hamiltonian constraint (7.9) and the use of the step function θ enables one to express the energy condition as a set of constraints

$$\theta(H^G(x)) = 0 \quad \forall x \in \Sigma \tag{7.14}$$

on the geometric variables. When these constraints are adjoined to the super-Hamiltonian and supermomentum constraints, the total system ceases to be first class:

$$\{\theta(H^G(x)), H(x')\} \not\approx 0 \ \textit{and}\ \{\theta(H^G(x)), \theta(H^G(x'))\} \not\approx 0\,. \tag{7.15}$$

The first inequality (7.15) tells us that the energy conditions are not preserved in time. The Poisson brackets (7.15) weakly vanish only if one imposes an additional constraint

$$P_k(x) = 0\,, \tag{7.16}$$

i.e., if one switches off the heat conduction and turns the Gaussian fluid into an incoherent dust. The constraint (7.16) tells us that the frame coordinates $Z^k(x)$ are no longer measurable.

When one tries to impose the energy conditions (7.14) as operator restrictions on the states,

$$\theta(\hat{H}^G(x))\ \Psi[X,g] = 0, \tag{7.17}$$

they become incompatible with the many-fingered time Schrödinger equation due to the commutator counterparts of Eq. (7.15). The consistency is restored for an incoherent dust, but the operator version of Eq. (7.16) makes the states independent of $Z^k(x)$. The interpretation of

$$|\Psi(x;\, T,g]|^2 \tag{7.18}$$

as the probability density for the geometry **g**(x) which is represented on the hypersurface T(x) by the metric $g_{ab}(x)$ to be found in the cell Dg centered about **g**(x) is hampered by the fact that the geometry operator **g**(x)× defined as a multiplication operator does not commute with the energy conditions. This does not mean that the probability density for the three-geometry **g**(x) does not exist, but that its identification requires a much more complicated and implicit procedure, analogous to the construction of the Newton-Wigner position operators for a relativistic particle.

It can be argued that the energy conditions (7.17) are unnecessarily strong, because they are imposed on energy *densities*, and these can become negative in a quantum field theory. One should keep in mind, however, that geometrodynamics is fundamentally different from an ordinary field theory: there, the classical energy density is positive-definite function of the field variables, while the classical expression $H^G(x)$ entering into Eq. (7.14) is indefinite.

• **Reference Fluids are Phenomenological Systems.** They can provide an interpretation of quantum gravity in "laboratory conditions," but one can scarcely expect to find such systems, say, in the early Universe.

▪ **Fundamental Fields as Matter Clocks.** One can counter both of these objections by saying that matter clocks are to be constructed from fundamental matter fields rather than from phenomenological fluids. The energy density of fundamental fields is supposed to satisfy at least the weak energy condition at the classical level. Such fields will always be present, even when phenomenological systems are destroyed by tidal forces. The idea of constructing a matter clock from a massless scalar field was tried in minisuperspace models.[96,13] (Incidentally, the massless scalar field is the reference fluid associated with the harmonic coordinate conditions.[132])

• **Shortcomings of Fundamental Matter Clocks.** However, when one starts exploring the fundamental matter fields as possible clocks, the advantages provided by a simple phenomenological system, like the Gaussian fluid, start quickly disappearing. The positivity of the classical energy density precludes it from being linear in the momentum conjugate to the field. When solving the super-Hamiltonian constraint with respect to such a momentum, the spectral analysis problem reappears in the quantum theory. Because the solution depends on $H^G(x)$, which is not positive-definite, the range of the geometric variables must usually be restricted for the solution to be real. If the matter clock is constructed from any other field then a spacetime scalar, one faces again the spacetime problem. By and large, identifying the embedding variables from among the fundamental matter fields is as complicated as looking for a geometric clock.

8. Unspecified Cosmological Constant and Cosmological Time

▪ **Unimodular Gravity.** In one special case, the reference fluid associated with a coordinate condition allows a geometric interpretation. This is the unimodular coordinate condition

$$| \det(\gamma_{KL}) |^{1/2} = 1 \tag{8.1}$$

fixing the spacetime volume element. The unimodular coordinates (8.1) were invoked by Einstein in the same paper in which he laid the foundations of the general theory of relativity.[142] By imposing the unimodular condition before rather than after the variation, one obtains the law of gravitation with an unspecified cosmological constant.[143] The reference fluid reduces in this case to the cosmological term. The law of gravitation with an unspecified cosmological constant was introduced by Einstein[144] and recently rediscovered, revived and reviewed by a number of authors.[145-148]

Unimodular gravity was cast into a canonical form by Henneaux and Teitelboim[149] and by Unruh.[143,150] The super-Hamiltonian constraint takes the familiar form

$$\lambda + |g|^{-1/2} H^G(x) = 0 \,, \tag{8.2}$$

except that the cosmological constant λ appears in the formalism as a canonical momentum conjugate to a coordinate τ , the "cosmological time." Heneaux and Teitelboim have shown that the increment of the cosmological time equals the four-volume enclosed between the initial and final hypersurfaces.[149] (In a related development, Sorkin used the four-volume time in a path-integral approach to quantum gravity.[151-152]) The action principle of general relativity with the cosmological term can be regarded as analogous to the Jacobi principle of classical mechanics.[153]

▪ **Cosmological Time and the Probabilistic Interpretation.** The appearance of the cosmological time in unimodular gravity raised hopes that this theory can solve the problem of time in quantum gravity. Because the constraint (8.2) is linear in the cosmological constant, its imposition on the state functional leads to an ordinary Schrödinger equation

$$i\partial_\tau \Psi(\tau; g] = (|g|^{-1/2} H^G(x; g,p])\hat{}\ \Psi(\tau; g] \,. \tag{8.3}$$

Unruh and Wald[76] suggested that any reasonable quantum theory should contain a parameter (which they called Heraclitian time) whose role is to set the conditions for measuring quantum variables and to provide the temporal ordering of such measurements. They surmised that the cosmological time of unimodular gravity is such a parameter. The Schrödinger equation (8.3) implies that the ordinary quadratic norm is conserved in τ . It is thus tempting to interpret $\Psi(\tau; g]$ as the probability amplitude for the outcomes of measurements carried out on geometric variables at a given instant of τ.

• **Cosmological Time is not a Functional Time.**[154] The problem with this suggestion is that the cosmological time is not in any obvious way related to the standard concept of time in the theory of relativity. The basic canonical variables $g_{ab}(x)$ and $p^{ab}(x)$, which unimodular gravity shares with geometrodynamics, are always supposed to be measured *on a single spacelike hypersurface* rather than *at a single value of the cosmological time*. To specify a hypersurface, one needs to use embedding variables, i.e., functions of three coordinates, rather than a single real parameter of Newtonian mechanics. It thus remains obscure in what sense the cosmological time "sets the conditions of quantum measurements."

To clarify this issue, one needs only to adjoin the unimodular condition to the action with a Lagrange multiplier and follow the algorithm developed in the last section. Through the parametrization process, the unimodular coordinates $X^K = (T, Z^k)$ appear in the action as embedding variables. However, the resulting constraint system is radically different from the single set of $(3+1)\,\infty^3$ constraints (7.9) of the Gaussian gravity. It splits into $3\infty^3 + (\infty^3 - 1)$ constraints on the embedding variables,

$$P_k(x) = 0 \quad \mathit{and} \quad (P_T(Z))_{,k} = 0 \tag{8.4}$$

the same number of constraints on the geometric variables,

$$H^G{}_a(x) = 0\ , \tag{8.5}$$

$$(|g(x)|^{-1/2}\ H^G(x))_{,a} = 0\ , \tag{8.6}$$

and one single constraint

$$\begin{aligned} &\int_\Sigma d^3Z\ P_T(Z)\ \Big/ \int_\Sigma d^3Z\ =:\ \lambda[X,P] \\ &= \lambda^G[g,p]\ :=\ -\int_\Sigma d^3x\ H^G(x)\ \Big/ \int_\Sigma d^3x\ |g(x)|^{1/2} \end{aligned} \tag{8.7}$$

coupling the geometric variables to the embedding variables.

When the embedding constraints (8.4) are imposed as operator restrictions on the states $\Psi[X,g]$, they ensure that Ψ depends on the unimodular embedding variables $X^K(x)$ only through a single quantity

$$\tau\ :=\ \int_\Sigma d^3Z\ T(Z)\ . \tag{8.8}$$

The parametrized unimodular condition tells us that this is the four-volume enclosed between a fiducial embedding $T(Z) = 0$ and the embedding $T(Z)$. The cosmological time τ thus labels the equivalence classes $\mathscr{E}_\tau$ of embeddings separated from the fiducial embedding by a given four-volume. Each equivalence class contains $4\infty^3 - 1$ embeddings, i.e., $\infty^3 - 1$ hypersurfaces, separated from each other by zero four-volume. The constraints (8.4) are to be interpreted as a statement that the individual embedding within each equivalence class are physically irrelevant.

The supermomentum constraint (8.5) ensures that the state Ψ depends only on the three-geometry $\mathbf{g}$: $\Psi = \Psi(\tau;\, \mathbf{g}]$. The constraint (8.7), which couples the embedding variables to the geometric variables, yields the Schrödinger equation

$$i\partial_\tau \Psi(\tau; g) = \int_\Sigma d^3x\ \hat{H}^G(x)\ /\ \int_\Sigma d^3x\ |\hat{g}(x)|^{1/2}\ \Psi(\tau; g]\ . \tag{8.9}$$

If this were the only equation on the state, one could interpret Ψ as the probability amplitude that the geometry g(x) has a given distribution at the time τ. However, Ψ must also satisfy the additional $\infty^3 - 1$ geometric constraints (8.6):

$$(|g(x)|^{-1/2}\ H^G(x)\ )^{\wedge}{}_{,a}\ \Psi(\tau; g] = 0\ . \tag{8.10}$$

The three-geometry operator does not commute with these constraints and the interpretation of Ψ as the probability amplitude becomes untenable. The $\infty^3 - 1$ geometric constraints (8.10) are actually a Wheeler-DeWitt equation which describes the evolution of states between embeddings of a single equivalence class $\mathscr{E}_\tau$. The cosmological time does not set the conditions of the measurement uniquely because it does not tell us on which one of the infinitely many hypersurfaces of the equivalence class the geometric variables are to be measured. The hypersurfaces with different values of τ are allowed to intersect and the cosmological time thus does not even provide the causal ordering required of the Heraclitian time. The fundamental reason why unimodular gravity does not and cannot solve the problem of time is simple: time in relativity is a collection of all spacelike hypersurfaces, and no single parameter, such as the cosmological time τ, is able to label uniquely so many instants.

9. Klein-Gordon Interpretation

The problem of identifying an internal time and handling the ensuing Schrödinger equation is beset with so many difficulties that one may start wondering if the net result is worth the trouble. I shall turn now to those interpretations of quantum gravity which give up the attempts to separate time from the dynamical degrees of freedom and rely instead on the Wheeler-DeWitt equation (2.2), (2.5).

The Wheeler-DeWitt equation is a second-order variational equation for the state functional $\Psi[g]$. One can view it as an infinitely-dimensional analogue of the Klein-Gordon equation for the relativistic particle. To illustrate the problems which arise when one tries to interpret the Wheeler-DeWitt equation, it is best to review the probabilistic interpretation of the Klein-Gordon particle.

▪ **The Hilbert Space Problem for a Relativistic Particle.** The super-Hamiltonian of a relativistic particle is given by the expression (3.6). In the Dirac constraint quantization, H is turned into the Klein-Gordon operator

$$\hat{H} = \frac{1}{2M}\left(-G^{AB}(Q)\nabla_A\nabla_B + M^2V(Q)\right) \tag{9.1}$$

and imposed as a restriction

$$\hat{H}\,\Psi(Q) = 0 \tag{9.2}$$

on the physical states. The main task is to endow the space of solutions $\mathscr{F}_0$ of Eq. (9.2) with an inner product and complete it into a Hilbert space. This constitutes the Hilbert space problem.

The construction of the inner product starts with the current

$$J^A_{12} := \frac{1}{2i}\,G^{AB}\left(\Psi^*_1 \overleftrightarrow{\partial}_B \Psi_2\right). \tag{9.3}$$

If Ψ_1 and Ψ_2 lie in $\mathscr{F}_0$, the current (9.3) satisfies the continuity equation

$$\nabla_A J^A_{12} = |G|^{-1/2}\,\partial_A(|G|^{1/2}J^A_{12}) = 0 \tag{9.4}$$

which implies that the integral

$$\langle\Psi_1|\Psi_2\rangle := \int_\Sigma d\Sigma_A\, J^A_{12}\;,$$
$$d\Sigma_A := \epsilon_{AA_1A_2A_3}\,d_{(1)}Q^{A_1}d_{(2)}Q^{A_2}d_{(3)}Q^{A_3} \tag{9.5}$$

does not depend on the choice of the hypersurface $\Sigma \to \mathcal{M}$. Unfortunately, the product (9.5) is not positive definite on $\mathscr{F}_0$ and hence is unsuitable to serve as a probability.

It is generally acknowledged that the Hilbert space problem for a single relativistic particle has no solution unless the spacetime background $G^{AB}(Q)$ and the potential $V(Q)$ are stationary.[155-158] In that case, there exists a timelike future-pointing Killing vector field $U^A(Q)$,

$$(L_U\,G)_{AB} = -\nabla_{(A}U_{B)} = 0\;, \tag{9.6}$$

which preserves $V(Q)$:

$$L_U\,V = U^A\,\partial_A V = 0\,. \tag{9.7}$$

These conditions imply that the energy

$$E := -\, U^A(Q)\, P_A \tag{9.8}$$

of the particle in the stationary reference frame is a constant of the motion,

$$\{E, H\} = 0\,. \tag{9.9}$$

Relativistic particles move along future-oriented timelike worldlines. Along such worldlines, the energy (9.8) is always positive, $E > 0$.

On quantization, the energy (9.8) is turned into the operator

$$\hat{E} := i(U^A\partial_A + \tfrac{1}{2}\,\mathrm{div}_G\, U) = iU^A\partial_A\,, \tag{9.10}$$

which, by virtue of Eqs. (9.6) and (9.7), commutes with the Klein-Gordon constraint operator:

$$[\hat{E}, \hat{H}] = 0\,. \tag{9.11}$$

This means that the eigenvalue equation for E,

$$\hat{E}\,\Psi = E\,\Psi, \tag{9.12}$$

and the Klein-Gordon constraint (9.2) have common solutions. One postulates that the quantum particle, like the classical one, must have positive energy. This amounts to saying that the physical space is not the whole space $\mathscr{F}_o$ of solutions of the Klein-Gordon constraint, but only a linear subspace of $\mathscr{F}_o$ spanned by the positive-energy solutions. Thus, if $\mathscr{F}^+$ is spanned by the positive-energy eigenfunctions of Eq. (9.12), and $\mathscr{F}_o$ is the solution space of Eq. (9.2), the physical space $\mathscr{F}_o^+$ is the intersection $\mathscr{F}_o^+ = \mathscr{F}_o \cap \mathscr{F}^+$. One can prove that on $\mathscr{F}_o^+$ the inner product (9.5) is positive definite. One can complete $\mathscr{F}_o^+$ in the norm given by Eq. (9.5). The statistical interpretation of the theory is then based on this inner product. One can show that this construction of the Hilbert space amounts to replacing the Klein-Gordon equation by a Schrödinger equation.[95]

If there is no Killing vector, there is no energy operator which commutes with $\hat{H}$, and the construction of the Hilbert space fails. Physically, one cannot maintain a one-particle interpretation of the theory because the dynamical background starts producing particles. Even if there is a Killing vector, one can prove the positivity of the inner product only if this vector is timelike and the potential V in the super-Hamiltonian (9.1) is non-negative. This sets the limits within which one knows how to resolve the Hilbert space problem for a relativistic particle.

■ **The Hilbert Space Problem in Geometrodynamics.** Proceeding by analogy, one can ask how much of this algorithm is applicable to quantum geometrodynamics. The analogue of the Klein-Gordon product (9.5) was introduced by DeWitt.[32] Let $\Psi_1[g]$ and $\Psi_2[g]$ be two solutions of the Wheeler-DeWitt equation and

$$d\Sigma^{ab}(x) := \epsilon^{ab\ a_1b_1\ \ldots\ a_5b_5} d_{(1)}g_{a_1b_1}(x)\ \ldots\ d_{(5)}g_{a_5b_5}(x) \tag{9.13}$$

the directed surface element of a hypersurface in superspace. Then

$$\langle\Psi_1|\Psi_2\rangle := \frac{1}{2i} \prod_{x\epsilon\Sigma} \int_\Sigma d\Sigma^{ab}\ G_{abcd}(x;\ g)\ (\Psi_1^*[g]\ \frac{\overleftrightarrow{\delta}}{\delta g_{cd}(x)}\ \Psi_2[g])\ . \tag{9.14}$$

At least formally, the inner product (9.14) does not depend on the choice of the hypersurface in superspace. As in particle dynamics, this product is not positive in the space $\mathcal{F}_0$ of solutions to the Wheeler-DeWitt equation. One must ask: Does the super-Hamiltonian (1.2) satisfy the conditions which would allow one to restrict $\mathcal{F}_0$ to a linear subspace $\mathcal{F}_0^+$ on which the product (9.14) is positive definite? Unfortunately, there are several reasons why the algorithm developed for the relativistic particle fails:

• **Super-Hamiltonian is not Stationary.**[33,159,160] To decide whether the supermetric $G_{abcd}(x;\ g)$ and the potential $-\ |g|^{1/2}R(x;\ g]$ admit an "isometry," one should look for a dynamical variable

$$E := -\int_\Sigma d^3x\ u_{ab}(x;\ g]p^{ab}(x) \tag{9.15}$$

that is linear in the momentum $p^{ab}(x)$ and has at least a weakly vanishing Poisson bracket with the super-Hamiltonian (1.2):

$$\{E,\ H(x)\} \approx 0. \tag{9.16}$$

Indeed, E should also have a vanishing Poisson bracket with the supermomentum (1.1), but this goal is easily met by requiring that the coefficient $u_{ab}(x;\ g]$ in Eq. (9.15) be a spatial tensor field.

It is quite difficult to prove whether such a variable exists or not (Kuchar[159]), the trouble being that the "Killing vector field" $u_{ab}(x;\ g]$ in the space of Riemannian metrics may in principle be a highly nonlocal functional of g. The proof relies on the fact that the super-Hamiltonian H(x) and the supermomentum $H_a(x)$, through which the weak equality in Eq. (9.16) is enforced, have quite definite locality and polynomial structures. Thus $H_a(x)$ is linear in the momentum and contains at most the first derivatives of the canonical variables $g_{ab}(x)$, $p^{ab}(x)$, while $H(x)$ is a sum of the "kinetic" part $H_{KIN}(x; g,p]$, which is an ultralocal quadratic form in $p^{ab}(x)$, and the "potential" part $H_{POT}(x; g]$, which is a scalar concomitant of $g_{ab}(x)$ of the second order. This enables one to prove first that $u_{ab}(x; g]$ must be local, i.e., a tensor concomitant of g of a finite order N, and then, by a glissade argument decreasing the order N step by step, end with the conclusion that $u_{ab}(x; g]$ must actually be ultralocal. The only ultralocal tensor concomitant of g is however (a constant multiple of) g itself, and hence the only candidate for the dynamical variable (9.15) is the integrated trace of p:

$$E = - \int_\Sigma d^3x \; g_{ab}(x)p^{ab}(x) \; . \tag{9.17}$$

The Poisson bracket of E with the kinetic part $H_{KIN}(x)$ of the super-Hamiltonian is proportional to $H_{KIN}(x)$:

$$\{E, H_{KIN}(x)\} = - \tfrac{3}{2} \; H_{KIN}(x) \; . \tag{9.18}$$

In other words, for each $x \in \Sigma$, $g_{ab}(x)$ is a conformal isometry (indeed, a homothetic motion) of the supermetric $G_{abcd}(x; g)$. Moreover, $g_{ab}(x)$ is a "timelike" supervector, i.e., $G^{abcd}(x)g_{ab}(x)g_{cd}(x) < 0$. Unfortunately, E does not scale the potential term $H_{POT}(x; g]$ in the super-Hamiltonian at the same rate as the kinetic term $H_{KIN}(x; g,p]$:

$$\{E, H_{POT}(x)\} = \tfrac{1}{2} \; H_{POT}(x). \tag{9.19}$$

The mismatch of the scaling factors in Eqs. (9.18) and (9.19) means that E does not weakly commute with the total super-Hamiltonian H(x), and hence our search for an isometry of the geometrodynamical constraint system fails. As a result, one cannot make the product (9.14) positive definite by restricting the solutions of the Wheeler-DeWitt equation to those with a positive "energy" (9.15). The situation in asymptotically flat (rather than spatially compact) spacetimes is discussed by Friedman and Higuchi.[161]

• **Potential Term is not Positive.** Even if the super-Hamiltonian were stationary, one would still be unable to prove that the Klein-Gordon product (9.14) restricted to the states $\mathcal{F}^+_\circ$ is positive definite because the potential $|g|^{1/2} R(x; g]$ can be both positive and negative. I have discussed this difference between geometrodynamics and the relativistic particle in Section 3.

• **States with Negative E are Physical.** A relativistic particle moves in *spacetime* along a future-oriented path, but a compact geometry moving in *superspace* can both expand and contract its total volume. The classical solutions of Einstein's equations can thus have either sign of E. The restriction of the quantum states to $\mathscr{F}_0^+$ would simply eliminate one half of entirely permissible states.

• **Spacetime Problem.** The Klein-Gordon product (9.5) is the same on any spacelike hypersurface T(Q) in spacetime; similarly, the DeWitt product (9.14) is the same on any "spacelike" hypersurface T(x; g] in superspace. I have already pointed out that the intrinsic time T(x; g] cannot be interpreted as labeling the spacetime events (see the discussion of the spacetime problem in Section 5). One should also remember that the signature of the supermetric has nothing to with the signature of the spacetime metric that is obtained by solving the classical Hamilton equations, and remains the same even in a Euclidean spacetime.[46,162] It is thus rather mysterious how the conservation of the DeWitt product in superspace is related to its conservation in spacetime.

Taken together, all of these points make it rather difficult to imagine how the space of solutions of the Wheeler-DeWitt equation can ever be turned into a Hilbert space.

10. Semiclassical Interpretation

Like the Klein-Gordon interpretation, the semiclassical interpretation tries to provide a dynamical interpretation of the Wheeler-DeWitt equation. It contends, however, that not necessarily all solutions of the Wheeler-DeWitt equation allow such an interpretation, but that time and quantum dynamics emerge only if a state Ψ becomes semiclassical. The semiclassical *interpretation* is this imposed only after the Wheeler-DeWitt equation has been subjected to a semiclassical *approximation.*[163]

The semiclassical approximation was developed to bridge the gap between the ill-defined quantum theory of gravity and the much better understood quantum field theory on a given background. In its simplest form, it represents a heuristic derivation of the latter theory from canonical quantum gravity. In this process, a functional *Schrödinger* equation for the field emerges from the *Wheeler-DeWitt* equation for the field coupled to gravity. It is this recovery of the Schrödinger equation from the second-order Wheeler-DeWitt equation which underlies the semiclassical interpretation.

▪ **The Particle Model.** The semiclassical approximation is best explained on a finite-dimensional model (3.9) in which a heavy particle M interacting with a light particle m is constrained to a stationary state.[164] The heavy particle is modeling the metric field and the light particle the matter field. For this purpose, it is suitable to assume that the metric G^{AB}, like the DeWitt supermetric (1.2), is indefinite, while the metric h^{kl} and the potential v, like the kinetic and potential energies of a field which satisfies the weak energy condition, are positive definite. In this case, one is allowed to put the total energy E equal to zero.

I adopt the Laplace-Beltrami factor ordering of the super-Hamiltonian H,

$$\begin{aligned}\hat{H} = &- \frac{1}{2M}|Gh|^{-1/2}\partial_A|Gh|^{1/2}G^{AB}\partial_B + \frac{1}{2}MV \\ &- \frac{1}{2m}|h|^{-1/2}\partial_k|h|^{1/2}h^{kl}\partial_l + \frac{1}{2}mv.\end{aligned} \tag{10.1}$$

The operator (10.1) is covariant under arbitrary transformations of the coordinates (Q^A, q^k). Under this factor ordering,

$$\hat{H} = -\frac{1}{2M}(|h|^{-1/2}\partial_A|h|^{1/2})\,G^{AB}\partial_B + \hat{H}_M + \hat{h}_m\,, \tag{10.2}$$

where

$$\hat{H}_M := -\frac{1}{2M}\Delta_G + \frac{1}{2}MV \quad \textit{and} \quad \hat{h}_m := -\frac{1}{2m}\Delta_h + \frac{1}{2}mv. \tag{10.3}$$

The first term in Eq. (10.2) arises because the volume measure $|h|^{1/2}$ depends on the position Q of the heavy particle.

The semiclassical approximation arises when one tries to solve the super-Hamiltonian constraint

$$\hat{H}\,\Psi(Q,q) = 0 \tag{10.4}$$

by expanding the state function $\Psi(Q,q)$ in the inverse powers of the mass parameter M:

$$\Psi(Q,q) = e^{iMS(Q,q)}\,\Phi(Q)\left(\psi(Q,q) + \Sigma_{n=1}^{\infty}\,\psi_{(n)}(Q,q)\,M^{-n}\right). \tag{10.5}$$

In relativity, the ansatz (10.5) corresponds to the weak-coupling expansion in powers of the Einstein constant of gravitation κ.

The leading term in $\hat{H}\Psi$ is proportional to M^2. In this order, Eq. (10.4) tells us that $h^{kl}\,S_{,k}S_{,l} = 0$. Because the metric h^{kl} is positive definite, S cannot depend on the coordinates q^k of the light particle: $S=S(Q)$. In the M^1 order, the constraint (10.4) subjects S(Q) to the Hamilton-Jacobi equation

$$G^{AB}(Q)S_{,A}(Q)S_{,B}(Q) + V(Q) = 0 . \tag{10.6}$$

In the M^0 order, one gets an equation for $\Phi(Q)\psi(Q,q)$. The split between the prefactor $\Phi(Q)$ and the state function $\psi(Q,q)$ is arbitrary. One can use this to simplify the equation for $\psi(Q,q)$ by choosing a prefactor $\Phi(Q)$ which satisfies the condition

$$\partial_A(|G|^{1/2}\Phi^2 S^A) = 0, \quad S^A := G^{AB}S_{,B} . \tag{10.7}$$

The M^0-order equation then becomes

$$i|h|^{-1/4}S^A\partial_A(|h|^{1/4}\psi) = h_m(Q; \hat{q},\hat{p})\, \psi. \tag{10.8}$$

Equations (10.6) - (10.8) constitute the semiclassical approximation. In the next order of the expansion, i.e., in the M^{-1} order, one gets an equation for $\psi_{(1)}(Q,q)$ which is of the form (10.8) except – and this is an important difference – the right-hand side of Eq. (10.8) is complemented by an inhomogeneous term constructed from the previously determined functions S(Q), $\Phi(Q)$, and $\psi(Q,q)$. This is the first term in the expansion which contains the second derivatives with respect to Q^A.

A particular solution S(Q) of the Hamilton-Jacobi equation (10.6) defines a congruence of classical trajectories.[7] These are the worldlines orthogonal to the hypersurfaces of constant S(Q). I shall label these worldlines by N – 1 "comoving" coordinates $y^a(Q)$,

$$S^A(Q)y^a{}_{,A}(Q) = 0. \tag{10.9}$$

The solution S(Q) also defines a time function t(Q): One chooses an arbitrary hypersurface transverse to the classical trajectories as a t = 0 hypersurface. By Lie dragging this hypersurface along the vector field $S^A(Q)$ one generates a foliation. The time function labeling the leaves of this foliation satisfies the condition

$$S^A(Q)t_A(Q) = 1, \quad t_A := t_{,A} . \tag{10.10}$$

There are as many time functions as there are choices of the initial hypersurface. A comparison of Eqs. (10.10) and (10.6) reveals that for a general V(Q), the time foliation necessarily differs from the S(Q) = *const* foliation.

In the coordinate system

$$Q^{A'} = (t(Q), y^a(Q)) \tag{10.11}$$

the remaining equations (10.7) and (10.8) simplify. The continuity equation (10.7) for the prefactor $\Phi(Q)$ takes the form

$$\partial_t (| G(t,y)|^{1/2} \Phi^2(t,y)) = 0 . \tag{10.12}$$

This equation has the solution

$$\Phi(t,y) = \varphi(y) \ | G(t,y)|^{-1/4}, \tag{10.13}$$

where $\varphi(y)$ is an arbitrary function of the comoving coordinates. One can choose $\varphi(y)$ such that the prefactor vanishes outside a thin worldtube centered on a selected classical worldline.

The equation (10.8) for the state function ψ reduces to

$$i|h(y; t,q)|^{-1/4} \partial_t(|h(y; t,q)|^{1/4} \psi(y; t,q)) = h_m(y; t,\hat{q},\hat{p}) \ \psi(y; t,q) . \tag{10.14}$$

This is an (N – 1) - parameter set of Schrödinger equations, one equation for each classical trajectory y. The Hamiltonian $\hat{h}_m$ differs from one classical trajectory to another, and it depends on the time t, because the metric $h^{kl}(Q,q)$ and the potential $v(Q,q)$ depend in general on Q. Under the factor ordering (10.3), each Hamiltonian $\hat{h}_m(y)$ is self-adjoint under the inner product

$$(\psi_1 | \psi_2) = \int d^nq \ |h(y; t,q)|^{1/2} \ \psi_1^*(y; t,q)\psi_2(y; t,q). \tag{10.15}$$

Moreover, this inner product does not depend on t by virtue of the Schrödinger equation (10.14).

In this scheme, the heavy particle follows a classical trajectory from the congruence determined by a fixed solution of the Hamilton-Jacobi equation (10.6), unaffected in its motion by the presence of the light particle. For each such trajectory, the motion of the light particle is quantized and described by the Schrödinger equation (10.14). This mixture of the classical and quantum worlds gave the semiclassical approximation its name.

▪ **The Klein-Gordon Product and the Schrödinger Product.** The Schrödinger equation (10.14) is a consequence of the Klein-Gordon constraints (10.4); accordingly, the Schrödinger norm (10.15) is induced by the Klein-Gordon product (9.5).[165]

For the block-diagonal metric of the two-particle system (10.1), the current (9.3) constructed from a solution Ψ of the constraint (10.4) has the components

$$J^A = \frac{1}{2i}\frac{1}{M} G^{AB} (\Psi^* \overleftrightarrow{\partial}_B \Psi) , \quad J^k = \frac{1}{2i}\frac{1}{m} h^{kl} (\Psi^* \overleftrightarrow{\partial}_l \Psi) , \qquad (10.16)$$

and it satisfies the continuity equation

$$|Gh|^{-½}\partial_A(|Gh|^{½}J^A) + |h|^{-½}\partial_k(|h|^{½}J^k) = 0. \qquad (10.17)$$

For the state function (10.5), the current (10.16) in the lowest, M^0, order of the M^{-n} expansion takes the form

$$j^A = |\Phi(Q)|^2 S^A(Q) |\psi(Q,q)|^2, \quad j^k = \frac{1}{2i}\frac{1}{m} h^{kl}(\psi^* \overleftrightarrow{\partial}_l \psi) . \qquad (10.18)$$

Because the prefactor $\Phi(Q)$ was chosen so that it satisfies its own continuity equation (10.7), the continuity equation (10.17) reduces in the lowest order to the equation

$$|h|^{-½} S^A\partial_A(|h|^{½}|\psi|^2 S^A) + |h|^{-½}\partial_k(|h|^{½}j^k) = 0 \qquad (10.19)$$

for the state function of the light particle. In the comoving coordinates (10.9) and with respect to the time function (10.10), the continuity equation for the prefactor assumes the form (10.12) and Eq. (10.19) reduces to

$$\partial_t (|h|^{½}|\psi|^2) + \partial_k (|h|^{½}j^k) = 0 . \qquad (10.20)$$

Equation (10.20) guarantees the conservation of the Schrödinger inner product (10.15). The form (10.18) of the current leads to the relation

$$\langle\Psi|\Psi\rangle = \int d^{N-1}y\, |G|^{½} |\Phi|^2 \cdot (\psi|\psi) \qquad (10.21)$$

between the Klein-Gordon product (9.5) and the Schrödinger product (10.15).

▪ **Geometrodynamics with Sources.** At least formally, canonical geometrodynamics with a field source can be subject to the same treatment as the particle model (Banks[164]). The gravitational field plays the role of the heavy particle and the source field of the light one. The expansion parameter M^{-1} is the

Einstein constant of gravitation 2κ. The three-metric $g_{ab}(x)$ replaces Q^A and DeWitt's supermetric $G_{abcd}(x;g)$ replaces $G^{AB}(Q)$. The source field $\phi(x)$ replaces q^k. In geometrodynamics, the metric $g_{ab}(x)$ enters as a coordinate Q^A both into the construction of the metric $h^{kl}(Q,q)$ and of the potential $v(Q,q)$ in the field space. For this reason, I took pain to explain how the model works for a Q-dependent metric h^{kl}, the point which is disregarded in the literature.

The gravitational field is largely unaffected by the source field because the gravitational coupling is weak. One seeks the solution of the Wheeler-DeWitt equation in the form

$$\Psi[g] = e^{i(2\kappa)^{-1}S[g,\phi]}\,\Phi[g]\left(\psi[g,\phi] + \Sigma_{n=1}^{\infty}\psi_{(n)}[g,\phi](2\kappa)^n\right) . \qquad (10.22)$$

As for the model system, one learns that the phase S cannot depend on ϕ and that it satisfies the vacuum Hamilton-Jacobi equation (1.8) - (1.9). One is again entitled to subject the prefactor to the continuity equation

$$\frac{\delta}{\delta g_{ab}(x)}\left(|G(x;\, g)|^{1/2}S_{ab}(x;\, g]\right) = 0, \quad S_{ab}(x;\, g] := G_{abcd}(x;\, g)\frac{\delta S[g]}{\delta g_{cd}(x)} . \qquad (10.23)$$

It then follows that the state functional $\psi[g,\phi]$ satisfies the first-order variational equation

$$i|h(x;\, g]|^{-1/4}S_{ab}(x;\, g]\,\frac{\delta}{\delta g_{ab}(x)}\left(|h(x;\, g)|^{1/4}\psi[g,\phi]\right) = h_m(x;\, g,\hat{\phi},\hat{\pi}]\,\psi[g,\phi] . \qquad (10.24)$$

As the Wheeler-DeWitt equation itself, the equation (10.24) is actually a system of ∞^3 equations, one per each point of Σ. The Jacobi principal function S[g] again enables one to introduce the comoving coordinates $Y^R(x;\, g]$, $R = 1, \ldots , 5$, and the intrinsic time function $T(x;\, g]$ which satisfies the equation

$$S_{ab}(x';\, g]\frac{\delta T(x;\, g]}{\delta g_{ab}(x')} = \delta(x,x') . \qquad (10.25)$$

In the new variables $T(x;\, g]$, $Y(x;\, g]$ on superspace, Eq. (10.24) takes the form of a functional Schrödinger equation in the intrinsic time $T(x;\, g]$:

$$i\,|\,h(x;\ Y,T]\,|^{-1/4}\ \frac{\delta}{\delta T(x)}\ (|\,h(x;\ Y,T]\,|^{1/4}\ \psi[Y;\ T,\phi])$$
$$= h_m(x;\ Y,T,\hat{\phi},\hat{\pi}]\ \psi[Y;\ T,\phi] \tag{10.26}$$

▪ **Back Reaction.** The semiclassical approximation has been modified to include the back reaction of the quantized source field on the evolution of the classical geometry by Brout et al.[166-168], Hartle,[169] Halliwell,[170] and Singh and Padhanabhan.[171,163] The aim of this development was to understand to what extent the classical background $\gamma_{\alpha\beta}(X)$ can be considered as being produced by the expectation value of the energy-momentum tensor of the quantized source field $\phi(x)$ according to the Einstein law

$$G^{\alpha\beta}(X;\ \gamma] = \kappa\,\langle\, T^{\alpha\beta}(X;\ \gamma,\phi]\,\rangle\ . \tag{10.27}$$

Equation (10.27) was originally proposed by Møller[172] and Rosenfeld[173] as an *exact* equation of a theory in which the metric field is always classical and only the ordinary matter fields are quantized. It was always felt that quantum gravity should yield Eq. (10.27) in a Hartree-Fock approximation. The development which I mentioned was aimed at getting the canonical version of Eq. (10.27) in the semiclassical approximation. For this purpose, Eq. (10.12) for the prefactor was proposed to be modified by the expectation value of a source term. One can trace a feeling of uneasiness in the literature about this procedure, the uneasiness which I happen to share. It seems to me that the semiclassical approximation in gravity is essentially a weak-field expansion in the powers of κ which should include the influence of the source field on geometry in the κ^1 order, rather than by tempering with the prefactor in the lower order. I shall thus largely concentrate on the issue of time in the simpler and less controversial framework provided by the Schrödinger equation (10.26) (or its model counterpart (10.14)) on the classical background uninfluenced by the quantum sources.

▪ **Decoherence.** It has been argued that the emergence of the classical behavior requires not only the WKB type of the wave function, but also the destruction of the interference terms (the decoherence) among possible classical alternatives. Such a destruction can be brought in by the interaction of the modes one is interested in with the modes of an "environment" which is subsequently traced out from the density operator. Decoherence in the semiclassical approximation to quantum gravity was discussed by Zeh,[174,175], Joos[176], Kiefer,[177] and Halliwell.[178]

▪ **Semiclassical Interpretation of the Model.** After explaining the formalism of semiclassical *approximation*, let me turn to the proposals which aim at using it as a tool for *interpreting* quantum gravity and, in particular, for resolving the problem of time. This approach can be traced back to DeWitt[32] and Misner[49]; it was elaborated by Lapchinski and Rubakov[179], by Banks[164], and most clearly stated by Vilenkin.[165] I explain it first on the particle model. The difficulties which emerge at this level indicate that semiclassical interpretation does not cope with the problems of time even in the simple transition from quantum gravity with sources to quantum field theory on a given classical background. Furthermore, the semiclassical interpretation must be modified before it can be applied to quantum gravity proper. I shall later discuss the additional problems brought in by such a modification.

The semiclassical interpretation of the particle model is based on Eqs. (10.6), (10.9) - (10.10), (10.14), (10.12) and (10.20), and (10.15), (10.21). A particular solution of the Hamilton-Jacobi equation (10.6) defines a congruence of classical trajectories of the heavy particle. Equation (10.9) labels the trajectories, and Eq. (10.10) introduces the time function of the congruence. In that time, the wave function ψ of the light particle satisfies the Schrödinger equation (10.14) along a fixed trajectory. The continuity equation (10.12) for the prefactor provides the measure in the congruence of classical trajectories. The continuity equation (10.20) yields the conserved norm (10.15) on which the probabilistic interpretation of the state of the light particle can be based. The statistical ingredients are summarized by Eq. (10.21) which relates the Klein-Gordon inner product to the Schrödinger inner product (10.15).

The semiclassical interpretation asserts that the Klein-Gordon constraint (10.4) has a probabilistic interpretation only if the system is in a semiclassical state (10.5). In that case,

$$|G(t,y)|^{1/2}\,\Phi^2(t,y)\,d^{N-1}y \;\cdot\; |h(y;\,t,q)|^{1/2}\,|\psi(y;\,t,q)|^2\,d^nq \qquad (10.28)$$

is the joint probability of finding at the time t the heavy particle moving along the worldline y = *const* in the cell $d^{N-1}y$ and the light particle localized in the cell d^nq about q. This interpretation is sometimes corroborated by evaluating the Wigner function of the semiclassical state (10.5) (Halliwell[180]).

The semiclassical interpretation has many weaknesses:

- **Global Time Problem.** The construction of the time function t(Q) presupposes that there exists a hypersurface which is transverse to the congruence of the classical trajectories generated by the Jacobi principal function S(Q).[165,170]

This begs the global time problem. The problem becomes even more severe when one tries to superimpose the semiclassical states with different principal functions.

• **Superposition Problem.** The probabilistic interpretation (10.28) was reached under the assumption that the total system (Q,q) is in a single semiclassical state (10.5). However, the linearity of the constraint (10.4) allows it to be in a superposition of two or more such states,

$$\Psi(q,Q) = \Phi_1(Q)e^{iMS_1(Q)}\psi_1(Q;q) + \Phi_2(Q)e^{iMS_2(Q)}\psi_2(Q;q) \ . \qquad (10.29)$$

When one calculates the current (10.16) for the state (10.29) to the lowest order M^0, one obtains the interference terms:

$$j^A = |\Phi_1|^2 S_1^A |\psi_1|^2 + |\Phi_2|^2 S_2^A |\psi_2|^2 \qquad (10.30)$$
$$+ \Phi_1\Phi_2\Big(\mathrm{Re}(\psi_1\psi_2^*)\cos M(S_2-S_1) + \mathrm{Im}(\psi_1\psi_2^*)\sin M(S_2-S_1)\Big)(S_1^A + S_2^A) \ .$$

The interference of states with the same $\Phi(Q)$ and $S(Q)$ but different $\psi_1(q)$, $\psi_2(q)$ is desirable: it corresponds to the quantum nature of the light particle. The interference of the WKB states of the heavy particle is a different matter. It raises the question whether the semiclassical interpretation can be consistently maintained. The following discussion explores the various facets of the superposition problem.

• **Complex Structure Problem.** The Klein-Gordon constraint does not involve imaginary coefficients and thus allows real solutions Ψ. The real and imaginary parts of a complex solution are not coupled by Eq. (10.4). On the other hand, the Schrödinger equation (10.14) contains the imaginary unit associated with the time derivative, and it thus couples the real and imaginary parts of ψ (Barbour and Smolin,[97] Barbour[181]). This highlights the fact that the Schrödinger equation cannot be possibly derived from the Klein-Gordon constraint without some further tacit assumption.

Indeed, the Schrödinger equation for $\psi(Q;q)$ follows from the M^0 order of the M^{-n} expansion only when the state Ψ is limited to a *single* WKB state (10.5). The derivation no longer holds for a *superposition* (10.29). Even if one assumes that S_1 and S_2 separately satisfy the Hamilton-Jacobi equation (10.6) (which does not *follow* from the M order of the expansion procedure, though it is *compatible* with it), and subject both prefactors Φ_1 and Φ_2 to the continuity equation (10.7), the M^0 order equation imposes only a *single* condition

$$(-i\, S_1^A |h|^{-\frac{1}{4}} \partial_A(|h|^{\frac{1}{4}} \psi_1) + \hat{h}_m \psi_1)\, \Phi_1\, e^{iMS_1}$$
$$+ (-i\, S_2^A |h|^{-\frac{1}{4}} \partial_A(|h|^{\frac{1}{4}} \psi_2) + \hat{h}_m \psi_2)\, \Phi_2\, e^{iMS_2} = 0 \quad (10.31)$$

on *two* state functions ψ_1 and ψ_2. Equation (10.31) has many more solutions than those in which ψ_1 and ψ_2 satisfy the separate Schrödinger equations. For example, one can put $\psi_1 = \psi_2 =: \psi$ and derive an equation for ψ,

$$i\, (\Phi_1 e^{iMS_1} + \Phi_2 e^{iMS_2})^{-1} (\Phi_1 e^{iMS_1} S_1^A + \Phi_2 e^{iMS_2} S_2^A)$$
$$|h|^{-\frac{1}{4}} \partial_A(|h|^{\frac{1}{4}} \psi) = \hat{h}_m \psi \,, \quad (10.32)$$

which is *not* a Schrödinger equation, because the vector field along which the directional derivative is taken is complex. In particular, for $S_1 = -S_2 =: S$, and $\Phi_1 = \Phi_2$, one gets the equation

$$-\tan MS \,.\, |h|^{-\frac{1}{4}} \partial_t (|h|^{\frac{1}{4}} \psi) = \hat{h}_m \psi \quad (10.33)$$

whose coefficients are all real. This clearly shows that the complex structure which one associates with the Schrödinger equation, and the equation itself, do not follow from the Klein-Gordon constraint by the M^{-n} expansion. An elucidating discussion of simple state functions of the type (10.33) is given in Barbour and Smolin.[97]

• **The Double Standard of Time.** Even those superpositions (10.29) in which the coefficients ψ_1 and ψ_2 separately satisfy the Schrödinger equations present problems. The Jacobi principal functions $S_1(Q)$ and $S_2(Q)$ lead to different time functions $t_1(Q)$ and $t_2(Q)$ and to different comoving coordinates $y_1^a(Q)$ and $y_2^a(Q)$. No single choice of the coordinate system will simultaneously reduce the first-order equations (10.8) to the Schrödinger form (10.14) and the continuity equations (10.7), (10.19) to the form (10.12), (10.20). On what time function should then one base the probabilistic interpretation of the pure state (10.29)?

One can counter this objection by pointing out that one does not need to commit oneself to any particular time function when evaluating the norm. The exact current (10.16) is exactly conserved, and the inner product (9.5) is thus the same on every hypersurface. When one replaces the exact current by the semiclassical current, the inner product is still the same on every hypersurface, up to the terms of order M^{-1}. This suggests that the integrand of

$$\langle\Psi|\Psi\rangle = \int d^{N-1}y \int d^n q \quad t_A j^A |G|^{1/2} \; |h|^{1/2} \tag{10.34}$$

can be interpreted as the probability density ρ to find the heavy particle in the cell $d^{N-1}y$ and the light particle in the cell $d^n q$ on *any* hypersurface $t(Q) = const.$

Some difficulties, however, still remain. One is not entitled to say that the heavy particle is in a definite state of motion. The global problem of time is aggravated because there may not exist any hypersurface which is *simultaneously* transverse to both congruences of classical trajectories. Most importantly, the probability density ρ does not need to be positive.

• **Interference Problem.** The interference terms in Eq. (10.30) indicate that it may not be possible to say that the heavy particle is in a definite state of motion. One can try to exorcise them by one of the two alternative methods:

I) One can choose the two prefactors Φ_1 and Φ_2 so that they do not overlap, concentrating, e.g., Φ_1 in a thin tube surrounding a classical trajectory y_1, and Φ_2 in a thin tube surrounding a classical trajectory y_2. As long as the two tubes do not intersect,

$$j^A = |\Phi_1|^2 |\psi_1|^2 S_1^A + |\Phi_2|^2 |\psi_2|^2 S_2^A \; . \tag{10.35}$$

This form of the current corresponds to a mixed state. This simultaneously solves the problem with the double standard of time, by using a time function t which coincides with t_1 along the tube y_1, and with t_2 along the tube y_2 . However, why should nature select for us the prefactors which kill the interference terms?

II) One can argue with Vilenkin[165] that for $S_1 \neq S_2$ and for a large M, the $\sin M(S_2 - S_1)$ and $\cos M(S_2 - S_1)$ are rapidly oscillating functions on a transverse hypersurface which, in the limit $M \to \infty$, average out to zero when integrated against the slowly varying functions $\Phi_1\Phi_2$ and $\psi_1\psi_2^*$. However, M, though large, is actually finite, and narrow interference fringes still remain, similar to the interference fringes of a massive particle in the double-slit experiment. Given a sufficiently fine resolution, the pure state (10.30) can be distinguished from the mixed state (10.35).

• **Negative Norms.** The greatest puzzle surrounding the semiclassical interpretation is how it can avoid the negative probabilities inherent in the Klein-Gordon product. This puzzle is highlighted by Eq. (10.21) whose left-hand side is indefinite and the right-hand side is positive definite.

A closer scrutiny reveals the trick. The semiclassical interpretation achieves the positivity in two steps: 1) by restricting the space of solutions of the Klein-Gordon equation to the states of the form (10.5) with a *single* Jacobi principal function S(Q), and 2) by choosing a hypersurface $t(Q) = 0$ transverse to the classical trajectories of S(Q) and orienting it in the direction of the vector field $S^A(Q) : t_A(Q)S^A(Q) > 0$.

The positivity of the norm is thus bought at the price of a serious infringement of the superposition principle. It is hard to see what physical mechanism selects only the states (10.5) with a single S(Q) and rejects all other possible solutions of the Klein-Gordon constraint. Characteristically, the discussion of the semiclassical interpretation of the state function of the Universe is often connected with cosmological speculation about the uniqueness of this state (Banks,[164] Vilenkin[182]). (An interpretation of an "arbitrary" single state was proposed by Padhanabhan[183] and Greensite[184].)

If one accepts an argument which removes the interference terms from Eq. (10.30), some superpositions (10.29) become permissible. However, one can never include all the states and still maintain the positivity. This is especially clear when one tries to superimpose classical states with the Jacobi functions S and – S.

If the Hamilton-Jacobi equation (10.6) has a solution S(Q), it necessarily also has a solution – S(Q). For a relativistic particle, one of those solutions may be excluded by the requirement that the particle must have a positive energy. No such requirement exists in vacuum geometrodynamics, where both S[g] and – S[g] are allowed, and stand to each other as an expanding versus a contracting Universe. I base the discussion on the time function (10.10) oriented in the direction of $S^A(Q)$.

It is easy to give examples of states with zero norm. The real state $\Psi = \Phi(e^{iMS} + e^{-iMS})\psi$, where ψ solves the real equation (10.33) instead of the Schrödinger equation, is such a state. The real state $\Psi = \Phi\psi e^{iMS} + \Phi\psi^* e^{-iMS}$, where ψ solves the Schrödinger equation associated with S and hence ψ^* solves the time-reversed Schrödinger equation associated with – S, is another such state. The state $\Phi\psi e^{iMS}$ taken by itself has a positive norm, the state $\Phi\psi^* e^{-iMS}$ has a negative norm. As long as the probabilistic interpretation of semiclassical theory is derived from the Klein-Gordon product, as in Eq. (10.21), one cannot allow the Hilbert space to be spanned by the semiclassical states corresponding to *all* possible Jacobi principal functions. At least one half of such states must be excluded. The semiclassical interpretation by itself does not give any hint what collection of states spans the Hilbert space and what states are to be left out.

• **Higher Corrections Problem**. The expectation that the semiclassical approximation solves the problem of time was raised by the emergence of the first-order Schrödinger equation from the second-order Klein-Gordon equation.

As pointed out by Barbour and Smolin,[97] the Schrödinger equation does not automatically follow in the M^0 order, and there are solutions to this order which satisfy an alternative first-order equation, Eq. (10.32). In the M^{-1} order, the situation gets worse. I pointed out that the equation for the first-order correction $\psi_{(1)}$ to the state function ψ contains an inhomogeneous term. This term appears because the Klein-Gordon constraint is a second-order, rather that the first-order, differential equation. The Schrödinger equation does not survive to the M^{-1} order. The quantum gravitational corrections to the Schrödinger equation for matter fields were discussed by Kiefer and Singh.[185]

• **Back Reaction and the Problem of Time.** The back reaction term was included into semiclassical approximation by amending the equation for the prefactor by the expectation value of the source term. Irrespective of whether such a procedure is or is not justified, it only further complicates the issue of time. It replaces the Schrödinger equation (10.26), or its model counterpart (10.14), by an integro-differential equation for the state ψ. This equation is nonlinear, and hence does not satisfy the superposition principle. The difficulties associated with such an equation were analyzed by several authors.[186-191]

▪ **Semiclassical Interpretation of Quantum Gravity.** The interpretation I have discussed so far is not directly applicable to quantum gravity. To maintain that quantum gravity has a probabilistic interpretation only if the Universe is in a semiclassical state (10.22) would literally mean that *quantum* gravity has a probabilistic interpretation only if it is *classical.* Obviously, the literal interpretation must be amended. This is achieved by the qualification that not all gravitational variables need to be classical; only those which represent an internal clock must become classical, while the true gravitational degrees of freedom should stay quantized. Unfortunately, this requires a split of the gravitational variables into those which represent time and become classical, and those which represent dynamics and remain quantum. Such a split carries to the semiclassical interpretation all the problems which beset the internal Schrödinger interpretation. I shall first discuss two major attempts to introduce a split into semiclassical interpretation, and then point out their difficulties.

▪ **Inhomogeneous Perturbations of a Homogeneous Universe.** The first attempt (Halliwell and Hawking[48]) assumes that the Universe is nearly homogeneous. Small inhomogeneities propagate on the homogeneous background and influence the evolution of the homogeneous mode, as in the fourth model of Section 3. Halliwell[170] then assumes that the homogeneous mode is in the WKB state while the inhomogeneous perturbations are quantized. The time variable is associated only with the homogeneous mode.

▪ **Quantum Perturbations of a Classical Background.** The second attempt (Vilenkin[165]) splits the quantum part $q_{ab}(x)$ of the spatial metric from the classical part $Q_{ab}(x)$ point by point, the quantum part being multiplied by the small parameter κ :

$$g_{ab}(x) = Q_{ab}(x) + 2\kappa q_{ab}(x) \ . \tag{10.36}$$

As one expands the super-Hamiltonian (1.2) with respect to 2κ, the quantum part $q_{ab}(x)$ and its conjugate momentum enter the expanded constraint in the same order (namely, $(2\kappa)^0$) as the source-field terms. They can thus formally be grouped with the field terms in the semiclassical expansion (10.22) of the state functional Ψ. Vilenkin talks about the $q_{ab}(x)$ variables as "corresponding to a small subsystem of the Universe" and justifies the split (10.36) by saying that "the semiclassical character of the Universe and the smallness of the subsystem are both due to the fact that the Universe is large."

• **Quantum Gravity and Time in the Small.** The Halliwell proposal, unlike the Vilenkin proposal, reduces the many-fingered time Schrödinger equation (10.26) to an ordinary Schrödinger equation in the "cosmological time" associated with the homogeneous mode. The privileged foliation labeled by this single time variable is specified by the assumption that the geometry is "almost homogeneous" on its leaves. As in linearized gravity, this still allows a small flexibility to wiggle the leaves. To fix the leaves completely, one must impose some "gauge conditions" on the inhomogeneous modes.

Such a specification of what is meant by time in quantum gravity would be justifiable if the Universe were almost homogeneous down to the scales at which one expects quantum gravity to become significant, i.e., down to the scales comparable with the Planck length. This is blatantly not true. The Universe is inhomogeneous even on much longer scales. To specify the foliation in a highly inhomogeneous Universe by the requirement that its leaves are almost homogeneous makes no sense, and Halliwell's method of choosing the mode which remains classical is no longer applicable. Halliwell's method remains what it probably was intended to be from the very beginning, namely, an interesting model, but it does not solve the problem of time in semiclassical interpretation of quantum gravity.

The proponents of the semiclassical interpretation may retort that in an inhomogeneous Universe the foliation should be entirely fixed by a slicing ("coordinate") condition. I criticized such a draconian solution in connection with the problem of functional evolution (Section 4). Moreover, because so far no slicing condition was found which would be privileged over the others, one must face the multiple choice problem.

• **Perturbative Canonical Quantization?** It is not at all clear in what sense the classical part of the split (10.22) corresponds to the Universe in the large and the quantum part to its small subsystem; intuitively, the behavior of the Universe in the large should be associated with some collective, integrated, variables (analogous to the center of mass of a system of particles) and its behavior in the small with the deviation from the collective variables. On the other hand, the split (10.22) is purely local, made point by point in Σ, and the "smallness" is brought in by the formal insertion of a small parameter κ. There is no physical principle which would unambiguously determine the split of small quantum perturbations from a large classical background. This suggests that either the whole metric field is to be treated classically, or the whole of it should be quantized (Singh and Padhanabhan[160]). In the first case, there is no quantum gravity, and in the second case, there is no semiclassical interpretation. The formal split (10.21) attempts at the canonical level what perturbative "covariant" quantization tried to do at the spacetime level. The failures of the perturbative covariant quantization are well known.

• **Classical Modes and Decoherence.** Even if one could defend a particular split, the question still remains why variables on one side of this split become classical while variables on the other side of the split remain quantum. At this point, one often hears a decoherence argument: A particular set of variables becomes classical because their interaction with the environment makes them to decohere. This type of argument was invoked by Kiefer[177] as an explanation why the homogeneous mode in the Halliwell-Hawking split becomes classical. In particular, Kiefer argued that the scale factor ("the radius") of the Friedmann universe driven by a massive scalar field is rendered classical by its interaction with inhomogeneous modes. The trouble with this kind of argument is that there seems to be no specific input which would make it to work for a given mode and to fail for other modes. One would thus expect that in a typical nonlinear field theory, when one chooses some higher, inhomogeneous mode, and treats all the other modes as an "environment", the interaction of the environment would again lead to decoherence. The decoherece arguments in their present form thus seem to be ill suited to provide a selection principle which determines the split of the classical modes from the quantum ones. A similar objection can be raised against Halliwell's argument[178] that the expanding and contracting modes in the state function of the universe decohere. It is thus difficult to dispute away the issue of the negative norms.

• **Spacetime Problem.** As in the Klein-Gordon interpretation, the many-fingered time variable (10.25) of the semiclassical interpretation is to be constructed from the spatial metric. No spacetime scalar can be built in this way.

In summary, the semiclassical interpretation does not solve the standard problems of time. It merely obscures them by the approximation procedure and, on the way, creates more problems.

11. Third Quantization

For a relativistic particle moving on a dynamical background, the failure to define the positive-energy sector of the solution space is taken as an indication that the one-particle interpretation of the theory cannot be consistently maintained because particles are produced and destroyed by the changing metric and/or the changing rest mass of the particle. This suggests that the quantum theory of a relativistic particle on a dynamical background makes sense only on an extended Hilbert space incorporating the states with an arbitrary number of particles. Formally, this is achieved by *second quantization*, in which the wave functions ψ satisfying the Klein-Gordon constraint are turned into quantum field operators. These operators can be thought of as acting on the state functionals $\Psi[\psi]$ of the fields. At least in simple cases, this representation is equivalent to the Fock space representation. Such a standpoint is, of course, experimentally corraborated in a closely related situation, namely, when the particle is an electron and the changing background is realized by external electromagnetic potentials.

The geometry $\mathbf{g}$ moving in superspace is analogous to a relativistic particle moving on a dynamical background. The role of the background is played by DeWitt's supermetric $G_{abcd}[g]$ and by the scalar curvature term $-|g|^{1/2}R[g]$; the dynamical nature of this background is demonstrated by a theorem proved by Kuchař.[159] By persuing the analogy with a relativistic particle, one can surmise that the dynamical background in superspace produces geometries in the same way that the dynamical background in spacetime produces particles. If so, the one-geometry interpretation of quantum geometrodynamics cannot be consistently maintained and one should work in an extended Hilbert space incorporating many-geometry states. Heuristically, one should turn the state functionals $\Psi[\mathbf{g}]$ which satisfy the Wheeler-DeWitt equation into quantum fields $\hat{\Psi}[\mathbf{g}]$ on superspace that act on some higher-order functionals $\tilde{\Psi}[\Psi[\mathbf{g}]]$.

A state functional $\Psi[\mathbf{g}]$ which satisfies the Wheeler-DeWitt equation already describes a state of the quantum gravitational *field*, i.e., of a second-quantized zero mass relativistic particle of spin 2. (Similarly, a state functional $\Psi[X,\mathbf{B}]$ that satisfies the constraint equations of parametrized Maxwell theory describes a state of the quantum electromagnetic *field*, i.e., of a second-quantized zero mass relativistic particle of spin 1.) The process in which the state functional $\Psi[\mathbf{g}]$ is turned into an operator thus deserves to be called *third quantization*. The logical possibility of third quantization was noticed by Kuchař[159] and taken as a serious physical option by Giddings and Strominger,[192] Coleman,[193] McGuigan,[194,195] and others.

The question which interests me here is whether third quantization can help to resolve the problems of time in quantum gravity. Unfortunately, very little work has been devoted to this question. The papers on the subject have developed the formalism rather than consider its interpretation. In this state of affairs, I cannot do much more than offer some rather general remarks about the conceptual difficulties of third quantization as an interpretational framework.

- **Does Third Quantization Solve the Problem of Time?** The expectation that third quantization solves the problem of time in quantum gravity is based on a belief that the second quantization solves the problem of time in quantum theory of a relativistic particle. Before going any further, it is advisable to check whether this belief is substantiated.

The second quantization approach to a quantum field theory is based on the construction of a Fock space. To build the Fock space, one takes a one-particle Hilbert space $\mathscr{F}_{(1)}$. From the direct product of the one-particle states $|\psi\rangle \in \mathscr{F}_{(1)}$, one builds the states

$$|\Psi_N\rangle = S\, |\psi_1\rangle \otimes |\psi_2\rangle \otimes \cdots \otimes |\psi_N\rangle \tag{11.1}$$

which span the N-particle sector $\mathscr{F}_{(N)}$. Symmetrization S under exchange of the particles is appropriate for Bose statistics. The Fock space $\mathscr{F}$ is the direct sum of such sectors,

$$\mathscr{F} = \mathscr{F}_{(0)} \oplus \mathscr{F}_{(1)} \oplus \mathscr{F}_{(2)} \oplus \cdots \, . \tag{11.2}$$

Here, $\mathscr{F}_{(0)}$ is spanned by the vacuum state $|\Psi_0\rangle$.

The norm of the special N-particle state (11.1) is induced by the norm of its one-particle constituents:

$$\langle\Psi_N|\Psi_N\rangle := \prod_{i=1}^{N} \langle\psi_i|\psi_i\rangle \, . \tag{11.3}$$

The norm of a state

$$|\Psi\rangle = |\Psi_0\rangle \oplus |\Psi_1\rangle \oplus |\Psi_2\rangle \oplus \cdots \tag{11.4}$$

in $\mathscr{F}$ is the sum

$$\langle\Psi|\Psi\rangle = \Sigma_{N=0}^{\infty} \langle\Psi_N|\Psi_N\rangle \tag{11.5}$$

of the norms of its N-particle components.

This short description makes it clear that one can define the Fock space $\mathscr{F}$ only if one knows the one-particle space $\mathscr{F}_{(1)}$ *as a Hilbert space*. This brings us back to the Hilbert space problem for a relativistic particle.

If the particle moves on a stationary background, the one-particle Hilbert space is spanned by the positive-energy solutions of the Klein-Gordon constraint (9.1) - (9.2), and the inner product is given by Eq. (9.5). If the particle moves on a dynamical background, it is not clear what one should take for the one-particle Hilbert space. If the background is stationary in the past or stationary in the future, one can construct the corresponding incoming and outgoing Hilbert spaces and answer scattering questions. However, one does not know how to pose meaningful questions in the dynamical region.

The absence of a privileged one-particle Hilbert space structure is the source of ambiguities in constructing a unique quantum field theory (even a free field theory) on a dynamical background. There are many ways of reformulating this problem, and in some of these the connection with the time problem (the Hilbert space problem) of the one-particle theory is far from obvious. It is then easy to start believing that the problem of time disappeared from the second-quantized theory and that one should merely fix some other technical problems. What I want to stress is that in one way of looking at things, the ambiguities of quantum field theory on a dynamical background are the direct consequence of the Hilbert space problem of the one-particle theory, and thus reflect the problem of time for the second-quantized system. The second quantization merely shifts the problem of time to a different level without really solving it. This is not a good omen for the attempts to solve the time problem in quantum gravity by third quantization.

• **How to Observe the Superfield?** Suppose that the ambiguities of a quantum field theory on a dynamical background were somehow resolved without having to solve the Hilbert space problem for the one-particle system. Let me further grant that the same scheme could be successfully applied to third-quantized gravity. How would one interpret such a theory?

The states of a quantum field can be given either in the *field representation* or in the *particle representation*. These representations reflect the choice of fundamental observables. The question is: How does one measure these observables?

The fundamental observables in the field representation are the spacetime field operators, the scalar field $\hat{\psi}(X)$, the electromagnetic field $\hat{F}_{\alpha\beta}(X)$, etc., or their canonical counterparts $\hat{\psi}(x)$, $\hat{\pi}(x)$; $\hat{\mathbf{E}}(x)$, $\hat{\mathbf{B}}(x)$, etc. The measurability of the electric and magnetic field operators played a major role in clarifying the conceptional foundations of quantum field theories.[121] The construction of appropriate thought experiments required a truly heroic effort.[196] I suspect that an equally heroic effort would be needed to understand how one should measure the scalar field $\hat{\Psi}[\mathbf{g}]$ on superspace. Indeed, the ordinary experimental devices operate in *spacetime* and are designed to measure spacetime quantities, like the average field strength in a spacetime region. Nobody as yet has imagined a piece of equipment which would be able to measure something like the mean value of the field $\hat{\Psi}[\mathbf{g}]$ in a region of superspace.

• **How Could a Many-Geometry State be Observed?** If it is difficult to imagine an apparatus which would measure the "field observable in superspace," perhaps it is easier to imagine an apparatus which would measure the "particle observable in superspace." A complete set of particle observables in spacetime is the number of particles and their positions at a given instant, i.e., on a given spacelike hypersurface. The apparatus which performs such a measurement is a collection of small Geiger counters densely distributed throughout space and brought into the state of readiness at a given instant of time. Notice that this is the *same* apparatus as that designed to measure the position of a *single* particle; the fact that the apparatus operates on a state in the Fock space rather than on a one-particle state manifests itself by more than one Geiger counter being triggered, i.e., in the *response* of the apparatus, rather than in its *construction*.

In third-quantized gravity, the many-particle states are replaced by many-geometry states. However, the analogy is not perfect. A three-geometry is not analogous to the position of a particle, because it contains also information about intrinsic time. This brings the problem of time back into the game. To be able to talk about measurements of the three-geometry rather than about measurements of a more complicated object (say, the conformal three-geometry), I can use the results of section 7, and couple gravity to a suitable material system, like a Gaussian reference fluid. (In a classical phenomenological discussion, one traditionally uses for the same purpose a system of small rigid rods dragged by the reference points,[197] or perfect clocks carried by the reference points together with the radar signals.[198]) It is fairly clear what response of such a material system considered as a classical apparatus corresponds to the detection of a definite three-geometry. If one takes the analogy with the Geiger counters seriously, the same material system should be able to detect also a many-geometry state. I must admit, however, that I am unable to imagine a classical response of the material systems like rods, clocks, signals, or a Gaussian reference fluid, which could be appropriately described by saying that on a single spacelike hypersurface one has detected two or more intrinsic geometries.

• **Changing Topology**. An alternative interpretation of the many-geometry state may be that the individual geometries $\mathbf{g}_1$ and $\mathbf{g}_2$ are not carried by a single hypersurface, but by *different* copies of Σ. Isham discusses the difficulties of this version of the interpretation.[96] First, the transition from a one-geometry state to a two-geometry state corresponds now to a topology change in which Σ bifurcates into two copies of itself. It is doubtful whether such a change is compatible with the third-quantized Wheeler-DeWitt formalism. Second, the Bose structure of the Fock space loses its operational significance. The symmetry of the two-geometry state should reflect the indistinguishability of the individual geometries. However, if the two copies of Σ are disjoint, there is presumably no causal connection between them, and it does not make sense to compare the two geometries with each other. (In the same way, it would not make much sense to talk about two particles as being indistinguishable if each of them moved in a separate space rather than in the same space.)

In summary, it seems that the third quantization does not really solve the problem of time, and that it is extremely hard to interpret its formalism operationally.

12. Naive Schrödinger Interpretation

Of all the proposals on how to interpret quantum gravity, the naive Schrödinger interpretation is certainly the most straightforward. It asserts that the square

$$|\Psi[\mathbf{g}]|^2 \tag{12.1}$$

of a solution $\Psi[\mathbf{g}]$ of the Wheeler-DeWitt equation (2.5) and the supermomentum constraint (2.4) is the probability density for finding a hypersurface with the intrinsic geometry $\mathbf{g}$. The designation "naive" (coined by Wald and Unruh[76]) distinguishes it from the internal Schrödinger representation, which never speaks about probabilities before splitting the dynamical variables from the many-fingered time. By denying the need for such a split, the naive Schrödinger interpretation is the first scheme which maintains that one does not need to know what is time before interpreting quantum gravity.

The naive Schrödinger interpretation was proposed by Hawking[199-203] and widely used by the members of his school. The conditional interpretation and the sum-over-histories interpretation, which I consider in the following two sections, may be considered as refinements of the naive Schrödinger interpretation. Here I shall discuss the naive Schrödinger interpretation in its unadulterated (and thus intuitively appealing) form.

The conceptual meaning of the naive Schrödinger interpretation is clarified when one applies it to a simple particle model. For a parametrized Newtonian or relativistic particle, the naive Schrödinger interpretation amounts to the statement that

$$|\Psi(T,Q)|^2 \tag{12.2}$$

is the probability density for finding an instant T with the particle in the position Q. The peculiar timeless nature of the interpretation is brought thereby into a sharp focus.

• **Naive Interpretation Lacks Dynamics.** A comparison of Eqs. (12.1) and (12.2) reveals that an attempt to normalize the state functional over the geometries **g** is analogous to an attempt to normalize the wave function of a particle over both the position and the time. Such an attempt must fail because, due to the Schrödinger or the Klein-Gordon equation, the wave function cannot be square integrable in the time direction. One is thus forced to talk only about relative probabilities.

I shall consider some typical probabilities which the interpretation (12.2) allows me to calculate for a Newtonian particle. I normalize $\Psi(T,Q)$, as in the standard quantum mechanics, to 1 at a given instant T:

$$\rho(T) := \int dQ \; |\Psi(T,Q)|^2 = 1. \tag{12.3}$$

The Schrödinger equation then implies that $\rho(T) = 1$ for all T. The naive Schrödinger interpretation allows me to reinterpret the expression (12.3) as the *relative probability to find an instant T*. One sees that in nonrelativistic quantum mechanics all instants are equally probable; to find one instant is as probable as to find any other. Similarly,

$$\rho(Q) := \lim_{T'\to\infty} \frac{1}{2T'}\int_{-T'}^{T'} dT \; |\Psi(T,Q)|^2 \tag{12.4}$$

is *the probability density to find the particle at Q irrespective of when*. Finally, I can write the expression

$$\langle\, Q \,\rangle := \lim_{T'\to\infty} \frac{1}{2T'}\int_{-T'}^{T'} dT \int dQ \; Q\,|\Psi(T,Q)|^2 \tag{12.5}$$

for *the mean position of the particle during the whole motion.*

One sees that the probabilistic scheme (12.3) - (12.5) is internally coherent. Thus, e.g., taking an ensemble of systems described by the state function Ψ, and measuring Q without measuring T, one is, according to Eq. (12.3), as likely to measure Q at one instant as at any other instant; by averaging all such Q's obtained at different instants T, one ends with the mean value formula (12.5).

However, one is missing the answer to a typical dynamical question which one can ask in the internal Schrödinger interpretation, namely "If one finds the particle at Q at the time T, what is the probability of finding it at Q′ at the time T′ > T?" Such questions make sense and can be realized by suitable experimental arrangements. Thus, the naive Schrödinger interpretation simply fails to ask and answer interesting dynamical questions.

Returning back from the particle model to quantum gravity, one sees that the naive Schrödinger interpretation can pose some questions, like "What is the mean value of this functional of the intrinsic geometry?", but these are not dynamical questions. A typical application of the naive Schrödinger interpretation is "To find the probability of this or that Universe" (leading to statements like "A Universe with a given density of mass is very likely flat and big") but *not* "To find the probability of this or that *evolution* of the same Universe." One of the main defects of the naive Schrödinger interpretation is that it does not describe how the Universe is changing.

- **Inappropriate for the Wheeler-DeWitt Equation.** The naive Schrödinger interpretation for a Newtonian particle is compatible with the internal Schrödinger interpretation, as exemplified by Eqs. (12.3) - (12.5). This can no longer be maintained for a relativistic particle, even on a Minkowskian background. Because the Schrödinger norm is no longer conserved by the Klein-Gordon evolution, Eq. (12.3) breaks down, and different instants of the Minkowskian time are not equally probable. Further, the multiplication operator $\hat{Q} = Q\times$ is not the position operator of a relativistic particle, and the expression (12.5) loses its meaning of the mean position. This shows that the naive Schrödinger interpretation is inappropriate for a relativistic particle. The Wheeler-DeWitt equation is certainly closer to the Klein-Gordon equation than it is to the Schrödinger equation. The naive Schrödinger probability (12.1) does not seem to be appropriate for it either.

- **Violation of the Constraints.** The interpretation of the expression (12.1) as the probability density of the intrinsic geometry $\mathbf{g}$ is consistent with the standard rules of quantum mechanics only if the geometry $\mathbf{g}$ is represented by the multiplication operator. However, the multiplication operator $\hat{\mathbf{g}}(x) = \mathbf{g}(x)\times$ throws the state $\Psi[\mathbf{g}]$ out of the physical space of states which satisfy the Wheeler-DeWitt equation. In other words, a measurement of $\mathbf{g}(x)$ violates the constraints. The naive Schrödinger interpretation avoids this issue because it refrains from making any statement about what is the state function of the Universe after one has found a hypersurface with the geometry $\mathbf{g}(x)$. If one believes that the intrinsic geometry is a carrier of information about time,[2] one should recall the old arguments why time should not be represented by a self-adjoint operator in quantum mechanics.

13. The Conditional Probability Interpretation

The conditional probability interpretation is an attempt to introduce dynamics into the naive Schrödinger interpretation. This requires an identification of time. However, the advocates of the interpretation hold that it is unnecessary to cast the constraints into the Schrödinger form with respect to any specific choice of time. Instead, they work within the Wheeler-DeWitt framework and assert that they can handle all choices of time at once without running into the multiple choice problem. They also feel that the lack of global time function is not necessarily a problem. In these respects, the conditional probability interpretation substantially differs from the internal Schrödinger interpretation.

The conditional probability interpretation was developed by Page and Wootters.[107,204-208] I sidestep those aspects of their discussion which relate the interpretation to the unobservability of the label time and the superselection rule for the conjugate energy.

Like the naive Schrödinger interpretation, the conditional probability interpretation works with the Hilbert space $\mathscr{F}$ of square-integrable functions on the configuration space(time) of the unconstrained system. The interpretation assigns probabilities to the measurements of the projection operators $\hat{A}$, $\hat{B}$, $\hat{C}$ on $\mathscr{F}$. It is well known that any projection operator

$$\hat{A}^2 = \hat{A} \tag{13.1}$$

has only two eigenvalues, 0 and 1. The eigenfunctions $|\Phi_0\rangle$ and $|\Phi_1\rangle$ to those eigenvalues span two mutually orthogonal linear subspaces of $\mathscr{F}$, $\mathscr{F}_{A=0}$ and $\mathscr{F}_{A=1}$: $\mathscr{F} = \mathscr{F}_{A=0} \otimes \mathscr{F}_{A=1}$. The measurement of A amounts to answering the yes - no question "Does $|\Psi\rangle$ lie in $\mathscr{F}_{A=1}$?" The eigenvalue A=1 means that the answer is "Yes", the eigenvalue A=0 that the answer is "No."

Let $|\Psi\rangle \epsilon\ \mathscr{F}_0 \subset \mathscr{F}$ be a physical state, i.e., a solution of the super-Hamiltonian constraint

$$\hat{H}|\Psi\rangle = 0\ . \tag{13.2}$$

The fundamental postulate of the conditional probability interpretation is the formula for the probability P(B|C) that the answer to the question $\hat{B}$ is yes if the answer to the question $\hat{C}$ is yes:

$$P(B|C) = \langle\Psi|\hat{C}\hat{B}\hat{C}|\Psi\rangle / \langle\Psi|\hat{C}|\Psi\rangle. \tag{13.3}$$

The connection with the time problem is established by constructing a projection operator $\hat{C}_{T'}$ corresponding to the question "Does an internal clock show the time T'?" (I chose the letter C for its two connotations: "the clock" and "the condition.")

The way in which the conditional probability interpretation works is best illustrated on simple models. The first model is the parametrized Newtonian particle moving in one dimension,

$$H := P_T + \frac{1}{2m} P_Q{}^2 + W(Q) = 0 . \tag{13.4}$$

The second model is a system of two non-interacting particles, each of them with a single degree of freedom, kept in a stationary state:

$$\begin{aligned} H &:= h_T + h_Q - E \\ &:= \frac{1}{2M} P_T{}^2 + V(T) + \frac{1}{2m} P_Q{}^2 + W(Q) - E = 0 . \end{aligned} \tag{13.5}$$

The position T of the first particle is cast in the role of the hand of a clock, and is thus called T.

The nature of the conditional probability interpretation is so clear from these examples that it is hardly necessary to spell out how the formalism looks in quantum gravity.

■ **Parametrized Newtonian Particle.** This model is ideal for showing how the naive Schrödinger interpretation follows from the conditional probability interpretation. Let $\mathcal{F}_T$ be the Hilbert space of square-integrable functions of T, and $\mathcal{F}_Q$ the Hilbert space of square-integrable functions of Q. One constructs the projection operator

$$\hat{C}_{T'} := |T'\rangle\langle T'| \tag{13.6}$$

in $\mathcal{F}_T$ and the projection operator

$$\hat{B}_{Q'} := |Q'\rangle\langle Q'| \tag{13.7}$$

in $\mathcal{F}_Q$. When acting on $\mathcal{F} = \mathcal{F}_T \otimes \mathcal{F}_Q$, $\hat{C}_{T'}$ acts as an identity operator on $\mathcal{F}_Q$, and

$\hat{B}_{Q'}$ as an identity operator on $\mathcal{F}_T$. Then $\hat{C}_{T'}$ is the questions "Is it T′ o'clock?" and $\hat{B}_{Q'}$ is the question "Is the particle at Q′?"

Let $|\Psi\rangle$ satisfy the constraint (13.2), i.e., let $\Psi(T,Q) := \langle T,Q|\Psi\rangle$ solve the Schrödinger equation. The conditional probability $P(B_{Q'}|C_{T'})$ that the particle is at Q′ at the time T′ can be calculated from the fundamental formula (13.3). The answer

$$P(B_{Q'}|C_{T'}) = |\Psi(T',Q')|^2 \Big/ \int_R dQ\ |\Psi(T',Q)|^2 \tag{13.8}$$

is, up to the normalization factor, the same as that given by the naive Schrödinger interpretation, Eq. (12.2).

For the future discussion, note that the yes answer to the question $\hat{C}_{T'}$ determines a unique state in $\mathcal{F}_T$ (i.e., the subspace $\mathcal{F}_{T\,\hat{C}_{T'}=1} \subset \mathcal{F}_T$ is one-dimensional), and similarly the yes answer to the question $\hat{B}_{Q'}$ determines a unique state in $\mathcal{F}_Q$. (Actually, due to the continuous spectrum of the operators $\hat{T}$ and $\hat{Q}$, these states are "improper" states which lie outside the respective Hilbert spaces $\mathcal{F}_T$ and $\mathcal{F}_Q$.) However, when $\hat{C}_{T'}$ is considered as an operator in $\mathcal{F} = \mathcal{F}_T \otimes \mathcal{F}_Q$, the yes answer to the question $\hat{C}_{T'}$ means that the system lies in the infinitely-dimensional subspace $\mathcal{F} = \mathcal{F}_T \otimes \mathcal{F}_Q$ of $\mathcal{F}$ because nothing is yet known about the position of the particle. The operators $\hat{C}_{T'}$ and $\hat{B}_{Q'}$ commute on $\mathcal{F}$.

▪ **Two Particles in a Stationary State.** One can apply the same strategy to the model (13.5). Let $\mathcal{F}$ be the Hilbert space of square-integrable functions on the configuration plane (T,Q) and $\mathcal{F}_o$ be the physical space of states satisfying the super-Hamiltonian constraint (13.2). The projection operators $\hat{C}_{T'}$ and $\hat{B}_{Q'}$ are defined as before, Eqs. (13.6) - (13.7). They represent the questions "Is the first particle at T′?" and "Is the second particle at Q′?" The conditional probability $P(B_{Q'}|C_{T'})$ that the second particle is at Q′ if the first particle is at T′ is again given by Eq. (13.8). This raises great expectations. The quadratic constraint (13.5) does not offer a privileged time like the Schrödinger constraint (13.4) does. The role of T and Q can easily be interchanged. Nevertheless, Eq. (13.8) seems to provide a probabilistic interpretation of the physical states $|\Psi\rangle \epsilon \mathcal{F}_o$. It is not necessary to say beforehand what is time. It is not necessary for T to be a global time variable. (Indeed, for many potentials V(T) and W(Q), as Hájíček's results indicate, it will not be a global time variable.) It is not necessary to cast the constraint (13.5) into a Schrödinger form before imposing it on the physical states. Equation (13.3) seems to be able to treat all conditions $\hat{C}$ (all times) at once and on the same footing. All of this seems to suggest that the problems of time are pseudoproblems.

■ **The Page-Wootters Treatment of the Model.** In my discussion of the system (13.5), I took the position of the first particle as a clock and the position of the second particle as a dynamical variable. Page and Wootters[205] generalize this procedure somewhat. To use the first particle as a clock, they choose a state $|\phi_0\rangle \in \mathscr{F}_T$ and define the projector

$$\hat{C}_0 := |\phi_0\rangle \langle\phi_0| \, . \tag{13.9}$$

By evolving the state $|\phi_0\rangle$ by the super-Hamiltonian $\hat{H}$, they arrive at a one-parameter family of projectors

$$\hat{C}_\tau := e^{-i\hat{H}\tau} \, \hat{C}_0 \, e^{i\hat{H}\tau} = e^{-i\hat{h}_T\tau} \, \hat{C}_0 \, e^{i\hat{h}_T\tau} \, . \tag{13.10}$$

They interpret $\hat{C}_0$ as the question "Does the clock show the time 0?", $\hat{C}_\tau$ as the question "Does the clock show the time τ ?", and the parameter τ as time.

Note that this interpretation is consistent only if $\hat{C}_0$ is *not* a constant of the motion:

$$[\hat{C}_0, \hat{H}] = [\hat{C}_0, \hat{h}_T] \neq 0 \, . \tag{13.11}$$

If $\hat{C}_0$ were a constant of the motion, Eq. (13.10) would yield $\hat{C}_\tau = \hat{C}_0$, i.e., the particle clock would not run.

The dynamical variable to be measured is a projector $\hat{B} = B(\hat{Q},\hat{P}_Q)$ constructed from the variables of the second particle. Take now a normalized state $|\Psi\rangle \in \mathscr{F}_O$ and calculate the probability $P(B|C_\tau)$ that the answer to $\hat{B}$ is yes at the clock time τ. Because $\hat{B}$ commutes with $\hat{C}_\tau$, $\hat{B}$ commutes with $\hat{h}_T$, $\hat{C}_\tau$ commutes with $\hat{h}_Q$, and $\hat{H} := \hat{h}_T + \hat{h}_Q - E$ annihilates the state $|\Psi\rangle$, one gets

$$P(B|C_\tau) = \langle\psi_0|B_\tau|\psi_0\rangle \, / \, \langle\psi_0|\psi_0\rangle \, , \tag{13.12}$$

where

$$|\psi_0\rangle := \langle\phi_0|\Psi\rangle \in \mathscr{F}_Q \tag{13.13}$$

describes the state of the second particle and

$$\hat{B}_\tau := e^{i\hat{h}_Q\tau} \, \hat{B} \, e^{-i\hat{h}_Q\tau} \tag{13.14}$$

satisfies the Heisenberg equation of motion

$$\partial_\tau \hat{B}_\tau = \frac{1}{i} [\hat{B}_\tau, \hat{h}_Q] \tag{13.15}$$

in the parameter τ. The conclusion drawn by Page and Wootters is that the conditional probability (13.12) depends on the time τ as if the dynamical variable $\hat{B}_\tau$ evolved according to Eq. (13.15) in spite of the fact that $|\Psi\rangle$ is a *stationary state* (13.2) of the total system.

The conditional probability interpretation does not really save the naive Schrödinger interpretation. As far as it entails the naive Schrödinger interpretation, it shares its difficulties. As far as it reaches beyond it, by making temporal statements possible, it brings its own defects from hiding into sharp focus.

- **Inappropriate for a Klein-Gordon System.** When applied to a parametrized relativistic particle on a Minkowskian background, the conditional probability interpretation makes the prediction (13.8) for the particle to be found at the position Q' at the Minkowski time T'. This prediction is different from the accepted expression for the probability of the localization of a relativistic particle.

- **Violation of the Constraints.** The formula (13.3), which the conditional probability interpretation takes as its fundamental postulate, is ordinarily derived from the rules governing measurements performed on a system by an outside apparatus. The ensemble $\mathscr{E}$ of systems is prepared in a state $|\Psi\rangle \epsilon \mathscr{F}_0$. First, $\hat{C}$ is measured on $\mathscr{E}$. Those systems for which the answer to the question $\hat{C}$ is yes are collected into a new ensemble $\mathscr{E}_{C=1}$. This ensemble is described by the state function

$$|\Psi_C\rangle = \hat{C}|\Psi\rangle \tag{13.16}$$

that is the projection of the initial state $|\Psi\rangle$ into the subspace $\mathscr{F}_{C=1} \subset \mathscr{F}$ spanned by the eigenfunctions of $\hat{C}$ to the eigenvalue 1. (The selection of the subensemble $\mathscr{E}_{C=1}$ of the original ensemble $\mathscr{E}$ is sometimes called - appropriately or not - the reduction of the wave packet.) Second, $\hat{B}$ is measured on $\mathscr{E}_{C=1}$. If the answer to the question $\hat{B}$ is yes, the system falls within the ensemble $\mathscr{E}_{B=1}$ described by the state function

$$|\Psi_B\rangle = \hat{B}|\Psi_C\rangle \, . \tag{13.17}$$

That is the projection of $|\Psi_C\rangle$ into the subspace $\mathscr{F}_{B=1} \subset \mathscr{F}$ spanned by the eigenfunctions of $\hat{B}$ to the eigenvalue 1. The conditional probability (13.3), defined as the ratio of the number of systems in $\mathscr{E}_{B=1}$ and the number of systems in $\mathscr{E}_{C=1}$, is given by the norm of $|\Psi_B\rangle$ divided by the norm of $|\Psi_C\rangle$.

From this description of the steps by which Eq. (13.3) is ordinarily derived it becomes clear that the conditional probability interpretation violates the constraints. The operators $\hat{C}$ and $\hat{B}$ act on $\mathscr{F}$ and they are not required to commute with the constraint $\hat{H}$. Indeed, I pointed out that $\hat{C}$ can play the role of a clock only if it does not commute with $\hat{H}$. Similarly, if the aim of the conditional probability interpretation is to describe how the Universe is changing, one must watch some $\hat{B}$'s which do not commute with $\hat{H}$, i.e., which are not constants of the motion.

However, if $\hat{C}$ does not commute with $\hat{H}$, the measurement of $\hat{C}$ will in general throw the state $|\Psi\rangle$ out of the physical space, $|\Psi_C\rangle \notin \mathscr{F}_0$. Even if the state $|\Psi_C\rangle$ were found to be in $\mathscr{F}_0$, the measurement of $\hat{B}$ would in general throw it out of the physical space, $|\Psi_B\rangle \notin \mathscr{F}_0$. The conditional probability interpretation is thus based on a postulate whose derivation violates the constraints.

Page counters this objection by asserting that the intermediate state (13.16) is "actually only a calculational tool useful for writing the conditional probability in the simple form" and that "the 'collapse' of the wavefunction ... should be merely thought of as part of one's computational process, not as a physical process, so it should not be surprising that the result is rather 'unphysical'."[206]

I fear that such an attempt to brush aside the violation of the constraint as a mere verbal artifice does not stand closer scrutiny. I shall try to show that this mechanism of violating the constraints is actually responsible for the fact that the conditional probability interpretation predicts *physically wrong* propagators for the simplest model system, namely, the parametrized Newtonian particle.

• **Wrong Propagators**. Curiously enough, the conditional probability interpretation was used by its proponents only to calculate the conditional probabilities of $\hat{B}$ at a *single* instant $\hat{C}_T$, but never applied to answering the fundamental *dynamical* question of the internal Schrödinger interpretation, namely, "If one finds the particle at Q′ at the time T′, what is the probability of finding it at Q″ at the time T″ > T′?"

Let me pose this question for a parametrized Newtonian particle within the framework of the conditional probability interpretation. Let $\hat{C}_{T'}$ and $\hat{B}_{Q'}$ be the projectors defined by Eqs. (13.6) - (13.7). Then their product

$$\hat{A}_{T'Q'} := \hat{B}_{Q'}\,\hat{C}_{T'} \tag{13.18}$$

is the projector expressing the question "Is the particle at Q′ at the time T′?" Because the projectors $\hat{B}_{Q'}$ and $\hat{C}_{T'}$ commute on $\mathscr{F}$, the definition (13.18) makes sense. Note that while the projectors $\hat{B}_{Q'}$ and $\hat{C}_{T'}$ taken separately generate multidimensional subspaces $\mathscr{F}_{B_{Q'}=1}$ and $\mathscr{F}_{C_{T'}=1}$ of $\mathscr{F}$, their product (13.18) generates a ray: $\mathscr{F}_{A_{T'Q'}=1}$ is a one-dimensional subspace of $\mathscr{F}$. (Of course, strictly speaking, none of the spaces $\mathscr{F}_{B_{Q'}=1}$, $\mathscr{F}_{C_{T'}=1}$, and $\mathscr{F}_{A_{T'Q'}=1}$ is a subspace of $\mathscr{F}$, because the operators $\hat{T}$ and $\hat{Q}$ on $\mathscr{F}$ have a continuous spectrum.)

The probability sought by the fundamental dynamical question is given in the conditional probability interpretation framework by the formula

$$P(A_{T''Q''}|A_{T'Q'}) = \langle\Psi|\hat{A}_{T'Q'}\ \hat{A}_{T''Q''}\ \hat{A}_{T'Q'}|\Psi\rangle\ /\ \langle\Psi|\hat{A}_{T'Q'}|\Psi\rangle\ . \qquad (13.19)$$

The physical answer to the fundamental dynamical question should, of course, be the square of the non-relativistic propagator. Instead, Eq. (13.19) yields

$$P(A_{T''Q''}|A_{T'Q'}) = |\delta(T'-T'')\delta(Q'-Q'')|^2\ . \qquad (13.20)$$

In brief, the conditional probability interpretation prohibits the absolute time to flow and the Newtonian particle to move!

The application of the fundamental formula to the Page-Wootters treatment of the two-particle model does not fare any better. For simplicity, take the initial state of the clock particle to be $|\phi_0\rangle = |T'\rangle$ and ask again about the position Q of the second particle. The projector

$$\hat{A}_{\tau'Q'} := \hat{B}_{Q'}\ \hat{C}_{\tau'}\ , \qquad (13.21)$$

with $\hat{C}_\tau$ given by Eq. (13.10), articulates the question "Is the second particle at the position Q′ at the clock time τ'?" The probability $P(A_{\tau''Q''}|A_{\tau'Q'})$ predicted by the conditional probability interpretation is

$$P(A_{\tau''Q''}|A_{\tau'Q'}) = |\langle T'|e^{-i\hat{h}_T(\tau''-\tau')}|T'\rangle|^2|\delta(Q'-Q'')|^2\ . \qquad (13.22)$$

Again, Eq. (13.22) prohibits the second particle to move. For $\hat{h}_T = \hat{P}_T$, Eq. (13.22) gives the old result (13.20).

One may wonder why Eq. (13.8) for the Newtonian particle seemed to replicate the standard quantum dynamics of Q′ with T′, while Eq. (13.20) miserably failed to do the same. The answer is that the measurements of $\hat{B}_{Q'}$ in Eq. (13.8) are done on an ensemble $\mathscr{E}_{C_{T'}=1}$ of particles whose time is known but the position Q′ is as yet undetermined, while the measurements of $\hat{A}_{T''Q''}$ in Eq. (13.20) are done on an ensemble $\mathscr{E}_{A_{T'Q'}=1}$ of particles that were already found to be at Q′ at the time T′. The measurement of $\hat{A}_{T'Q'}$ with the result 1 is a complete measurement which uniquely fixes the state $|T',Q'\rangle$ of the ensemble $\mathscr{E}_{A_{T'Q'}=1}$ reducing thereby the original physical state $|\Psi\rangle \epsilon \mathscr{F}_{\circ}$ to an unphysical state $|T',Q'\rangle \notin \mathscr{F}_{\circ}$ that has nothing to do with the original $|\Psi\rangle$. The unphysical result (13.20) is thus a direct consequence of the violation of the constraint by the first measurement, the violation which Page tried to dispute away as a purely computational device.

I regard Eq. (13.20) as a reduction ad absurdum of the conditional probability interpretation. If the interpretation does not yield an expected physical prediction for a parametrized Newtonian particle, how can one rely on it in quantum gravity? Of course, one can try to modify the conditional probability interpretation, say, by projecting the state back into the physical space $\mathscr{F}_{\circ}$ each time the measurement of a projector $\hat{A}$, $\hat{B}$, $\hat{C}$, ... brings it out of the physical space. I better abstain from analyzing the shortcomings of such a scheme before someone seriously proposes it.

14. Sum-Over-Histories Interpretation

In canonical quantization of a Newtonian system, path integrals are tools for calculating the propagator which determines the change of the state in the Hilbert space from one instant of time to another. The idea pursued by the sum-over-histories interpretations is that path integrals somehow make sense even outside the Hilbert space framework and provide an interpretation for quantum systems which do not automatically carry with them the notions of time and of the Hilbert space.

Such an approach to quantum gravity was advocated by Teitelboim who based on it a statistical interpretation of the scattering process (in proper time) in which the Universe starts in a velocity-dominated singularity and, encountering the scalar potential of the Wheeler-DeWitt equation, ends in a velocity-dominated singularity.[209-211] A more elaborate sum-over-histories interpretation of quantum gravity was offered by Hartle. Hartle's interpretation is a part of an ambitious program aiming at connecting quantum mechanics with the theory of initial conditions of the Universe.[75,130,131,169,212-217] I shall restrict my discussion strictly to those aspects of the program which deal with the problem of time.

At the technical level, Hartle used path integrals for reinterpreting ordinary nonrelativistic quantum mechanics,[212] and then applied the same scheme to the toy universe of two particles kept in a stationary state.[75] At a more heuristic level he argued that the main ideas of his interpretation can be generalized from the toy model to quantum gravity.[130,213,214,216] I shall summarize the essential features of Hartle's interpretation.

▪ **Sum-Over-Histories Interpretation for Two Model Systems.** The model systems studied by Hartle are the parametrized Newtonian particle (13.4), and two noninteracting particles, with coordinates T and Q, kept in a stationary state (13.5). The second model is thus the same as that used by Page and Wootters, and it is a specialization of the model (3.9) considered in the semiclassical interpretation.

The fundamental rules of the sum-over-histories interpretation are the following:

I) *Observables.* The basic observables are projectors associated with regions of the configuration space (T,Q). A detector measuring an observable registers if the system crosses the region at least once but possibly many times, and it does not register if the system avoids the region.

II) *Probability Amplitudes.* These are to be calculated by Feynman's standard prescriptions in which the super-Hamiltonian is treated as an ordinary Hamiltonian and the state is propagated in a label time τ. (For the super-Hamiltonian (13.4), the label time flows at the same pace as the internal Newtonian time T; for the model (13.5), it corresponds to the Newtonian time of an outside observer.)

III) The *label time* should be treated as an *unobservable label.* This means that the amplitudes must be summed over τ *before* being squared to yield joint probabilities. Both positive and negative τ's are allowed; because the label time order is unobservable, the amplitudes corresponding to the positive and negative τ must also be summed over before the probabilities are determined.

To extract from these rules a probabilistic interpretation, one *prepares a state* $|\Phi\rangle$ by imposing *conditions* which the *histories* (the paths of the system in the configuration space (T,Q)) must satisfy. The conditions assert that the histories must pass through given regions; the preparation of the state means that the detectors associated with those regions must register. After that, one fixes an *"exhaustive and exclusive" set of observations* consistent with the conditions. This can be done by removing from the configuration space the regions which defined the conditions, and breaking the remainder into a collection of non-overlapping regions. For definiteness, let there be only three such regions, 1, 2, and 3. The exhaustive and exclusive set of observations corresponds then to the following possible combinations of the responses of the three detectors:

(0): None of the three detectors registers.
(1): One and only one of the detectors registers.

(1;1): This is the detector 1.
(1;2): This is the detector 2.
(1;3): This is the detector 3.
(2): Two and only two of the detectors register.
(2;12): These are the detectors 1 and 2.
(2;13): These are the detectors 1 and 3.
(2;23): These are the detectors 2 and 3.
(3): All three detectors register.

By performing these exhaustive and exclusive observations on an ensemble of systems each of which is prepared to satisfy the same conditions, one can experimentally determine the relative frequencies of the possible outcomes. The sum-over-histories interpretation predicts what these relative frequencies ought to be by following the rules I to III. The results are obtained by evaluating path integrals, but they can be summarized in the language of operators on an auxiliary Hilbert space $\mathscr{F}$ of square integrable functions of T and Q. I shall write down two such probabilities, one from the class (1),[75]

$$P(1;3|\Phi) = \mathcal{N}\int_3 dTdQ\ |\langle TQ|\hat{K}_+|\Phi\rangle + \langle TQ|\hat{K}_-|\Phi\rangle|^2$$
$$= \mathcal{N}\int_3 dTdQ\ |\langle TQ|\hat{K}|\Phi\rangle|^2 \tag{14.1}$$

and another one from the class (2),[75]

$$P(2;31|\Phi) = \mathcal{N}\int_1 dTdQ\ |\langle TQ|\hat{K}_+\hat{\Pi}_3\hat{K}_+|\Phi\rangle + \langle TQ|\hat{K}_-\hat{\Pi}_3\hat{K}_-|\Phi\rangle|^2 + (1 \leftrightarrow 3). \tag{14.2}$$

The common normalization factor $\mathcal{N}$ is fixed by the requirement that the probabilities of all the listed exhaustive and exclusive options must sum to 1. The operators $\hat{K}_\pm$ are the forward and backward propagators

$$\hat{K}_\pm = \frac{1}{2\pi}\int_{-\infty}^{\infty} d\tau\ \Theta(\pm\tau)e^{-i\hat{H}\tau} \tag{14.3}$$

obtained by integrating the familiar forward and backward Schrödinger propagators over the positive (negative) label time τ, as indicated by the step function Θ. The operator $\hat{\Pi}_3$ is the projector

$$\hat{\Pi}_3 := \int_3 dTdQ\ |TQ\rangle\langle TQ| \tag{14.4}$$

into the region 3. The state vector $|\Phi\rangle$ describes an ensemble of systems prepared in the state satisfying the given conditions.

The construction of the probabilities (14.1) and (14.2) allows an intuitive interpretation. The probability $P(1;3|\Phi)$ that in the state $|\Phi\rangle$ one and only one detector registers, and that this is the detector 3, is obtained by squaring the sum of two probability amplitudes, the first one for the state $|\Phi\rangle$ to be propagated forward to a point T,Q of the region 3, the second one for the state to be propagated backward. The result is the same as if the state were propagated by

$$\hat{K} := \hat{K}_+ + \hat{K}_- = \frac{1}{2\pi}\int_{-\infty}^{\infty} dt\ e^{-i\hat{H}\tau} = \delta(\hat{H}). \tag{14.5}$$

The probability $P(2;31|\Phi)$ that in the state $|\Phi\rangle$ two and only two detectors register, and that these are the detectors 1 and 3, is obtained by summing up two *probabilities*. Each of them is the square of the sum of two *probability amplitudes*. The first probability amplitude is for the state $|\Phi\rangle$ to propagate forward to the region 3, to project to that region, and the projected state to propagate forward to a point of the region 1. The second probability amplitude corresponds to the same trajectory being traversed backwards. These two probability amplitudes yield the first probability. The second probability is constructed in the same way, but the regions 1 and 3 are now interchanged, the region 1 becoming an intermediate region, and the trajectory ending at a point of the region 3. The integrations over τ which yield the forward and backward propagators, and the sum over these two kinds of propagations , both the integration and the sum taking place in the probability *amplitudes*, are the manifestation of the rule III) about the unobservability of the label time.

A major part of Hartle's analysis is devoted to the recovery of the Schrödinger–Heisenberg quantum dynamics from the sum-over-histories interpretation. For this purpose, a system of thin-slab detectors is arrayed to cover a hypersurface in the (T,Q) plane. By analyzing the joint response of such detectors, Hartle draws a number of conclusions:

- In general, there is no Hilbert space formulation associated with the hypersurface. This is because the histories do not in general intersect the hypersurface once and only once.

- Consider only the hypersurfaces T = *const.* For a parametrized

Newtonian particle, there are states for which the clock variable T is forever precisely correlated with the label time τ. In such a state, the response of the thin-slab detectors can be exactly described in terms of a state evolving in accordance with the Schrödinger equation. For the model with two particles in a stationary state, there are states ("good clock states") in which T and τ are correlated for a physically interesting interval of time. The response of the detectors is then at least approximately obtained from a Schrödinger equation.

The sum-over-histories interpretation is clearly preferable to the semiclassical interpretation because *every* solution of the constraint equation, not only a "good clock state" solution, has a probabilistic interpretation. It is also preferable to the naive Schrödinger interpretation because it is able to describe correlations among responses of two or more detectors. However, it does not, in my opinion, escape the issues of time:

- **Multiple Choice Problem.** The predictions of the sum-over-histories interpretation depend on which one of the classically equivalent super-Hamiltonians is chosen in the canonical path integral. The interpretation thus requires a commitment to one definite form of the constraints. The presumption seems to be that in canonical gravity one should prefer the standard quadratic form (1.2) of the super-Hamiltonian, i.e., work within the Wheeler-DeWitt framework. I take this to be a satisfactory answer to the multiple choice problem, as long as one sticks to it all the way through and does not switch the super-Hamiltonian somewhere in the middle of the argument (e.g., use the extrinsic time super-Hamiltonian when constructing a "good clock state").

- **The Functional Evolution Problem.** In Hartle's models, the label time τ is a single parameter. In geometrodynamics, τ is replaced by an external many-fingered time: each choice of τ corresponds to a choice of the lapse and shift multipliers, i.e., to a choice of the foliation. When evaluating the path integral, one must make a definite choice. The resulting integral can be expected to depend on the choice of the foliation. (The formal arguments that path integrals are "gauge independent" are hardly reliable at this level of discussion.) This is how the functional evolution problem reappears in the sum-over-histories framework.

- **Violation of the Constraints.** Hartle repeatedly emphasizes that there is no Hilbert space associated with a hypersurface in the (T,Q) plane. There is, however, a Hilbert space behind the formalism, namely, the Hilbert space $\mathscr{F}$ of square-integrable functions over the whole (T,Q) plane. Note that Eqs. (14.1) - (14.5) which determine the detector responses involve operations in $\mathscr{F}$. One should pay close attention to the role of $\mathscr{F}$ in the sum-over-histories interpretation.

When discussing the naive Schrödinger interpretation and its conditional probabilities refinement, I pointed out that the Hilbert space of square-integrable functions over the entire configuration space(time) does not seem to be a natural arena for interpreting a constrained system. The sum-over-histories interpretation, by sharing this Hilbert space with the two previous interpretations, shares with them also the difficulties. To find what they are, I analyze Eqs. (14.1) - (14.5) from the standpoint of canonical quantum mechanics.

When only one of the detectors registers, the probability that it is a particular detector (say, the detector 3) replicates the probability postulated by the naive Schrödinger interpretation. Note that the state $|\Phi\rangle \in \mathscr{F}$ prepared by the conditions of the experiment does not need to satisfy the super-Hamiltonian constraint. However, the state $|\Psi\rangle = \hat{K}|\Phi\rangle$ which enters into Eq. (14.1) does satisfy the constraint,

$$\hat{H}|\Psi\rangle = \hat{H}\hat{K}|\Phi\rangle = \hat{H}\delta(\hat{H})|\Phi\rangle = 0 , \tag{14.6}$$

because the forward and backward propagators $\hat{K}_+$ and $\hat{K}_-$ combine into the projector $\hat{K} = \delta(\hat{H})$ into the physical space $\mathscr{F}_0$. Equation (14.1) thus directly corresponds to Eq. (12.2).

The naive Schrödinger interpretation does not say what happens to the state $|\Psi\rangle$ when the system is found in the region 3, and what would then be the probability of another observation. It is to the credit of the sum-over-histories interpretation that it does tell one the probability with which another detector registers. However, its answer, Eq. (14.2), does not correspond to the idea that the state of the system satisfies the super-Hamiltonian constraint between observations. This is brought in by the fact that the forward and backward propagators no longer neatly combine into the projector $\hat{K}$. The constituents of Eq. (14.2) no longer satisfy the constraint: $|\Phi\rangle$ was not prepared to do so, and $\hat{K}_+$ does not project it into the physical space; even if it did, $\hat{\Pi}_3$ would kick the vector $\hat{K}_+|\Phi\rangle$ again off the physical space, and the subsequent propagator $\hat{K}_+$ would do nothing to bring it back. The same remarks apply to the remaining three terms in Eq. (14.2). It is thus profoundly puzzling what connection the sum-over-histories interpretation has with the space $\mathscr{F}_0$ of states which satisfy the super-Hamiltonian constraint. Hartle devotes considerable attention to setting up the path integral formulation so that the constraints are satisfied for the state function that determines the response of a single detector. On the other hand, he evidently regards as unimportant whether the constraints are satisfied by the intermediate states connecting two or more detectors. From the point of view of the canonical formalism, the sum-over-histories interpretation has essentially the same defect as the conditional probability interpretation: the structure of Eq. (14.2) indicates that the observations are not performed so that the state is always kept in the physical space $\mathcal{F}_0$.

This conclusion had to be expected from the beginning. The detectors designed to measure "Q *and* T" must be at odds with the constraints. Effectively, the constraints prohibit an independent determination of Q and T; they allow only for the measurements of "Q *at* T". This strongly suggests that the probabilistic interpretation of a constrained system should be based on *fluxes through hypersurfaces* rather than on *densities in space*.

• **Why Should one Believe in Probability Amplitudes**? For a relativistic particle on a stationary background, the Hilbert space $\mathcal{F}_0^+$ of positive-energy one-particle states can be constructed from the wave function $\Psi(T,Q)$ and its normal rate of change (restricted by the positive-energy requirement) on an arbitrary spacelike hypersurface. However, the inner product of two wave functions on a given hypersurface is given by the Klein-Gordon rule (9.5) rather than by the "square of the amplitude" rule postulated by the sum-over-histories interpretation. The "square-of-the-amplitude" rule is appropriate only if the state Ψ is given in the Schrödinger position representation and if it is propagated according to the Schrödinger equation. Neither of these conditions is satisfied for the wave function $\Psi(T,Q)$. As a result, the rules I) – III) do not lead to the one-particle propagator in the Hilbert space $\mathcal{F}_0^+$ of positive-energy solutions of the Klein-Gordon constraint.

This example illustrates the point that there is no reason to believe that the Feynman rule which treats the state function as the probability amplitude provides an appropriate framework for quantizing a generic constrained system. In particular, why should it work in quantum gravity?

15. Frozen Time Formalism and Evolving Constants of the Motion

In a gauge theory, only those dynamical variables that are left unchanged by gauge transformations have a physical meaning. Such variables are called *observables*. Thus, the electric and magnetic field strengths in electrodynamics are observables, while the scalar and vector potentials are not.

In canonical formalism, gauge transformations are generated by constraints that are linear in the momenta. These constraints move a point in the phase space along an orbit of the gauge group. Two points on the same orbit are physically indistinguishable; they represent merely two equivalent descriptions of the same physical state. An observable should not depend on what description of the state one chooses. This means that it must be the same along the orbit, i.e., that its Poisson bracket with the constraints must vanish.

General relativity is a constraint theory par excellence: All its physical content is encoded in the constraints. Previous experience with ordinary gauge theories led to the idea that the observables F in general relativity are those dynamical variables that have vanishing Poisson brackets with all the constraints:

$$\{F, H_a(x)\} = 0 = \{F, H(x)\} . \tag{15.1}$$

It is generally agreed that the supermomentum constraint (1.1) generates a gauge, namely, the group of diffeomorphisms DiffΣ of spatial coordinates, and that any observable in general relativity must be invariant under DiffΣ. However, the super-Hamiltonian constraint plays a different role: It generates the dynamical change of the geometrodynamical variables from one hypersurface to another. Any dynamical variable which commutes with the super-Hamiltonian must be the same on every hypersurface $\Sigma \to \mathcal{M}$, i.e., it must be a *constant of the motion.* To maintain that the only observable quantities are those that commute with all the constraints seems to imply that the Universe can never change. For this reason, this standpoint on observables was dubbed the *frozen time formalism.*[218-223] The frozen time formalism never successfully explained the evolution that we see all around us.

A similar difficulty appears already in simple parametrized dynamical systems, as the Newtonian particle. An observable defined as a dynamical variable that has the vanishing Poisson bracket with the super-Hamiltonian constraint (1.11) must necessarily be a constant of the motion. Nevertheless, in ordinary Newtonian dynamics one happily works with dynamical variables that change with time, like the position of the particles, and assumes that such variables can be measured. This indicates that the transformations generated by the super-Hamiltonian should not be interpreted as a gauge transformations: Two points on the same orbit of the super-Hamiltonian are two events in the dynamical evolution of the system. Such events are physically distinguishable rather than being two descriptions of the same physical state. For this reason, I took the position in the rest of this paper that observables are to be defined as those dynamical variables that have vanishing Poisson brackets with the gauge constraints, but not necessarily with the super-Hamiltonian constraint (see Section 6, Eq. (6.3)).

Recently, Rovelli proposed a reinterpretation of the frozen time formalism that allows one to describe within it the dynamical change of a system.[224-226] He introduced the scheme on a model of two harmonic oscillators in a stationary state[225] and then attempted to generalize it to an arbitrary finite-dimensional system whose dynamics is generated by a single super-Hamiltonian constraint.[226] His main objective was to show that one does not need a global time function for interpreting quantum dynamics. Dynamics is introduced by the concept of an *evolving constant of the motion.*

▪ **Evolving Constants of the Motion.** I shall explain how the scheme is expected to work for a system with the super-Hamiltonian H(Q,P) that admits a global time function T(Q,P). This assumption means that each dynamical trajectory intersects every hypersurface

$$T(Q,P) = t = const \tag{15.2}$$

once and only once, i.e., that

$$\{T, H\} \neq 0 \,. \tag{15.3}$$

(Throughout his work, Rovelli does not distinguish the weak equations from the strong ones, and I shall not try to fix this deficiency.)

Let F(Q,P) be an arbitrary dynamical variable that is not necessarily a constant of the motion:

$$\{F, H\} \neq 0 \,. \tag{15.4}$$

Choose a fixed value of t and define a phase-space function $F_t(Q,P)$ by the requirements that (I) $F_t(Q,P)$ coincide with F(Q,P) when restricted to the hypersurface (15.2), and that (II) $F_t(Q,P)$ be constant along each classical trajectory:

$$\{F_t, H\} = 0 \,. \tag{15.5}$$

If T is a global time function, Eq. (15.3), these two requirements determine $F_t(Q,P)$ uniquely. When repeated for each value of t, the procedure replaces a dynamical variable F(Q,P) by a one-parameter family $F_t(Q,P)$, $t\in(-\infty,\infty)$ of observables. The first requirement amounts then to the statement that

$$F_{t=T(Q,P)} = F(Q,P) \,. \tag{15.6}$$

Dynamics does not tell one how a single $F_t(Q,P)$ changes when evolved by H, because each $F_t(Q,P)$ is a constant of the motion and hence does not change at all, but it tells one how the observables $F_t(Q,P)$ depend on the parameter t. By taking the Poisson bracket of Eq. (15.6) with the super-Hamiltonian, one learns that

$$\left.\frac{dF_t(Q,P)}{dt}\right|_{t=T(Q,P)} = \{T,\ H\}^{-1}\ \{F,\ H\}\ . \qquad (15.7)$$

This means that the observables $F_t(Q,P)$ change with t in the same rate at which the dynamical variable F evolves in the internal time T.

Rovelli calls the one-parameter family of observables $F_t(Q,P)$ an *evolving constant of the motion*. For a classical system with a global time function, the evolution of dynamical variables can be described in terms of evolving constants of the motion.

▪ **Quantization.** Rovelli suggests that the algorithm used to define the evolving constants of the motion can be repeated at the quantum level. One simply turns the classical variables H(Q,P), T(Q,P) and F(Q,P) into operators $H(\hat{Q},\hat{P})$, $T(\hat{Q},\hat{P})$ and $F(\hat{Q},\hat{P})$ and defines the *quantum evolving constant of the motion* by the requirements

$$[\hat{F}_t,\ \hat{H}] = 0 \qquad (15.8)$$

and

$$F_{t=T(\hat{Q},\hat{P})}(\hat{Q},\hat{P}) = F(\hat{Q},\hat{P}) \qquad (15.9)$$

that mimic the classical equations (15.5) - (15.6). The change of the quantum evolving constant of the motion $F_t(\hat{Q},\hat{P})$ with t describes the quantum dynamics of the system.

Rovelli conjectures that this procedure enables one to solve the Hilbert space problem. For a generic super-Hamiltonian $\hat{H}$, the physical space $\mathcal{F}_0$ of state functions $\Psi(Q)$ that satisfy the quantum constraint

$$\hat{H}\Psi(Q) = 0 \qquad (15.10)$$

does not have any obvious Hilbert space structure. Rovelli proposes to choose a "complete set" of dynamical variables $F_k(\hat{Q},\hat{P})$ and turn all of them into corresponding quantum constants of the motion $F_{k\,t}(\hat{Q},\hat{P})$. His conjecture is that there exists a unique inner product in $\mathcal{F}_0$ such that all $F_{k\,t}(\hat{Q},\hat{P})$ operators are self-adjoint under this inner product.

I want to show that Rovelli's procedure is ambiguous and that, even if its ambiguities were somehow resolved, it is not likely to solve any one of the outstanding problems of time.

• **Quantum Evolving Constants of the Motion are Ambiguous.**

Rovelli's procedure starts by turning the classical variables H(Q,P), T(Q,P) and F(Q,P) into operators. Without some further guiding principle the factor ordering of the operators $\hat{Q}$ and $\hat{P}$ is arbitrary and so are the resulting operators $\hat{H}$, $\hat{T}$ and $\hat{F}$. Even if these factor ordering problems were resolved, there is an additional problem of writing down Eq. (15.9) (Hájíček[227]). The time operator $\hat{T} = T(\hat{Q},\hat{P})$ does not commute with $\hat{Q}$ and $\hat{P}$, and the replacement of the classical parameter by the operator $\hat{T}$ is thus ambiguous. Even if this problem were resolved, the commutator condition (15.8), unlike its classical counterpart (15.5), is not sufficient to determine the operator $\hat{F}_t$ uniquely. Equation (15.8) determines the quantum evolving constant of the motion only if the super-Hamiltonian has the Schrödinger form (1.11) with respect to the time function T.

In view of all these factor-ordering problems, it seems best to drop out the requirement (15.9) and to formulate the quantum problem somewhat differently. One can first, at the classical level, construct the classical evolving constant of the motion $F_t(Q,P)$ which satisfies the classical requirements (15.5) - (15.6). One can then pose the problem of factor ordering this one-parameter family of classical observables so that they remain observables even in the quantum theory. Unfortunately, one lacks an algorithmic procedure for achieving this aim. This is a manifestation of the problem of time: In the Schrödinger case, the algorithm depends on the identification of the true Hamiltonian as the variable canonically conjugate to the time. The evolving constants of the motion are then constructed by means of the standard evolution operator that requires one to decide on a factor ordering of a single quantity, namely, the true Hamiltonian of the system.

• **The Hilbert Space Problem.** I want to illustrate why Rovelli's conjecture on resolving the Hilbert space problem is likely to fail. To do that, I choose a simple model with a global time function, namely, a relativistic particle moving on a dynamical background.

I select an intrinsic time function T(Q) and label the worldlines of the reference frame that is orthogonal to the T-foliation by the comoving coordinates $q^A(Q)$, as in Eq. (6.22). Let $p_a(Q,P)$ be the momentum canonically conjugate to q^a. The dynamical variables $q^a(Q)$, $p_a(Q,P)$ form a complete set of fundamental canonical variables. Their Poisson brackets are

$$\{q^a, p_b\} = \delta^a_b, \quad etc. \tag{15.11}$$

The variables q^a, p_a are not constants of the motion, but following Rovelli's algorithm one can replace each of them by the corresponding evolving constant of the motion, i.e., by the one-parameter family of constants of the motion q_t^a, $p_{t\,a}$. For each value of t, these constants of the motion have the same Poisson brackets as the fundamental canonical variables that they represent:

$$\{q_t^a, p_{t\,b}\} = \delta_b^a , \quad etc. \tag{15.12}$$

One knows from the vanHove theorem[228] that one cannot preserve the whole classical Poisson algebra upon quantization. A crucial decision which one faces is the selection of a suitable subalgebra that is taken into quantum theory without modification. Though Rovelli does not explicitly state this, I presume (as it is common in quantum mechanics of particle systems), that he wants to take over into quantum theory the algebra (15.12) of the complete set of observables. If so, he must factor order the variables $q_t^a(\hat{Q})$ and $p_{t\,a}(\hat{Q},\hat{P})$ so that they both preserve their classical algebra,

$$\frac{1}{i} [q_t^a(\hat{Q}), p_{t\,b}(\hat{Q},\hat{P})] = \delta_b^a , \quad etc., \tag{15.13}$$

and remain the constants of the motion,

$$[q_t^a(\hat{Q}), H(\hat{Q},\hat{P})] = 0 = [p_{t\,a}(\hat{Q},\hat{P}), H(\hat{Q},\hat{P})] . \tag{15.14}$$

This is a non-trivial task because $q_t^a(Q)$, and $p_{t\,a}(Q,P)$ are complicated functions of the classical arguments Q^A and P_A (thus, e.g., $p_{t\,a}(Q,P)$, unlike $p_a(Q,P)$, is no longer linear in the momenta P_A). I already pointed out that Rovelli's algorithm for achieving (15.14) does not in general work. This is not, however, my present concern: *Assume* that, by hook or by crook, one finds a factor ordering such that Eqs. (15.13) - (15.14) hold for the quantum operators.

Rovelli's conjecture is that one can endow $\mathscr{F}_0$ with an inner product such that $\hat{q}_t^a$ and $\hat{p}_{t\,a}$ are self-adjoint operators. Because the Klein-Gordon operator $\hat{H}$ is a second-order differential operator, the solution space $\mathscr{F}_0$ is quite big: it contains solutions corresponding to both the positive and negative energies. It is quite possible that there does not exist an inner product in the whole of $\mathscr{F}_0$ under which the operators q_t^a and $p_{t\,a}$ are self-adjoint. The crux of the Hilbert space problem, however, lies elsewhere:

Choose another time function, $\bar{T}(Q)$, and repeat the whole procedure. Get thereby another set, $\hat{\bar{q}}_t^a$ and $\hat{\bar{p}}_{t\,a}$, of quantum observables. Keep repeating this for all possible intrinsic time functions T(Q). Now, against Rovelli's conjecture, let me pose a counterconjecture: It is impossible to turn all the operators $\hat{q}_t^a$, $\hat{p}_{t\,a}$ corresponding to all possible intrinsic time functions T(Q) into self-adjoint operators by a *single* choice of an inner product in $\mathscr{F}_0$.

If it were possible, one would have a consistent one-particle interpretation of the Klein-Gordon theory on a dynamical background, the interpretation that would be independent of the chosen foliation. Nobody has so far succeeded in constructing such an interpretation. This is the challenge which the relativistic particle puts to Rovelli's proposal. The problem that I just spelled out is a modified version of the multiple choice problem for the relativistic particle discussed in Section 6.

• **Multiple Choice Problem.** Before going any further, let me note that Rovelli's formalism has yet another multiple choice problem: It depends on which one of the classically equivalent super-Hamiltonians one turns into an operator $\hat{H}$ that defines the quantum theory. I presume that in quantum geometrodynamics most people would be inclined to solve this problem by insisting that the super-Hamiltonian constraint be imposed in its familiar quadratic form.

• **Global Time Problem.** When there is no global time function, the replacement (15.5) - (15.6) of a dynamical variable F by an evolving constant of the motion F_t breaks down. Nevertheless, Rovelli claims that one can give a dynamical interpretation of such systems without a global time function. He bases his claim on a discussion of a single highly special model, namely, a two-oscillator system. Hájíček[227] pointed out an error in Rovelli's construction of an allegedly self-adjoint observable based on a nonglobal time function in this model. In response to Hájíček's criticism, Rovelli modified the factor ordering of his observable.[229] In my opinion, the physical interpretation of Rovelli's observable remains problematic. Moreover, the two-oscillator model has such a high degree of symmetry that drawing any general conclusions from its treatment is clearly premature. Hartle[230] raised the question whether Rovelli's interpretation based on a non-global time function is consistent even for a parametrized Newtonian particle.

• **Functional Evolution Problem.** Rovelli never got so far as to discuss the evolving constants of the motion for a field-theoretical system with a many-fingered time. Hartle pointed out that the generalization of Rovelli's procedure to such systems may present difficulties.[230]

Rovelli's procedure enables one to discuss classical dynamics of systems with a global time function in terms of constants of the motion. However, it does not seem to be able to solve a single one of the outstanding problems of time. Indeed, Rovelli himself did not develop his proposal to the point at which it becomes clear where the problems of time lie.

16. Summary

In general relativity, dynamics is entirely generated by constraints. The dynamical data do not explicitly include a time variable. I identified several major problems which this situation causes in quantum gravity. A classical obstacle to interpreting the quantum formalism is the possibility that there is *no global time function* on the phase space of the theory, and the associated *sandwich problem*. Even if a global time function exists, quantum theory may depend on its particular choice; this constitutes the *multiple choice problem*. Moreover, even when one sticks to a definite choice of the many-fingered time, one still faces the *problem of functional evolution*: the propagation of the state functional may depend on the foliation connecting the initial hypersurface with the final hypersurface. Those interpretations of quantum gravity that do not try to separate time from dynamical variables trade the multiple choice problem for the *Hilbert space problem*: how does one endow the space of solutions of the quantum constraints with a Hilbert space structure?

I classified ten major attempts at interpreting quantum gravity into three classes: I) *An internal time framework*, II) *the Wheeler-DeWitt framework*, and III) *quantum gravity without time*.

The first class of interpretations tries to identify time and bring the super-Hamiltonian constraint into the Schrödinger from prior to quantization. I reviewed various attempts to construct time from the geometric variables in the *internal Schrödinger interpretation* (the intrinsic and extrinsic times of minisuperspace and midisuperspace models, of linearized gravity, and the mean extrinsic curvatures time of full geometrodynamics). Such attempts suffer from the multiple choice problem, the spacetime problem (it is difficult to find a many-fingered internal time which is a spacetime scalar), and from the spectral analysis problem (the Hamiltonian of the theory is typically defined only implicitly by spectral analysis). Internal times proposed so far and likely to be contemplated in the future make the spectral analysis problem so involved and the true Hamiltonian so cumbersome that I find it hard to believe that the internal Schrödinger interpretation of full quantum gravity is a viable option.

The attempts to find an internal time among the matter variables have a hard time to cope with the energy conditions. The *matter time* provided by Gaussian reference dust comes as close to accomplishing the goal of the internal time framework as anyone can reasonable hope. It must be ultimately dismissed as a fundamental solution to the time problem only because it relies on a highly phenomenological system.

The last candidate for an internal time, the cosmological time of *unimodular gravity*, is not a many-fingered time variable, and hence it leaves an infinite number of remaining super-Hamiltonian constraints without dynamical interpretation.

The interpretations within the Wheeler-DeWitt framework suffer from the Hilbert space problem. All of them are also beset by the sandwich problem and the spacetime problem. The *Klein-Gordon interpretation* fails to construct the Hilbert space because the super-Hamiltonian is not stationary in superspace and the potential term is not positive. The *semiclassical interpretation* hides the problem of time behind an approximation procedure. It claims that the Wheeler-DeWitt equation for a semiclassical state approximately reduces to the Schrödinger equation, and the Klein-Gordon norm reduces to the Schrödinger norm. Unfortunately, it achieves the positivity of the norm at an unacceptable price of suspending the superposition principle. Its attempts to exorcise the interference terms, by decoherence arguments or otherwise, are unconvincing. When one allows superpositions, one can show that the Schrödinger equation with its complex structure actually does not follow from the real Klein-Gordon equation, and one can easily exhibit states with a negative norm. Last but not least, when applying the semiclassical interpretation to quantum gravity proper, one again faces the problem of separating the classical modes that define time from the quantum modes.

A closer look at *third quantization* reveals that it does not really solve the problem of time and that its formalism resists an operational interpretation.

The last class of interpretations maintains that one does not need time to interpret quantum gravity. The *naive Schrödinger interpretation*, the *conditional probability interpretation*, and the *sum-over-histories interpretation* share the common Hilbert space, that of square-integrable functions on the configuration space(time) before the imposition of the constraints. With it they also share the same basic difficulties. The main one is that each of these interpretations in its own way violates the constraints. Further, none of them reproduces the standard results when applied to a Klein-Gordon system. In the case of the sum-over-histories interpretation, this casts doubts on the universal applicability of the probability amplitude rule. Finally, while the naive Schrödinger interpretation lacks dynamics, the conditional probability interpretation predicts wrong dynamics for well understood systems, like the parametrized Newtonian particle.

The *frozen time formalism* stands apart from the other three interpretations in this class. Its difficulty has always been how to explain quantum dynamics in terms of constants of the motion. Rovelli's proposal on how to achieve this goal is ambiguous. When its consequences are analyzed, it becomes apparent that it faces all the familiar difficulties, especially the Hilbert space problem and the multiple choice problem.

In my opinion, none of us has so far succeeded in proposing an interpretation of quantum gravity that would either solve or circumvent the problems of time. He who made eternity out of years remains beyond our reach. His ways remain unscrutable because He not only plays dice with matter but also with time.

17. Acknowledgments

This report on the problems of time in canonical quantum gravity stems from the talks which I gave at the 10th Conference of Czechoslovak Physicists at Brno, Czechoslovakia, August 27 - 31, 1990, and at the 4th Canadian Conference on General Relativity and Relativistic Astrophysics at Winnipeg, Manitoba, Canada, May 16 - 18, 1991. I am grateful to many colleagues with whom I discussed the subject over the years, especially to Julian B. Barbour, Petr Hájíček, James B. Hartle, Christopher J. Isham, Carlo Rovelli, Charles G. Torre, Claudio Teitelboim, William G. Unruh, Robert M. Wald, John A. Wheeler, and James W. York.

The paper was written during my stay at Imperial College, London, in lively interaction with Christopher J. Isham. Julian B. Barbour kindly softened its Slavic accent. The work was partially supported by a SERC Fellowship and by the NSF grant PHY-897973 to the University of Utah.

18. References

1. A. Gryphius, *Gesamtausgabe der deutschsprachigen Werke*, eds. M. Szyrocki and H. Powell (Niemeyer, Tübingen, 1964), Vol. 2, pp. 182-3.
2. R.F. Baierlein, D.H. Sharp and J.A. Wheeler, *Phys. Rev.* **126** (1962) 1864. "Three-Dimensional Geometry as Carrier of Information About Time."
3. K.V. Kuchař, *J. Math. Phys.* **13** (1972) 768. "Bubble-Time Canonical Formalism for Geometrodynamics."
4. A. Peres, *Nuovo Cimento* **26** (1962) 53. "On the Cauchy Problem in General Relativity II."
5. U. Gerlach, *Phys. Rev.* **177** (1969) 1929. "Derivation of the Ten Einstein Field Equations from the Semiclassical Approximation to Quantum Geometrodynamics."
6. C. Lanczos, *The Variational Principles of Mechanics*, 4th ed. (Univ. of Toronto Press, Toronto, 1970), Ch. VIII. 7, pp. 264-276.
7. Ref. 6, Ch. VIII. 6, pp. 262-264.
8. Ref. 6, Ch. XI. 10, pp. 185-192.
9. P. Hájíček, *Phys. Rev.* **D34** (1986) 1040. "Origin of Non-Unitarity in Quantum Gravity."
10. P. Hájíček, *J. Math. Phys.* **30** (1989) 2488. "Topology of Parametrized Systems."
11. P. Hájíček, *Class. Quantum Grav.* **7** (1990) 871. "Dirac Quantization of Systems with Quadratic Constraints."

12. M. Schön and P. Hájíček, *Class. Quantum Grav.* **7** (1990) 861. "Topology of Quadratic Super-Hamiltonians."
13. P. Hájíček, *Lecture Notes on Quantum Cosmology* (University of Bern, Bern, 1990); unpublished.
14. C. Carathéodory, *Variationsrechnung und partielle Differentialgleichungen erster Ordnung I* (Teubner, Leipzig, 1956).
15. J.L. Synge, in *Handbuch der Physik*, Vol. III/1, ed. S. Flügge (Springer, Berlin, 1960). "Classical Dynamics. 88. Generating Functions."
16. A. Peres and N. Rosen, *Nuovo Cimento* **13** (1959) 430. "On Cauchy's Problem in General Relativity."
17. A. Peres, *Nuovo Cimento* **18** (1960) 32. "Invariants of Relativity II. - Classification of Spaces."
18. J.A. Wheeler, in *Relativity, Groups, and Topology*, eds. C. and B. DeWitt (Gordon and Breach, New York, 1964). "Geometrodynamics and the Issue of the Final State."
19. J.A. Wheeler, in *Gravitation and Relativity*, eds. H -Y. Chiu and W.F. Hoffmann (Benjamin, New York, 1964). "Mach's Principle as Boundary Condition for Einstein's Equations."
20. E.P. Belasco and H. Ohanian, *J. Math. Phys.* **10** (1969) 1503. "Initial Conditions in General Relativity: Lapse and Shift Formulation."
21. P.G. Bergmann, in *Relativity*, eds. M. Carmeli, S. Fickler and L. Witten (Plenum, New York, 1980). "The Sandwich Conjecture."
22. A. Komar, *J. Math. Phys.* **11** (1970) 820. " General Relativistic Thin-Sandwich Theorem."
23. C.M. Pereira, *J. Math. Phys.* **14** (1973) 1498. "Thin-Sandwich Conjecture and a Fifth Initial Value Equation for the Gravitational Field."
24. D. Christodoulou and M. Francaviglia, in *Enrico Fermi Proceedings*, ed. J. Ehlers (North Holland, New York, 1979). "The Geometry of the Thin-Sandwich Problem.
25. C.M. Pereira, *J. Math. Phys.* **22** (1980) 1064. "Nonglobal Proof of the Thin-Sandwich Conjecture."
26. P.A.M. Dirac, *Lectures on Quantum Mechanics* (Yeshiva Univ., New York, 1964).
27. P.A.M. Dirac, *Can. J. Math.* **3** (1951) 1. "The Hamiltonian Form of Field Dynamics."
28. K.V. Kuchař and C.G. Torre, *Phys. Rev.* **D43** (1990) 419. "Gaussian Reference Fluid and Interpretation of Quantum Geometrodynamics."
29. C. W. Misner, *Rev. Mod. Phys.* **29** (1957) 497. "Feynman Quantization of General Relativity."
30. P.W. Higgs, *Phys. Rev. Lett.* **1** (1958) 373. "Integration of Secondary Constraints in General Relativity."

31. J.A. Wheeler, in *Batelle Rencontres: 1967 Lectures in Mathematics and Physics*, eds. C. DeWitt and J.A. Wheeler (Benjamin, New York, 1968). "Superspace and the Nature of Geometrodynamics."
32. B.S. DeWitt, *Phys Rev.* **160** (1967) 1113. "Quantum Theory of Gravity I."
33. K.V. Kuchař, in *Quantum Gravity 2: A Second Oxford Symposium*, eds. C.J. Isham, R. Penrose and D.W. Sciama (Clarendon, Oxford, 1981). "Canonical Methods of Quantization."
34. C.J. Isham, in *Relativity, Groups and Topology II: Les Houches 1983*, eds. B.S. DeWitt and R. Stora (North-Holland, Amsterdam, 1984). "Topological and Global Aspects of Quantum Theory."
35. K.V. Kuchař, *Phys. Rev.* **D4** (1971) 955. "Canonical Quantization of Cylindrical Gravitational Waves."
36. R.H. Gowdy, *Ann. Phys. (N.Y.)* **83** (1974) 203. "Vacuum Spacetimes with Two-Parameter Spacelike Isometry Groups and Compact Invariant Hypersurfaces: Topologies and Boundary Conditions."
37. C.W. Misner, *Phys. Rev.* **D8** (1973) 3271. "A Minisuperspace Example - The Gowdy T^3 Universe."
38. B. Berger, *Phys. Rev.* **D11** (1975) 2770. "Quantum Cosmology: Exact Solution for the Gowdy T^3 Model."
39. B. Berger, *Phys. Lett.* **B108** (1982). "Singularity Avoidance in the Semiclassical Gowdy T^3 Model."
40. B. Berger, *Ann. Phys. (N.Y.)* **156** (1984) 155. "Quantum Effects in the Gowdy T^3 Cosmology."
41. V. Husain, *Class. Quantum Grav.* **4** (1987) 1587. "Quantum Effects on the Singularity of the Gowdy Cosmology."
42. M.O. Green, J. Schwarz and E. Witten, *Superstring Theory* (Cambridge Univ. Press, Cambridge, 1987).
43. L. Brink and M. Henneaux, *Principles of String Theory* (Plenum, New York, 1988).
44. K.V. Kuchař and C.G. Torre, in *Conceptual Problems of Quantum Gravity*, eds. A. Ashtekar and J. Stachel (Birkhäuser, Boston, 1991). "Strings as Poor Relatives of General Relativity."
45. P.A.M. Dirac, *Can. J. Math.* **3** (1951) 1. "The Hamiltonian Form of Field Dynamics."
46. K.V. Kuchař, *J. Math. Phys.* **17** (1976) 792. "Kinematics of Tensor Fields in Hyperspace."
47. K.V. Kuchař, *J. Math. Phys.* **17** (1976) 801. "Dynamics of Tensor Fields in Hyperspace."
48. J.J. Halliwell and S.W. Hawking, *Phys. Rev.* **D31** (1985) 1777. "Origin of Structure in the Universe."

49. C.W. Misner, in *Magic Without Magic: John Archibald Wheeler, A Collection of Essays in Honor of his 60th Birthday*, ed. J. Klauder (Freeman, San Francisco, 1972). "Minisuperspace."
50. M. Ryan, *Hamiltonian Cosmology* (Springer, Berlin, 1972).
51. M.A.H. MacCallum, in *Quantum Gravity*, eds. C.J. Isham, R. Penrose and D. W. Sciama (Clarendon, Oxford, 1975). "Quantum Cosmological Models."
52. J.J. Halliwell, *Int. J. Mod. Phys.* **A5** (1990) 2473. "A Bibliography of Papers on Quantum Cosmology."
53. C.W. Misner, *Phys. Rev. Lett.* **22** (1969) 1071. "Mixmaster Universe."
54. C.W. Misner, *Phys. Rev.* **186** (1969) 1319, 1328. "Quantum Cosmology."
55. C.W. Misner, in *Relativity*, eds. M. Carmeli, S. Fickler and L. Witten (Plenum, New York, 1970). "Classical and Quantum Dynamics of a Closed Universe."
56. J. Abbot, S. Giddings and K.V. Kuchař, *Gen. Rel. Grav.* **16** (1984) 751. "Einstein's Theory in a Three-Dimensional Spacetime."
57. E. Witten, *Nucl. Phys.* **B311** (1988) 46. "2+1 Dimensional Gravity as an Exactly Soluble System."
58. E. Witten, *Nucl. Phys.* **B323** (1989) 113. "Topology-Changing Amplitudes in 2+1 Dimensional Gravity."
59. V. Moncrief, *J. Math. Phys.* **30** (1989) 2907. "Reduction of the Einstein Equations in 2+1 Dimensions to a Hamiltonian System over Teichmüller Space."
60. A. Hosoya and K. Nakao, *Class. Quantum Grav.* **7** (1990) 163. "(2+1)-Dimensional Pure Gravity for an Arbitrary Closed Initial Surface."
61. S. Carlip, *Phys. Rev.* **D42** (1990) 2647. "Observables, Gauge Invariance, and Time in (2+1)-Dimensional Quantum Gravity."
62. S. Carlip, *Preprint* UCDPHYS-PUB-23-90. "Time in 2+1 Dimensional Quantum Gravity."
63. G. Nordström, *Phys. Z.* **13** (1912) 1126. "Relativitätsprinzip und Gravitation."
64. G. Nordström, *Ann. Phys. (Leipzig)* **42** (1913) 533. "Zur Theorie der Gravitation vom Standpunkt des Relativitätsprinzips."
65. A. Einstein and A.D. Fokker, *Ann. Phys. (Leipzig)* **44** (1914) 321. "Die Nordströmsche Gravitationstheorie vom Standpunkt des Absoluten Differentialkalküls."
66. P. Hájíček and K.V. Kuchař, *Phys. Rev.* **D41** (1990) 1091. "Constraint Quantization of Parametrized Relativistic Gauge Systems in Curved Spacetimes."

67. K.V. Kuchař, *Phys. Rev.* **D22** (1980) 1285. "Gravitation, Geometry, and Nonrelativistic Quantum Theory."
68. J.B. Hartle and K.V. Kuchař, *J. Math. Phys.* **25** (1984) 57. "Path Integrals in Parametrized Theories: Newtonian Systems."
69. J.B. Barbour, *Nature* **249** (1974) 328. "Relative-Distance Machian Theories."
70. J.B. Barbour, *Nature* **250** (1974) 606. "Erratum."
71. J.B. Barbour, *Nuovo Cimento* **B26** (1975) 16. "Forceless Machian Dynamics."
72. J.B. Barbour and B. Bertotti, *Nuovo Cimento* **B38** (1977) 1. "Gravity and Inertia in a Machian Framework."
73. J.B. Barbour and B. Bertotti, *Proc. Roy. Soc. London* **A382** (1982) 295. "Mach's Principle and the Structure of Dynamical Theories."
74. W. Pauli, in *Handbuch d. Physik V*, ed. S. Flugge (Springer, Berlin, 1958). "Die allgemeinen Prinzipien der Wellenmechanik."
75. J.B. Hartle, *Phys. Rev.* **D38** (1988) 2985. "Quantum Kinematics of Spacetime II. A Model Quantum Cosmology with Real Clocks."
76. W.G. Unruh and R.M. Wald, *Phys. Rev.* **D40** (1989) 2598. "Time and the Interpretation of Canonical Quantum Gravity."
77. K.V. Kuchař, *Phys. Rev.* **D39** (1989) 1579. "Parametrized Scalar Field on $R \times S^1$: Dynamical Pictures, Spacetime Diffeomorphisms and Conformal Isometries."
78. K.V. Kuchař, *Phys. Rev.* **D39** (1989) 2263. "The Dirac Constraint Quantization of a Parametrized Field Theory by Anomaly-Free Operator Representations of Spacetime Diffeomorphisms."
79. K.V. Kuchař, in *Highlights in Gravitation and Cosmology*, eds. B.R. Iyer et al. (Cambridge Univ. Press, Cambridge, 1988). "Canonical Quantization of Generally Covariant Systems."
80. G. Torre, *Class. Quantum Grav.* **8** (1991) 1895. " A Complete Set of Observables for Cylindrically Symmetric Gravitational Fields."
81. K.V. Kuchař and C.G. Torre, *J. Math. Phys.* **30** (1989) 1769. "World Sheet Diffeomorphisms and the Canonical String."
82. D.G. Boulware and S. Deser, *J. Math. Phys.* **8** (1967) 1468. "Stress-Tensor Commutators and Schwinger Terms."
83. J. Anderson, *Phys. Rev.* **114** (1959) 1182. "Factor Sequences in Quantized General Relativity."
84. J. Anderson, in *Proceedings of the First Eastern Theoretical Physics Conference, October 26-27, 1962*, ed. M.E. Ross (Gordon and Breach, New York, 1962). "Q-Number Coordinate Transformations and the Ordering Problem in General Relativity."

85. J. Schwinger, *Phys. Rev.* **132** (1963) 1317. "Quantized Gravitational Field II."
86. P.A.M. Dirac, in *Contemporary Physics: Trieste Symposium 1968*, V. 1, ed. L. Fonda (IAEA, Vienna, 1969). "The Quantization of the Gravitational Field."
87. A. Komar, *Phys. Rev.* **D19** (1979) 2908. "Constraints, Hermiticity, and Correspondence."
88. A. Komar, *Phys. Rev.* **D20** (1979) 830. "Consistent Factor Ordering of General-Relativistic Constraints."
89. A. Ashtekar, *Phys. Rev.* **D36** (1987) 1587. "New Hamiltonian Formulation of General Relativity."
90. N.C. Tsamis and R.P. Woodard, *Phys. Rev.* **D36** (1987) 3641. "The Factor-Ordering Problem Must be Regulated."
91. J.L. Friedman and I. Jack, *Phys. Rev.* **D37** (1988) 3495. "Formal Commutators of the Gravitational Constraints are not Well-Defined: A Translation of Ashtekar's Ordering to the Schrödinger Representation."
92. J.L. Friedman and I. Jack, in *Conceptual Problems of Quantum Gravity*, eds. A. Ashtekar and J. Stachel (Birkhäuser, Boston, 1991). "Formal Commutators of the Gravitational Constraints are not Well-Defined."
93. P. Hájíček and K.V. Kuchař, *J. Math. Phys.* **31** (1990) 1723. "Transversal Affine Connection and Quantization of Constrained Systems."
94. P.A.M. Dirac, *Phys. Rev.* **114** (1959) 924. "Fixation of Coordinates in the Hamiltonian Theory of Gravitation."
95. K.V. Kuchař, in *Conceptual Problems of Quantum Gravity*, eds. A. Ashtekar and J. Stachel (Birkhäuser, Boston, 1991). "The Problem of Time in Canonical Quantization of Relativistic Systems."
96. C.J. Isham, *Imperial College Preprint* TP/90-91/14. "Conceptual and Geometrical Problems in Quantum Gravity. Lectures Presented at the 1991 Schladming Winter School."
97. J.B. Barbour and L. Smolin, *Yale University Preprint April 8, 1988.* "Can Quantum Mechanics be Sensibly Applied to the Universe as a Whole?"
98. J.J. Halliwell, *Phys. Rev.* **D43** (1991) 2590. "Global Spacetime Symetries in the Functional Schrödinger Picture."
99. R. Arnowitt, S. Deser and C.W. Misner, in *Gravitation: An Introduction to Current Research*, ed. L. Witten (Wiley, New York, 1962). "The Dynamics of General Relativity."
100. R. Arnowitt, S. Deser and C.W. Misner, *Phys. Rev.* **117** (1960) 1595. "Canonical Variables for General Relativity."
101. K.V. Kuchař, *J. Math. Phys.* **11** (1970) 3322. "Ground State Functional of the Linearized Gravitational Field."

102. K.V. Kuchař, in *Relativity, Astrophysics and Cosmology*, ed. W. Israel (Reidel, Dordrecht, 1973). "Canonical Quantization of Gravity."
103. J.W. York, *Phys. Rev. Lett.* **28** (1972) 1082. "Role of Conformal Three-Geometry in the Dynamics of Gravitation."
104. J.W. York, in *Sources of Gravitational Radiation*, ed. L. Smarr (Cambridge Univ. Press, Cambridge, 1979). "Kinematics and Dynamics of General Relativity."
105. Y. Choquet-Bruhat and J.W. York, in *General Relativity and Gravitation the Hundred Years After the Birth of Albert Einstein*, eds. P. Bergmann, J. Goldberg and A. Held (Plenum, New York, 1980). "The Cauchy Problem."
106. I.M. Singer, *Commun. Math. Phys.* **60** (1978) 7. "Some Remarks on the Gribov Ambiguity."
107. D.N. Page, in *Quantum Concepts in Space and Time*, eds. R. Penrose and C.J. Isham (Clarendon, Oxford, 1986). "Hawking's Wave Function for the Universe."
108. Y. Choquet-Bruhat, *C.R. Acad. Sci. (Paris)* **280A** (1975) 169. "Sous variétés maximales, ou à courbure constant, de variétés lorentziennes."
109. D. Brill and F. Flaherty, *Commun. Math. Phys.* **50** (1976) 157. "Isolated Maximal Surfaces in Spacetime."
110. A.J. Goddard, *Commun. Math. Phys.* **54** (1977) 279. "Foliations of Space-Times by Spacelike Hypersurfaces of Constant Mean Curvature."
111. L.L. Smarr and J.W. York, *Phys. Rev.* **D17** (1978) 2529. "Kinematical Conditions in the Construction of Spacetime."
112. D.M. Eardley and L.L. Smarr, *Phys. Rev.* **D19** (1979) 2239. "Time Functions in Numerical Relativity: Marginally Bound Dust Collapse."
113. J.E. Marsden and F.J. Tipler, *Physics Rep.* **66** (1980) 109. "Maximal Hypersurfaces and Foliations of Constant Mean Curvature in General Relativity."
114. J. Isenberg and J.E. Marsden, *J. Geom. and Phys.* **1** (1984) 85. "The York Map is a Canonical Transformation."
115. W. Blyth and C.J. Isham, *Phys. Rev.* **D11** (1975) 768. "Quantization of a Friedman Universe Filled with a Scalar Field."
116. K.V. Kuchař, *J. Math. Phys.* **17** (1976) 777, Section 10. "Geometry of Hyperspace."
117. K.V. Kuchař, *J. Math. Phys.* **17** (1976) 792, Section 3. "Kinematics of Tensor Fields in Hyperspace."
118. A. Einstein, *Über die spezielle und die allgemeine Relativitätstheorie* (Vieweg, Braunschweig, 1920), §28, p.67; English transl. R.W. Lawson, *Relativity: The Special and the General Theory*, 17th ed. (Crown, New York, 1961).

119. D. Hilbert, *2 Mitt.*, *Nachr. Ges. Wiss. Göttinger* **53** (1917). "*Grundlagen der Physik.*" Included in *David Hilbert gessammelte Abhandlungen*, Vol. **3**, 2nd ed. (Springer, Berlin, 1970).

120. B.S. DeWitt, in *Gravitation: An Introduction to Current Research*, ed. L. Witten (Wiley, New York, 1962). "The Quantization of Geometry."

121. N. Bohr and L. Rosenfeld, *Mat.-Fys. Medd. Dan. vid. Selsk.* **12** No. 8 (1933). "Zur Frage der Messbarkeit der elektromagnetischen Feldgrossen." English transl. in *Selected Papers by Léon Rosenfeld*, eds. R.S. Cohen and J. Stachel, *Boston Studies in the Philosophy of Science* **21** (Reidel, Dordrecht, 1978). "On the Question of the Measurability of Electromagnetic Field Quantities."

122. C. Rovelli, *Sissa Preprint* 101/89/ep. (1989). "What is Observable in Classical and Quantum Gravity."

123. P.G. Bergmann and A. Komar, *Int. J. Theor. Phys.* **5** (1972) 15. "The Coordinate Group of Symmetries of General Relativity."

124. D.C. Salisbury and K. Sundermeyer, *Phys. Rev.* **D27** (1983) 740. "Realization in Phase Space of General Coordinate Transformations."

125. J. Lee and R.M. Wald, *J. Math. Phys.* **31** (1990) 725. "Local Symmetries and Constraints."

126. R.M. Wald, *Univ. of Chicago Preprint* (1991). "The Relationship between Local Symmetries and Constraints in General Relativity and Other Classical Lagrangian Field Theories."

127. C.J. Isham and K.V. Kuchař, *Ann. Phys. (N.Y.)* **164** (1985) 288. "Representations of Spacetime Diffeomorphisms. I. Canonical Parametrized Field Theories."

128. C.J. Isham and K.V. Kuchař, *Ann. Phys. (N.Y.)* **164** (1985) 316. "Representations of Spacetime Diffeomorphisms. II. Canonical Geometrodynamics."

129. K.V. Kuchař, *Found. Phys.* **16** (1986) 193. "Canonical Geometrodynamics and General Covariance."

130. J.B. Hartle, in *Proceedings of the Fifth Marcel Grossmann Meeting*, eds. D. Blair and M. Buckingham (World Scientific, Singapore, 1989). "Time and Prediction in Quantum Cosmology."

131. J.J. Halliwell and J.B. Hartle, *Phys. Rev.* **D43** (1991) 1170. "Wave Functions Constructed from Invariant Sums-Over-Histories Satisfy Constraints."

132. K.V. Kuchař and C.G. Torre, *Phys. Rev.* **D44** (1991); in print. "Harmonic Gauge in Canonical Gravity."

133. C.L. Stone and K.V. Kuchař, submitted to *Class. Quantum Grav.* "Representations of Spacetime Diffeomorphisms in Canonical Geometrodynamics under Harmonic Coordinate Conditions."

134. K.V. Kuchař, submitted to *Phys. Rev.* **D.** "Extrinsic Curvature as a Reference Fluid in Canonical Gravity."

135. C.G. Torre, *Phys. Rev.* **D40** (1989) 2588. "Hamiltonian Formulation of Induced Gravity in Two Dimensions."

136. K.K. Tam and J. O'Hanlon, *Nuovo Cimento* **62B** (1969) 351. "Relativistic Magnetohydrodynamics of a Gravitating Fluid."

137. B.F. Schutz, *Phys. Rev.* **D4** (1971) 3559. "Hamiltonian Theory of a Relativistic Perfect Fluid."

138. C. Eckart, *Phys. Rev.* **58** (1940) 919. "The Thermodynamics of Irreversible Processes III. Relativistic Theory of the Single Fluid."

139. G.A. Kluitenberg and S.R. deGroot, *Physica* **21** (1955) 148. "Relativistic Thermodynamics of Irreversible Processes. IV. Systems with Polarization and Magnetization in an Electromagnetic Field."

140. S.W. Hawking and G.F.R. Ellis, *The Large Scale Structure of Space-Time* (Cambridge Univ. Press, Cambridge, 1973) Section 4.3.

141. Ref 28, Sections 8 and 10.

142. A. Einstein, *Ann. Phys. (Leipzig)* **49** (1916) 769; see pp. 801 and 812. "Die Grundlage der allgemeinen Relativitätstheorie."

143. W.G. Unruh, *Phys. Rev.* **D40** (1989) 1048. "Unimodular Theory of Canonical Quantum Gravity."

144. A. Einstein, *Sitzunsber. Preuss. Acad. Wiss., Phys. - Math. Kl.* (1919). English transl. in *The Principle of Relativity* (Dover, New York, 1952). "Do Gravitational Fields Play an Essential Part in the Structure of the Elementary Particles of Matter?"

145. J.J. Van de Bij, H. van Dam and Y.J. Ng, *Physica* **A116** (1982) 307. "The Exchange of Massless Spin-Two Particles."

146. F. Wilczek, *Phys. Rep.* **104** (1984) 143. "Foundations and Working Pictures in Microphysical Cosmology."

147. W. Buchmüller and N. Dragon, *Phys. Lett.* **B207** (1988) 292. "Einstein Gravity from Restricted Coordinate Invariance."

148. S. Weinberg, *Rev. Mod. Phys.* **61** (1989) 1. "The Cosmological Constant Problem."

149. M. Henneaux and C. Teitelboim, *Phys. Lett.* **B222** (1989) 195. "The Cosmological Constant and General Covariance."

150. W.G. Unruh, in *Proceedings of the Fourth Seminar on Quantum Gravity*, eds. M.A. Markov, V.A. Berezin and V.P. Frolov (World Scientific, Singapore, 1988). "Time and Quantum Gravity."

151. R.D. Sorkin, in *History of Modern Gauge Theories*, eds. M. Dresden and A. Rosenblum (World Scientific, Singapore, 1990). "On the Role of Time in the Sum-Over-Histories Framework for Gravity."

152. R.D. Sorkin, in *Conceptual Problems of Quantum Gravity*, eds. A. Ashtekar and J. Stachel (Birkhäuser, Boston, 1991). "Problems with Causality in the Sum-Over-Histories Framework for Quantum Mechanics."
153. J.D. Brown and J.W. York, *Phys. Rev.* **D40** (1989) 3312. "Jacobi's Action and the Recovery of Time in General Relativity."
154. K.V. Kuchař, *Phys. Rev.* **D43** (1991) 3332. "Does an Unspecified Cosmological Constant Solve the Problem of Time in Quantum Gravity?"
155. A. Ashtekar and A. Magnon, *Proc. Roy. Soc. London* **A346** (1975) 375. "Quantum Fields in Curved Spacetimes."
156. A. Magnon-Ashtekar, *Thesis* (Département de Mathématiques Pures, Université de Clermont, Clermont, 1976); unpublished.
157. A. Ashtekar and A. Magnon-Ashtekar, *C.R. Acad. Sci. (Paris)* **A286** (1978) 531. "Remark on Quantum Field Theory in Curved Space-Times."
158. B.S. Kay, *Commun. Math. Phys.* **62** (1978) 55. "Linear Spin-Zero Quantum Fields in External Gravitational and Scalar Fields."
159. K.V. Kuchař, *J. Math. Phys.* **22** (1981) 2640. "General Relativity: Dynamics Without Symmetry."
160. K.V. Kuchař, *J. Math. Phys.* **25** (1982) 1647. "Conditional Symmetries in Parametrized Field Theories."
161. J. Friedman and A. Higuchi, *Phys. Rev.* **D41** (1990) 2479. "Symmetry and Internal Time on the Superspace of Asymptotically Flat Geometries."
162. K.V. Kuchař, *J. Math. Phys.* **18** (1977) 1589. "Geometrodynamics with Tensor Sources."
163. T.P. Singh and T. Padhanabhan, *Ann. Phys. (N.Y.)* **196** (1989) 296. "Notes on Semiclassical Gravity."
164. T. Banks, *Nucl. Phys.* **B249** (1985) 332. "TCP, Quantum Gravity, the Cosmological Constant, and All That."
165. A. Vilenkin, *Phys. Rev.* **D39** (1989) 1116. "The Interpretation of the Wave Function of the Universe."
166. R. Brout, *Found. Phys.* **17** (1987) 603. "On the Concept of Time and the Origin of the Cosmological Temperature."
167. R. Brout and G. Venturi, *Phys. Rev.* **D39** (1989) 2436. "Time in Semiclassical Gravity."
168. R. Brout, G. Horowitz, and D. Wiel, *Phys. Lett.* **B192** (1987) 318. "On the Onset of Time and Temperature in Cosmology."
169. J.B. Hartle, in *Gravitation in Astrophysics, Cargese, 1986*, eds. B. Carter and J.B. Hartle (Plenum, New York, 1986). "Prediction and Observation in Quantum Cosmology."
170. J.J. Halliwell, in *Quantum Cosmology and Baby Universes*, eds. S. Coleman, J. Hartle, T. Piran and S. Weinberg (World Scientific, Singapore, 1991). "Introductory Lectures on Quantum Cosmology."

171. T. Padhanabhan, *Class. Quantum Grav.* **6** (1989) 533. "Semiclassical Approximations to Gravity and the Issue of Back-Reaction."
172. C. Møller, in *Les Théories Relativisties de la Gravitation*, eds. A. Lichnerowicz and M.A. Tonnelat (CNRS, Paris, 1962). "The Energy-Momentum Complex in General Relativity and Related Problems."
173. L. Rosenfeld, *Nucl. Phys.* **40** (1963) 353. "On Quantization of Fields."
174. H.D. Zeh, *Phys. Lett.* **A116** (1986) 9. "Emergence of Time From a Universal Wave Function."
175. H.D. Zeh, *Phys. Lett.* **A126** (1988) 311. "Time in Quantum Gravity."
176. E. Joos, *Phys. Lett.* **A116** (1986) 6. "Why do We Observe a Classical Spacetime?"
177. C. Kiefer, *Class. Quantum Grav.* **4** (1987) 1369. "Continuous Measurement of Mini-Superspace Variables by Higher Multipoles."
178. J.J. Halliwell, *Phys. Rev.* **D39** (1989) 2912. "Decoherence in Quantum Cosmology."
179. V. Lapchinski and V.A. Rubakov, *Acta Phys. Pol.* **B10** (1979) 1041. "Canonical Quantization of Gravity and Quantum Field Theory in Curved Spacetime."
180. J.J. Halliwell, *Phys. Rev.* **D36** (1987) 3626. "Correlations in the Wave Function of the Universe."
181. J.B. Barbour, *Preprint 1991*. "Time, Gauge Fields, and the Schrödinger Equation in Quantum Cosmology."
182. A. Vilenkin, *Phys. Rev.* **D37** (1988) 888. "Quantum Cosmology and the Initial State of the Universe."
183. T. Padhanabhan, *Pramana J. Phys.* **35** (1990) L199. "A Definition for Time in Quantum Cosmology."
184. J. Greensite, *Nucl. Phys.* **B342** (1990) 409. "Time and Probability in Quantum Cosmology."
185. C. Kiefer, *Zürich University Preprint* ZU-TH 10/91. To appear in the *Proceedings of the Journées Relativistes, Cargèse, 2.5.-4.5 1991*. "How does Quantum Gravity Modify the Schrödinger Equation for Matter Fields?" C. Kiefer and T.P. Singh, *Preprint 1990*. "Quantum Gravitational Corrections to the Functional Schrödinger Equation."
186. G.T. Horowitz and R.M. Wald, *Phys. Rev.* **D21** (1980) 1462. "Quantum Stress Energy in Nearly Conformally Flat Spacetimes."
187. S. Danjbar-Daemi, B.S. Kay and T.W.B. Kibble, *Phys. Lett.* **91B** (1980) 417. "Renormalization of Semiclassical Field Theories."
188. T.W.B. Kibble, in *Quantum Gravity 2: A Second Oxford Symposium*, eds. C.J. Isham, R. Penrose and D.W. Sciama (Clarendon, Oxford, 1981). "Is Semiclassical Theory of Gravity Viable?"

189. M.J. Duff, in *Quantum Gravity 2: A Second Oxford Symposium*, eds. C.J. Isham, R. Penrose and D. W. Sciama (Clarendon, Oxford, 1981). "Inconsistency of Quantum Field Theory in Curved Space-Time."
190. W-M. Suen, *Phys. Rev. Lett.* **62** (1989) 2217. "Minkowski Spacetime is Unstable in Semiclassical Gravity."
191. W-M. Suen, *Phys. Rev.* **D40** (1989) 315. "Stability of the Semiclassical Einstein Equation."
192. S. Giddings and A. Strominger, *Nucl. Phys.* **B321** (1988) 481. "Baby Universes, Third Quantization and the Cosmological Constant."
193. S. Coleman, *Nucl. Phys.* **B310** (1988) 643. "Why is There Something Rather Than Nothing: A Theory of the Cosmological Constant."
194. M. McGuigan, *Phys. Rev.* **D38** (1988) 3031. "Third Quantization and the Wheeler-DeWitt Equation."
195. M. McGuigan, *Phys. Rev.* **D39** (1989) 2229. "Universe Creation From the Third Quantized Vacuum."
196. L. Rosenfeld, in *Niels Bohr and the Development of Physics* (Pergamon, London, 1955). "On Quantum Electrodynamics."
197. C. Møller, *The Theory of Relativity*, 2nd ed. (Clarendon, Oxford, 1972). "8.8. Space and Time Measurements in an Arbitrary System of Reference."
198. L.D. Landau and E.M. Lifshitz, *Teorija polja* (Gosudarstvennoe izdatel'stvo fizikomatematiceskoj literatury, Moskva, 1962). English transl. M. Hamermesh, *The Classical Theory of Fields* (Pergamon, Oxford, 1962). "84. Distances and Time Intervals."
199. J.B. Hartle and S.W. Hawking, *Phys. Rev,* **D28** (1983) 2960. "Wave Function of the Universe."
200. S.W. Hawking, *Nucl. Phys.* **B239** (1984) 257. "The Quantum State of the Universe."
201. S.W. Hawking, in *Relativity, Groups and Topology II*, *Les Houches Session XL*, eds. B. DeWitt and R. Stora (North Holland, Amsterdam, 1984). "Lectures on Quantum Cosmology."
202. S.W. Hawking and D.N. Page, *Nucl. Phys.* **B264** (1986) 185. "Operator Ordering and the Flatness of the Universe."
203. S.W. Hawking and D.N. Page, *Nucl. Phys.* **B298** (1988) 789. "How Probable is Inflation?"
204. D.N. Page, *Phys. Rev.* **D34** (1986) 2267. "Density Matrix of the Universe."
205. D.N. Page and W.K. Wootters, *Phys. Rev.* **D27** (1983) 2885. "Evolution Without Evolution: Dynamics Described by Stationary Observables."
206. D.N. Page, *ITP Preprint* NSF-ITP-89-18 (1989). "Time as an Inaccessible Observable."

207. D.N. Page, in *Conceptual Problems of Quantum Gravity*, eds. A. Ashtekar and J. Stachel (Birkhäuser, Boston, 1991). "Interpreting the Density Matrix of the Universe."

208. W.K. Wootters, *Int. J. Theor. Phys.* **23** (1984) 701. "Time Replaced by Quantum Correlations."

209. C. Teitelboim, *Phys. Lett.* **96B** (1980) 77. "Proper Time Approach to the Quantization of the Gravitational Field."

210. C. Teitelboim, *Phys. Rev.* **D25** (1982) 3159. "Quantum Mechanics of the Gravitational Field."

211. M. Henneaux and C. Teitelboim, *Ann. Phys.* **143** (1982) 127. "Relativistic Quantum Mechanics of Supersymmetric Particles."

212. J.B. Hartle, *Phys. Rev.* **D37** (1988) 2818. "Quantum Kinematics of Spacetime I: Non-Relativistic Theory." *Phys. Rev.* **D43** (1991) 1434. "Erratum."

213. J.B. Hartle, *VCSB Preprint*. "Quantum Kinematics of Spacetime III: General Relativity."

214. J.B. Hartle, in *The Proceedings of the 60th Birthday Celebration for Murray Gell-Mann, Pasadena, CA, January 27, 1989*, eds. J. Schwarz and F. Zachariasen (Cambridge Univ. Press, Cambridge, 1990). "Excess Baggage."

215. M. Gell-Mann and J.B. Hartle, in *Complexity, Entropy, and the Physics of Information, Santa Fe Institute Studies in the Sciences of Complexity, VIII*, ed. W.H. Zurek (Addison-Wesley, Reading, 1990). "Quantum Mechanics in the Light of Quantum Cosmology."

216. J.B. Hartle, in *Quantum Cosmology and Baby Universes*, eds. S. Coleman, J. Hartle, T. Piran and S. Weinberg (World Scientific, Singapore, 1991). "The Quantum Mechanics of Cosmology."

217. J.B. Hartle, *Preprint UCSBTH91-02*. "Spacetime Coarse Grainings in Non-Relativistic Quantum Mechanics."

218. P.G. Bergmann, *Nuovo Cimento* **3** (1956) 1956. "Introduction of 'True Observables' into the Quantum Field Equations."

219. P.G. Bergmann, *Helv. Phys. Acta Suppl.* **4** (1956) 1956. "Quantisierung allgemeinkovarianter Feldtheorien."

220. E. Newman and P.G. Bergmann, *Rev. Mod. Phys.* **29** (1957) 443. "Observables in Singular Theories by Systematic Approximation."

221. A. Komar, *Phys. Rev.* **111** (1958) 1182. "Construction of a Complete Set of Independent Observables in the General Theory of Relativity."

222. B.S. DeWitt, Ref. 32, Section 4, p. 160.

223. D.R. Brill and R.H. Gowdy, *Rep. Progress Phys.* **33** (1970) 413. "Quantization of General Relativity. 7.1. The Frozen Formalism Problem of Generally Covariant Theories."

224. C. Rovelli, in *Conceptual Problems of Quantum Gravity*, eds. A. Ashtekar and J. Stachel (Birkhäuser, Boston, 1991). "Is There Incompatibility Between the Ways Time is Treated in General Relativity and in Standard Quantum Mechanics?"

225. C. Rovelli, *Phys. Rev.* **D42** (1990) 2638. "Quantum Mechanics Without Time: A Model."

226. C. Rovelli, *Phys. Rev.* **D43** (1991) 442. "Time in Quantum Gravity: An Hypothesis."

227. P. Hájíček, *Phys. Rev.* **D44** (1991) 1337. "Comment on 'Time in Quantum Gravity: An Hypothesis'."

228. L. Van Hove, *Acad. Roy. Belg. Bull. Cl. Sci.* **37** (1951) 610. "Sur le probléme des relations entre les transformations unitaires de la mécanique quantique et les transformations canoniques de la mécanique classique."

229. C. Rovelli, *Phys. Rev.* **D44** (1991) 1339. "Quantum Evolving Constants. Reply to 'Comment on Time in Quantum Gravity: An Hypothesis'."

230. J.B. Hartle, *Private communication* (1991).

COMPUTER ALGEBRA IN GRAVITY – PAST, PRESENT AND FUTURE

M.A.H. MacCALLUM[1]
Laboratoire de Physique Théorique, Institut Henri Poincaré, 11, rue Pierre et Marie Curie, 75231 Paris cedex 05, FRANCE.

1 Introduction

General relativity was a testing ground of many early computer algebra (CA) systems. This was essentially because of the large size of the Einstein equations expressed in terms of a general metric in coordinate form: there are 4 equations with 9 990 terms and 6 with 13 280 terms. Such large expressions stretched early systems to their limits, under the storage and time constraints of those days, and enabled system designers to discover where they could improve the efficiency of their algorithms. However, most of this activity did not really feed back into research in gravity, being directed to checking already known results.

Moreover, a multiplicity of different systems arose, each essentially used only by its creator and his or her group. (D'Inverno's review [1] listed these systems.) Such bespoke systems also exist in celestial mechanics, where one leading practitioner, Deprit, has likened them to a toothbrush, something one uses every day but would not lend to anybody else. Unfortunately, they were not treated like a toothbrush in another way – they were not replaced with a newer and better one at regular intervals. Part of the reason for this was the slow dissemination of information about the comparative efficiency of different algorithms (for some relevant comments see [2]), and other techniques: much of this knowledge is still not recorded or discussed in the open literature.

The other factor inhibiting the growth of computer algebra systems for gravity was the storage requirement, which I feel has also been the main technological determinant of the history of computer algebra in general. (One can therefore criticise my history of the subject as having a rather Marxist flavour – arguing that material factors determine cultural and social evolution.) But before proceeding to explain this aspect, I would like to say a few words about why CA is a good thing.

The first advantage is that it is accurate: well-written computer algebra packages do not lose factors of 2 or get the sign wrong. For example, in an investigation of a sample of results in integral tables Klerer and Grossman [3] showed that even such well-respected tables as those of Gradsteyn and Rhyzik contained, as of 1960, (with 95% confidence) errors in more than 4% of the entries. (In fact, new tables being prepared now use computer algebra to prepare the entries!) Computer algebra is

[1]Permanent address: Astronomy Unit, School of Mathematical Sciences, Queen Mary and Westfield College, Mile End Road, LONDON E1 4NS, U.K. E-mail: mm@maths.qmw.ac.uk

not only accurate but fast: the classic example is Delaunay's calculations of lunar motion in the last century, which took him 20 years, but have since been repeated in 9 months programming and a few minutes CPU time. (For the most recent developments in this particular area see [4].) An example in relativity is given by the famous Bondi metric, where the initial calculations for [5] took 6 months and still contained errors, whereas by 1967 d'Inverno's ALAM could repeat the calculation (correctly!) in 4 minutes and the computation time is now down to 30s or less.

A third advantage is the removal of the tediousness of repeated almost identical calculations. Both this and the final remark I would make, which is that the power of the systems opens up projects which would be unthinkable by hand, sometimes with unexpected results, are especially relevant to my own research, and to other applications in gravity theory.

2 The requirements of computer algebra systems

In a numerical program, the storage required for a particular variable is predictable in advance: for double precision floating point numbers in FORTRAN, typically 64 bits. Thus the programmer can work out in advance the total storage requirement of a program, and find out whether it will run on a simple PC or needs a supercomputer. The same applies to estimating computation times.

In CA, this is no longer the case. The algebraic expression which is the value of a variable may be of any length, and even such a simple expression as $2x^3 + y^2$ converted to the usual input-output format as 2*x**3+y**2 and stored as 8-bit ASCII code would already take more than the 64 bits of typical double precision FORTRAN. This basic technical difference between symbolic and numerical calculation has a number of side-effects.

One is that the overall storage requirement tends to be high. A megabyte is now the minimum for most gravity calculations, which is why the repeated requests to workers on CA in gravity to supply systems for IBM PCs with a maximum 640K have tended to fall on deaf ears. (However, the MuTensor system of Harper and Dyer [6] is capable of quite advanced calculations inside these restrictions, and other small systems, e.g. GRG [7], exist.) 4 Mb machines can accommodate routine problems comfortably, though requirements can of course rise substantially beyond that. (For instance, on 6 November 1990 one of my department's machines essentially came to a standstill while a pure mathematics colleague ran a 21Mb symbolic calculation.)

In fact one has to be careful in advertising the extent to which CA can handle large problems. It can indeed tackle expressions with hundreds of thousands or even millions of terms (though the value of the output if the expressions do not simplify seems rather doubtful), but combinatorial explosions quickly outrun the capacity of any system. For instance, the number of terms in the determinant of an $n \times n$ matrix each of whose entries is a sum of 2 terms is $2^n n!$. This grows rapidly with n, and most general purpose algebra systems on an average machine become unable to cope at about $n = 10$, which often surprises unwary users who reason that since they can handle $n = 3$ easily and are not daunted by $n = 4$, a machine should be able to do much more. A reasonable estimate, due to Fritz Schwarz, of the improvement in the size of expressions which can be handled by using CA instead of "hand calculation" is a factor of 10^4.

A second effect comes from "intermediate expression swell". As an example, consider the multiplication of 2 polynomials of 10th degree, each having 11 terms. Taking all possible cross-products gives 121 terms, while the final answer must be a 20th degree polynomial with at most 21 terms. So in the middle of this calculation one has 100 more terms than at the end. This has two consequences: one of them is to again boost the overall storage requirement, but the new feature is the importance of "garbage collection", where storage used but no longer required can be re-allocated for a new use. It is not impossible to design garbage collection in FORTRAN, but it is not a standard feature, whereas it is standard in LISP and the more recently available C. That is why those languages are the most common bases of CA systems, LISP being used for systems like MACSYMA and REDUCE whose development started long ago, and C for more recent systems like MAPLE and *Mathematica*.

These language choices however acted, and to some extent still act, as an inhibiting factor in the spread of CA. Users are reluctant to learn a new language, and still more a new programming style. In particular, recursive algorithms, which as yet are impossible in FORTRAN, are almost essential in CA as a way to cope with expressions whose length is *a priori* unknown. It has been said, by my collaborators in Stockholm, that one can learn LISP programming in 3 weeks unless one has learnt FORTRAN first, in which case one must double the time.

The third effect is the need for simplification. There is a theorem showing that even apparently elementary classes of simplification problems are not formally decideable (this result is due to Richardson and Caviness: see e.g. [8]), but more importantly in practice there may be simplifications special to the problem which it would be uneconomic to program into simplification routines for general use, since in most cases the time spent testing their applicability would just be wasted. Therefore CA systems (while they vary in the amount of automatic simplification) usually provide a mechanism for the user to introduce problem-specific substitutions and simplification rules. To control and test these rules, an interactive operating system is highly desirable (one has to be really dedicated to use CA systems in a batch environment). Such operating systems only became commonplace with microcomputers (and their slow introduction by most mainframe manufacturers may explain the lead held for many years by Digital Equipment machines as a base for algebra systems).

The last effect of the data dependence of the performance of algebra systems is seen in execution times. Complexity estimates in CA depend on the data much more than their analogues for numerical programs. Prediction is difficult. As an example, consider the first two lines of Table 1, an edited version of the table in [9]. The standard REDUCE test file took about 15 times as long on an Atari as on a Sun 3/260, but the specific example of the Nielsen-Pedersen metric took only 1.5 times as long. This variation, with nominally the same algebra system (REDUCE) though using different LISP implementations, to me illustrates that one should disbelieve any sales talk for CA systems and pay attention to actual performance on problems of the sort one really wants to run.

Table 1: Comparison of times for verification of a metric of Nielsen and Pedersen. Times are given as (CPU time for calculation) + (garbage collection time). All times are in seconds. The line labelled "Barton-Fitch" refers to the segment of the standard REDUCE test program which concerns relativity. The SHEEP lines labelled "CORD" and "FRAME" refer to coordinate and tetrad facilities in SHEEP respectively.

Input	Atari	Orion	Sun 3
REDUCE test file	275	80	18.5
Barton-Fitch D	766 (+126)	355 (+116)	< 592
EXCALC coordinates (original)	>15000+7200	13140	890
EXCALC coordinates (modified)	2384 (+490)	1271 (+460)	< 421
EXCALC tetrad	202 (+23)	104 (+40)	< 21
SHEEP CORD	206 (+58)	115 (+42)	
SHEEP FRAME	25	10.6+0	

The table also shows other interesting features. One is that even in a good package (which EXCALC is) a poor choice of algorithm can have dramatic effects, as shown in the first two lines for EXCALC. (In fact SHEEP [20] on the Atari ran 6 times *faster* than this algorithm on a Cray X-MP.)

A second is that a specialized system (which SHEEP is, for relativity and tensor theories) may be worth having if one repeatedly performs calculations of a restricted type, rather than using a package based on a general purpose system. (The drawback of course lies in the limited features and the sometimes convoluted means used to overcome the related shortcomings: for example, the way in which clever substitutions in SHEEP are used to overcome the lack of polynomial division or factorization [10].)

Finally, the table illustrates the principle first clearly stated in [11], that tetrad (equivalently, vierbein, moving frame, or Newman-Penrose) calculations are generally much more efficient than coordinate methods (in fact the speed-up in this example, a factor of 8-10, is greater than usual).

Returning to the main theme of this section, the storage needs of CA in GR, the good news, which we all know, is that availability of store on desktop machines is now much greater than it was on mainframes 10 years ago. When I began using CA in GR (not counting a early experience in the late 1960s), I could, on sufferance, get use of 0.5 Mb on a DEC or IBM mainframe. Now at home I have a 4Mb machine which can run Maple, REDUCE and SHEEP, and the availability and cheapness of such machines is bringing CA within everybody's reach. Moreover, these machines run faster than those old mainframes. (As examples, starting from the cheapest, one can cite the Atari ST, IBM 386 PCs, Macintosh 2, Acorn R140, Sun SLC, DECstation 3100 and so on.)

Again this should not be overstated. Students, brought up in the new age of plenty, sometimes give the impression of being hard done-by without a 32Mb 20Mips machine of their own, but readers should perhaps ponder whether it is still cheaper to pay 6 months salary while a more efficient system is written rather than just throw silicon at a problem.

3 CA packages for gravity theory

Packages for gravity theory are of two main types, those which calculate components of tensors and those which handle indexed objects in the standard textbook manner. I will refer to these as component calculations and indicial calculations. Component calculation packages are the more common. To clarify what the outputs might look like here is some printout from a component calculator (SHEEP's CLASSI package [21])

```
                2 -2       -2              -1
R     = -BCFR U    -2BFU    +BG +1/2CFU
 1313


             -2   -2
R     = FGR    -R
 2323
```

and an indicial calculator (SHEEP's STENSOR package)

```
 i            i     n       i     n       i         i
R     = -GAM   GAM    +GAM   GAM    -GAM       +GAM
  klm          nm    kl      nl    km      kl,m      km,l
```

With modern high-resolution screens and output routines it is possible to produce output even closer to textbook styles, using Greek characters, variable vertical spacing, and so on, but these refinements, which are icing on the cake, are not yet generally available (though most systems do have some interface to TeX).

As for capabilities, the systems of the past were, in many cases, limited to calculation of the curvature and Ricci tensors from a metric in terms of coordinates. We would now expect, of a good system, the following facilities

- a much wider built-in range of tensor definitions
- a variety of tetrad formalisms
- spinor formalisms
- variable dimension, non-metric connections, and auxiliary fields (e.g. Yang-Mills)
- easy methods for the user to extend the system, e.g. add new tensors
- ability to handle commutation rules
- both indicial and coordinate calculations, and full integration of the two modes

No current system completely meets all these requirements, but several provide a significant fraction, and they are still developing. The major general purpose systems all offer packages for gravity theory, or at least for general relativity.

The oldest of the still current general purpose systems are REDUCE and MACSYMA. REDUCE is a highly portable system available on a very wide range of machines and is sufficiently cheap to have become widespread in most parts of the world (it is probably the most widely used in a geographical sense). Its main disadvantage until recently has been the rather small range of auxiliary packages for applied mathematics, but this is improving rapidly with the availability of contributed programs through electronic mail (to reduce-netlib@rand.org: an initial mail should contain the one line 'send index'), and their inclusion with the distribution tapes of the just released version 3.4.

The exterior calculus program EXCALC based on REDUCE is now distributed with REDUCE. For some details of its capabilities, which can be summarized as the ability to handle differential forms in a notation very close to that of a textbook, see [12]. Some other REDUCE programs for relativity, for example those of Dautcourt and McCrea, are described in [14] and [15]. One very useful recent development is a symbiotic system containing REDUCE and SHEEP (or CLASSI or STENSOR) developed by Dr. J.E.F. Skea in my department [10]. Another notable REDUCE package for gravity theory is REDTEN, developed by Dyer and Harper [16] at Toronto, which has some indicial calculation features at least on input, and is a REDUCE implementation of a system also available under MuMATH [6] and Maple.

MACSYMA has for many years been the leader in terms of auxiliary packages. It is a very large system (e.g. the running image for a Sun is 13 Mb), widely used in the U.S. and Europe. There are now three versions: one from Symbolics, Inc., one named ALJABR from Fort Pond Research, and one called PARAMACS from Paradigm Associates (this last was formerly sold as DoE-Macsyma). Because the Symbolics group has been led by two people with Ph.D.'s in gravity theory (Richard Pavelle and Richard Petti), this version's facilities in our area have probably been developed the most.

The MACSYMA CTENSOR and ITENSOR programs for component and indicial manipulation were recently corrected and improved. In particular both were extended to provide handling of frame fields and of affine torsion and nonmetricity. New programs, CARTAN for exterior calculus and ATENSOR for tensor algebra, are now available. CARTAN provides handling of Lie derivatives and contractions of vectors with forms as well as the basic exterior algebra and differentiation. ATENSOR covers multilinear operators, symplectic, Clifford and Grassmannian algebras, and Lie enveloping algebras.

The small system MuMATH, first distributed as MuMATH-79 for Z80-based microcomputers, and rewritten as MuMATH-83 for the 16-bit IBM PCs, has not been widely used in relativity, perhaps because it seems to use too small a data space to be able to cope with the size of the Einstein equations. The Dyer and Harper package MuTensor, already referred to above, turns out, however, to be surprisingly effective. (It should be noted that MuMATH is no longer distributed and has been replaced by DERIVE, for which programmability has only just been implemented. However, MuMATH has been rewritten for the Atari ST and TT with 32-bit address pointers, allowing a much larger data space, and with a windowing interface: this version, whose manual is currently available only in German, is called RIEMANN.)

More important for relativity, probably, are the newer systems Maple and *Mathematica*, both written in C rather than the Lisp which underlies the older systems. Maple now rivals MACSYMA in its applications packages, and its structure makes it an excellent system for teaching purposes (although I personally have some reservations about its use in research). It is of course a Canadian product, from the University of Waterloo.

The gravity and relativity programs available for Maple also come mainly from Waterloo, though as mentioned above there is a Maple version of the Dyer and Harper package. The Waterloo group producing these programs is led by Ray McLenaghan. The capabilities provided are now quite extensive. They include: computation of connection and curvature in coordinate, general moving frame and complex null bases; determination of the Petrov type of the Weyl tensor; computer-aided integration of the Einstein equations in Newman-Penrose and orthonormal bases; computation of integrability conditions; classification of Bianchi group type from structure constants; and some differential form capabilities. One small criticism is that at the present stage these programs are not unified into one system so that data can readily be passed from one to another package. (During the Winnipeg meeting the programs were demonstrated by Prof. McLenaghan and Dr. Fee and Dr. Fee also presented a paper on the differences between algorithms for computing curvature.) One useful new Maple program (by Czapor and Mclenaghan) can take spinor equations in indicial form and generate the corresponding dyad equations, i.e. write out all components of the equations using Newman-Penrose notation.

Mathematica is still new. It is being very aggressively marketed and its features seem to be designed with the physicist in mind (to a degree where some mathematicians have complained about the system in much the same way they complain about physicists' mathematics in general). Initially, its unique strengths were in its integration of numerical and graphical facilities with algebraic ones, and its very nice user interface. In these respects it put other systems, whose progress in these areas, while not non-existent, had been leisurely, to shame, but recently the other systems have been catching up fast, especially on the graphics. (I still find it odd to see algebra systems sold on the basis of their graphics, however.) Workers on other computer algebra systems have been hesitant to give an opinion of the algebraic part after the rather acrimonious controversy [17] about the previous algebra system SMP also designed by *Mathematica*'s principal designer, Stephen Wolfram, but an critical review by Richard Fateman is now in press [18].

Stephen Christensen and Leonard Parker have written a system MathTensor in *Mathematica*, which they aim to market commercially. (This, incidentally, is a new development in computer algebra for relativity, the programs for which, though not the underlying systems, have up-to-now been free.). The structure of this system is based on a kernel of functions and rules, the most basic being simplification functions capable of coping with long expressions containing products and contractions of Riemann and other tensors, to which a knowledge base for specific tensors is added. This makes the system easy to extend and suitable for pedadogic purposes. The system recognises the summation convention, and 'knows' for example that commuting covariant derivatives leads to terms involving the Riemann tensor. As far as I am aware, it is still at the "beta-test" phase and under very active development.

The other general purpose systems are either little used or at least little used for gravity. IBM's SCRATCHPAD II system, which has been under development for many years now, has at last been given commercial release by NAG under the name Axiom. It is a large and powerful, and very high-level, system which may be of interest for symbolic calculation in gravity because of the very abstract approach to datatypes, based on category theory and an object-oriented programming approach, which it embodies. Another newly-available system is Jos Vermaseren's FORM (for Atari STs), which has been specially designed to handle very large expressions of the type arising in some quantum field calculations, and in particular to exploit the use of temporary disk storage (also used in STENSOR).

As machine facilities have grown the value of small specialized systems has declined and most of them have become defunct. Of the specialized systems which were in use among relativists, Camal and Ortocartan have undergone little or no development in recent years, and Schoonschip remains confined to a smallish number of people working with quantum field theory problems. The GRATOS system for relativistic celestial mechanics is still under development and in use by the Leningrad group led by Brumberg, and there is the useful small system, GRG, with some exterior calculus facilities, already referred to. There will always be a place for specialized systems for specific problem domains where efficiency, or availability of specialized data structures, is very important: for examples in our field one can cite the programs of Hoenselaers or Fulling (the latter of which was presented in another talk at Winnipeg).

The only other major specialized system is SHEEP, with its extensions CLASSI and STENSOR, of which I have been an intensive user for some years. This system offers component and indicial calculus, and still has, in my opinion, the widest range of features for gravity (or tensor handling in general), integrated in a unified system, of any of the current packages (though there are also aspects where other systems are definitely better). A full description is given in [10].

In summary, the programs available for gravity theorists are still under active development. The main trend is for increased abstraction: more and more programs handle differential forms (e.g. EXCALC, Maple's *difforms*, MACSYMA's CARTAN) or tensors with symbolic indices (e.g. STENSOR, MathTensor, Excalc) and the "indicial" manipulators can all interface to component-wise computations. The answer to "Which CA system is best?" has to be the question "What for?", but I hope the remarks above give readers some help in starting to form a judgement.

4 Applications of CA in gravity theory

To relativists, algebraic computing is primarily a tool, rather than an end in itself, although interesting and new problems can arise in designing algorithms for relativity. The number of researchers using the tool is rising rapidly, and I want to briefly outline its uses.

In investigations of non-linear theories, in particular of gravity, there are four approaches one could adopt.

A One could attempt to create a new theory, or a new variant of an old theory. As the literature shows, it is not hard to do this, but it is very hard to do it successfully.

B One can attempt to prove general theorems about a theory or a group of theories. As an example, one can take the singularity theorems of Hawking, Penrose and others. Results of this kind are very hard to obtain.

C One can work, as most theoretical physicists do, on the investigation of properties of theories using approximation schemes. For example, perturbative quantum field theory or numerical fluid dynamics fall into this category. Not only are the computations far from trivial, but in the case of non-linear theories it is very hard to justify them, i.e. to prove convergence, estimate error bounds and so on.

D Finally, one can turn to exact solutions. These are necessarily very special, in complicated theories, but offer a different perspective to other approaches in that they are guaranteed to fully respect the non-linearity. Here it is hard to obtain useful results (though one can readily obtain more or less trivial results), but it is possible.

All of these approaches can be helped by CA [22]. In the area of exact solutions, applications include the most traditional, checking solutions, as well as use as an aid in finding new solutions, or, more systematically, applying the generating techniques for solutions with commuting Killing vectors [23]. The systems can also be used in studying properties of solutions, e.g their classification (my own interest), the investigation by examples of alternative gravity theories to check experimental viability, the properties of singularities and horizons, and the behaviour of auxiliary classical and quantum fields. In the area of approximations, one can study perturbations, and one of the major uses is in numerical relativity [24, 25], where CA is used to generate the input expressions for numerical programs. Finally, CA can be used to investigate general equations in classical and quantum theories of gravity, for example, supergravity, higher dimensional theories, and computations in Ashtekar variables [12, 26, 28, 13].

To digress a little, let me comment further on the interaction of CA and numerical work. CA can be used to generate numerical code, which is its main use in numerical relativity. This aspect has been extensively explored: for example, among REDUCE's auxiliary packages are one for generating FORTRAN, RATFOR or C (GENTRAN) and one for optimizing the resulting code (SCOPE). CA can also be used in an approach where part of the problem is handled symbolically and the rest numerically. An example is given in [29], where some calculations relevant to betatrons were speeded up by factors up to 90. This is in principle especially useful where the symbolic code can be used to make a step which would numerically be ill-conditioned. Finally CA can be used to analyse numerical schemes: for example, Eastwood and Arter [30] used it to show that some numerical simulations of tokamaks showed numerical rather than real physical instabilities.

The scale of use can vary quite a bit. At the simplest level, a sequence of top-level commands was used in an investigation of representations of superalgebras [31]: the REDUCE input for this application took me less than one hour to develop. At an intermediate scale, in terms of development time, come such applications as Shaker-Jomaa's use of STENSOR in Kaluza-Klein theory [32]. The largest scale is exemplified by the programs developed in CLASSI for the classification of exact solutions in general relativity, which have taken about 13 years so far to develop, and necessitated theoretical as well as programming innovations.

As I have recently given an extended review of this work elsewhere [33], I will not repeat that here but just summarize some main points. The essential difficulty is the "equivalence problem" of deciding whether two metrics represent the same geometry in different coordinates, or classifying a metric in a completely invariant way. (One can then compare metrics by comparing their invariant classifications. It should be noted that the classification and comparison is local: global topology is not determined, and even non-overlapping coordinate patches of the same manifold would not in general be identified.) For the case of metrics without symmetry this was solved by Christoffel, but in an impractical way. Important improvements in the theory were made by Cartan, and Brans considered implementation of these ideas, but the practical version of the theory was introduced by Karlhede [34].

By considering the example of comparing plane waves and flat space, one can see that calculating scalar polynomial invariants in the curvature, which would be most relativists' first idea, is not sufficient. From the work of Christoffel and Cartan we find that to produce an invariant classification one has to compute the Riemann tensor and a finite number of its covariant derivatives, considered as quantities on a suitable restricted frame bundle. By putting the tensors into canonical form, Karlhede showed one could significantly reduce the frame bundle and limit the number of quantities to be computed. There were a number of steps in the underlying procedure which needed detailed development, for example: Petrov classification of the Weyl tensor, Segre classification of the Ricci tensor, determination of the canonical frames, specification of a minimal set of components of the covariant derivatives, and computation of the Lie algebra structure of the spacetime isometries. All of these have been studied in one or more papers, and are described in [33].

5 Recent developments in computer algebra

There are quite a number of areas in which CA has recently made substantial strides whose existence is not yet known outside the CA developers' community. Some of these involve new algorithms of interest even for hand calculation, and all potentially could be useful to gravity theorists, so I want to briefly indicate them here.

One very important but tricky area is in the handling of algebraic numbers, that is to say numbers which are solutions of some polynomial equation with integer coefficients. That this is not just of interest to pure mathematicians can be seen by considering the fact that even to integrate such expressions as $1/(x^2-2)$ one must introduce an algebraic number (in this case $\sqrt{2}$). What makes algebraic numbers tricky to handle is that in principle any finite set of them can be expressed in terms of one basic one, but the equation this satisfies may be of very high degree and hence hard to compute with [35]. However, one has to take the result into account, and not just treat all the algebraic numbers as independent, or one may overlook cancellations between terms. Various strategies for overcoming these difficulties are in use (for example, REDUCE has at least three different algebraic number packages) and the problems are not yet resolved. They extend also to algebraic functions (such as $\sqrt{x-1}$).

A second, but more satisfactorily solved, problem is that of indefinite integration of transcendental and algebraic functions. Risch gave the first practical algorithm, for the case of elementary functions (these are combinations of rational, exponential and logarithmic functions) in 1969, and it was implemented in MACSYMA and Maple: a description of this is in [35]. REDUCE uses a different procedure, more efficient but less complete, due to Risch and Norman (described in [36]). Davenport [37] solved the algebraic function case, but this was only implemented (in REDUCE) for the case of square roots. An improved formulation was given by Trager, and then in recent years the final steps to a workable algorithm for all cases have been made, principally by Bronstein. The resulting procedure is described in [38], and has been implemented in Axiom, and, I think, Maple, and a REDUCE implementation is well advanced.

We all know how to solve simultaneous linear equations, but fewer are aware there are systematic procedures for solving simultaneous polynomial equations. One technique, known as Gröbner bases, was initiated by Buchberger in 1965 and is implemented in most general-purpose systems. It has been extended to more general cases than polynomials with rational coefficients, including (though not fully) systems of non-linear differential equations, and its efficiency has been investigated and improved, though it remains computationally intensive: for details see [39, 40]. Carminati, Czapor and McLenaghan (see McLenaghan's review in this conference) have used it to investigate the integrability conditions of Huygens' principle. A second technique, which is more directed to solving equations than Gröbner bases, which also have other uses, is due to Wu [41] and has particularly been used in proving theorems in Euclidean geometry: it is implemented in REDUCE, at least.

The algorithmic solution of ordinary differential equations has also made great strides in recent years. For first-order equations, the situation is, and perhaps always will be, somewhat messy, but work based on ideas similar to those in indefinite integration has proved quite successful [42, 43]. For linear equations, methods based on differential algebra and the differential Galois group have enabled a complete answer by Kovacic to the problem of finding solutions in "quadratures" of second-order linear ordinary differential equations with rational function coefficients (or the related problem for first-order Riccati equations) [44]. This result however seems to be not widely appreciated yet. Ulmer [45] has developed the similar method for third-order equations. Singer [46] has given in principle a procedure for more general coefficients and equations of any order. Recently Ulmer and Bronstein have given a series of improvements in the effectiveness of these methods [47, 48].

For non-linear ordinary differential equations of order greater than 1 the main tool is the use of Lie symmetries [49, 50]. Many programs for finding Lie symmetries of both ordinary and partial differential equations have been written, using different strategies for the parts where no fully-developed algorithm is known, but these are not yet integrated into a general ODE solver. It turns out that for second order equations, ansätze for contact symmetries also play an important part (Thomas Wolf, 1991, unpublished).

For partial differential equations the problems are even less completely solved (not surprisingly), but useful programs for integrable systems, systems obtainable from Lagrangians, Lie-Bäcklund transformations, self-similar solutions, and so on do exist and could be more widely used.

As well as algorithmic developments, CA systems are developing new interfaces. I have already outlined some interfaces to numerical computation, but there are also the systems permitting the use of routines written in other languages, interfaces to the NAG and other FORTRAN libraries (notably from AXIOM, REDUCE and the SENAC system), and large programs for generating code for systems of differential equations [51]. As well as *Mathematica*'s integrated system, there are similar developments for Maple and Axiom, and Maple is also used as a part of the MathStation system sold for computer-aided design and engineering.

Graphical output has been stimulated by the success of *Mathematica*, and its initial lead is now challenged by graphics for Maple, REDUCE, Derive and Axiom (and, for all I know, the other systems). The user interface for input is less well-developed: in particular there is no system yet combining the nice features of packages like Milo and Theorist (which enable usual mathematical layout for input on the screen, and have an efficient window, mouse and pointer interface) with the power of a fully-developed CA program. It has been widely rumoured that "active book" computers will soon appear: these are computers in the shape of a book with a screen on top on which one writes with a special pen. There is no keyboard, and the computer has to be trained in character recognition for an individual's writing. The first developments of similar facilities for mathematical input are already under way.

6 The future

One can foresee quite a large extension of the present usage of CA in gravity just from the more general exploitation of existing systems and the wider dissemination of knowledge about the already available algorithms and systems. Hardware will continue to get smaller, faster, and cheaper. But there are also some further developments one would like to see, and can expect to see.

In computer algebra systems in general, these include improved definite integration (a tricky problem in general: see [52]), better systems for ordinary differential equations using the methods mentioned above (this is an area I am working on), better handling of infinite series and their summation, developments in application of "sum-substitution" (where one tries to make optimal use of identities like $\sin^2 x + \cos^2 x = 1$), and systems with fully integrated numerical, symbolic, and graphical output *and input*.

In computer algebra for gravity, one would like to see systems meeting simultaneously all the objectives outlined at the start of §3 above, so providing a reliable and flexible fully-developed combined indicial-component calculator, and better user interfaces. I am sure they are on their way.

References

[1] R.A. d'Inverno, "Algebraic computing in general relativity", *Gen. Rel. Grav.*, **6**, 567 (1976).

[2] R.G. McLenaghan, "Maple applications in general relativity" in *Algebraic computing in general relativity (Proceedings of the first Brazilian school on computer algebra, vol. 2)*, ed. M.J. Rebouças and W.L. Roque, Oxford University Press (1991).

[3] Klerer, M. and Grossman, F., "Error rates in tables of indefinite integrals", *Industrial Mathematics (J. Ind. Math. Soc.)*, **18**, 31-62 (1968).

[4] Deprit, A. and Deprit, E., "Processing Poisson series in parallel", *J. Symb. Comp.*, **10**, 179-201 (1990).

[5] Bondi, H., van der Burg, M. G. J. and Metzner, A. W. K., "Gravitational waves in general relativity VII. Waves from axi-symmetric isolated systems", *Proc. Roy. Soc. A*, **269**, 21 (1962).

[6] Dyer, C.C., Allen, S. and Harper, J.S., "The MuTENSOR computer system for general relativity", in *Abstracts of the 11th international conference on general relativity and gravitation, Stockholm, Sweden*, ed. B. Laurent and K. Rosquist (1986).

[7] I.G. Obukhova, S.I. Tertychniy and V.V. Zhytnikov, "GRG – computer algebra program for gravity and classical field theory", Moscow State Pedagogical Institute preprint (1991).

[8] B. Buchberger and R. Loos, "Algebraic simplification", in *Computer algebra: symbolic and algebraic computation*, ed. B. Buchberger, G.E. Collins and R. Loos, pp. 11-44, Springer-Verlag, Wien (2nd edition) (1983).

[9] MacCallum, M.A.H., "Comments on the performance of algebra systems in general relativity and a recent paper by Nielsen and Pedersen", *ACM SIGSAM Bulletin*, **23**, 22-25 (1989).

[10] M.A.H. MacCallum and J.E.F. Skea, "SHEEP: a computer algebra system for general relativity", in *Algebraic computing in general relativity (Proceedings of the first Brazilian school on computer algebra, vol. 2)*, ed. M.J. Rebouças and W.L. Roque, Oxford University Press (1991).

[11] S.J. Campbell and J. Wainwright, "Algebraic computing and the Newman-Penrose formalism in general relativity", *Gen. Rel. Grav.*, **8**, 987 (1977).

[12] E. Schrüfer, F. Hehl and J.D. McCrea, "Application of the REDUCE package EXCALC to the Poincare gauge field theory", *Gen. Rel. Grav.*, **19**, 197 (1987).

[13] A. Giannopoulos and V. Daftardar, "Stensor direct evaluation of the Ashtekar variables for any metric", *Abstracts of contributed papers, 12th International conference on general relativity and gravitation*, 297, Boulder (1989).

[14] M.A.H. MacCallum, "Symbolic computation in relativity theory", in *EUROCAL 87 (Proceedings of the European Conference on Computer Algebra, Leipzig, 1987)*, ed. J.H. Davenport, (Lecture Notes in Computer Science, **378**), pp. 34-43, Springer-Verlag, Berlin and Heidelberg (1989).

[15] J.D. McCrea, "REDUCE in general relativity and Poincaré gauge theory", in *Algebraic computing in general relativity (Proceedings of the first Brazilian school on computer algebra, vol. 2)*, ed. M.J. Rebouças and W.L. Roque, Oxford University Press (1991).

[16] J.F. Harper and C.C. Dyer, "The REDTEN reference manual, version 2.41", University of Toronto report (1986).

[17] M. Monagan, G. Gonnet and B. Char, "Symbolic mathematical computation", *Communications of the ACM*, **29**, 680-682 (1986).

[18] R.J. Fateman, "A Review of Mathematica", Berkeley preprint, submitted to *J. Symb. Comp.* (1991).

[19] A. Krasinski, "The program ORTOCARTAN for applications in Einstein's relativity theory", *Lect. Notes Comp. Sci.* **204**, 159-160, Springer-Verlag, Berlin and Heidelberg (1985).

[20] I. Frick, "The Computer Algebra System SHEEP, what it can and cannot do in general relativity", University of Stockholm Institute of Theoretical Physics report 77-14 (1977).

[21] J.E. Åman, "Manual for CLASSI: classification programs in general relativity (third provisional edition)", University of Stockholm Institute of Theoretical Physics report (1987).

[22] M.A.H. MacCallum, "Computer-aided classification of exact solutions in general relativity", in *Gravitational collapse and relativity*, ed. T. Nakamura and H. Sato, World Scientific, Singapore (1986).

[23] M. Bradley and A. Curir, "SOLITON", Universities of Stockholm and Turin preprint (1989).

[24] T. Nakamura, "Numerical relativity", in *General Relativity and Gravitation: proceedings of the 11th International conference*, ed. M.A.H. MacCallum, Cambridge University Press, London and New York (1987).

[25] C. Evans, L.S. Finn and D. Hobill (eds.) *Frontiers in numerical relativity*, Cambridge University Press, London and New York (1989).

[26] H. Caprasse, J. Demaret, and E. Schrüfer, "Can EXCALC be used to investigate high-dimensional comological models with non-linear Lagrangians?", pp. 116-124 in *Proceedings of ISSAC-88* ed. P. Gianni, Springer-Verlag, Berlin and Heidelberg (1989).

[27] M.A.H. MacCallum, "The GR12 Algebraic computing workshop", in *General Relativity and Gravitation, 1989*, ed. N. Ashby, D.F. Bartlett and W. Wyss, Cambridge University Press, Cambridge, New York and Melbourne, pp. 113-120 (1990).

[28] S.A. Fulling, "The analytic approach to recursion relations", *J. Symb. Comp*, **9**, 73-85 (1990).

[29] B. Autin and J.A. Bengtsson, "Symbolic evaluation of integrals occurring in accelerator orbit theory", *J. Symb. Comp*, **7**, 183 (1989).

[30] J.W. Eastwood and W. Arter, "Interpretation of disruptions in tokamaks", *Phys. Rev. Lett.*, **57**, 2528-2531 (1986).

[31] J.W.B. Hughes and R.C. King, "A conjectured character formula for atypical irreducible modules of the Lie superalgebra $sl(m/n)$", *J. Phys. A*, **20**, L1047-52 (1987).

[32] K. Shaker-Jomaa, "Computing Kaluza-Klein ansatz", *Abstracts of contributed papers, 11th International conference on general relativity and gravitation*, ed. B. Laurent and K. Rosquist, Stockholm (1986), and Ph. D. thesis, University of London (1985).

[33] M.A.H. MacCallum. "Computer-aided classification of exact solutions in general relativity", in *Proceedings of the Italian Relativity Meeting 1990*, ed. B. Marmo, World Scientific, Singapore (1991, in press).

[34] A. Karlhede, "A review of the geometrical equivalence of metrics in general relativity", *Gen. Rel. and Grav.*, **12**, 693 (1980).

[35] J.H. Davenport, Y. Siret and E. Tournier, *Computer algebra: systems and algorithms for algebraic computation*, Academic Press, New York and London (1988).

[36] M.A.H. MacCallum and F.J. Wright, *Algebraic computing with REDUCE*, (Proceedings of the first Brazilian school on computer algebra, vol. 1, ed. M.J. Rebouças and W.L. Roque), Oxford University Press (1991).

[37] J.H. Davenport, *On the integration of algebraic functions*, Lect. Notes in Comp. Sci. **102**, Springer-Verlag, Berlin and Heidelberg (1981).

[38] M. Bronstein, J.H. Davenport and B.M. Trager, "Symbolic integration is algorithmic", in *Computers and Mathematics 1989*, ed. E. Kaltofen, (1989).

[39] B. Buchberger, "Gröbner bases: an algorithmic method in polynomial ideal theory", In *Progress, directions and open problems in multidimensional systems theory*, ed. N.K. Bose pp. 184-232, Reidel, Dordrecht (1985).

[40] T. Becker and V. Weisspfenning (with H. Kredel), *Gröbner bases*, Springer-Verlag, Berlin (1991).

[41] Wu Wen-tsun, "A zero structure theorem for polynomial-equations-solving," *Mathematics-mechanization preprints*, **1**, 2-12, Institute of Systems Science, Academia Sinica, Beijing (1987).

[42] M.J. Prelle and M.F. Singer, "Elementary first integrals of differential equations", *Trans. Amer. Math. Soc.* **279**, 215 (1983).

[43] M. Bronstein, "The Risch differential equation on an algebraic curve", in ISSAC '91: Proceedings of the 1991 International Symposium on symbolic and algebraic computation, Bonn, ed. S.M. Watt, ACM, New York (1991).

[44] J.J. Kovacic, "An algorithm for solving second order linear homogeneous differential equations", *J. Symb. Comp.* **2**, 3-43 (1986).

[45] F. Ulmer, "On Liouvillian solutions of third order homogeneous linear differential equations", Karlsruhe preprint 16/90 (1990).

[46] M.F. Singer, "Liouvillian solutions of linear differential equations with Liouvillian coefficients", *J. Symb. Comp.* **11**, 251 (1991).

[47] M. Bronstein, "On solutions of linear ordinary differential equations in their coefficient field", Zurich preprint ETH-152 (1991).

[48] F. Ulmer, "On algebraic solutions of linear differential equations with primitive unimodular Galois group", Karlsruhe preprint (1991).

[49] P.J. Olver, *Applications of Lie groups to differential equations*, Springer-Verlag, Berlin and Heidelberg (1986).

[50] H. Stephani, *Differential equations - their solution using symmetry*, ed. M.A.H. MacCallum, Cambridge University Press, New York and London (1989).

[51] G.O. Cook, Jr., "ALPAL, a program to generate physics simulation codes from natural descriptions", *Int. J. Mod. Phys. A*, to appear (1990). [Available as Lawrence Livermore preprint UCRL-102076 (1989).]

[52] V.S. Adamchik and O.I. Marichev, "The algorithm for calculating integrals of hypergeometric type functions and its realization in REDUCE system", pp. 212-224 in *ISSAC '90 Symbolic and algebraic computation*, ed. S. Watanabe and M. Nagata, ACM, New York (1990).

BACKGROUND RADIATION FROM DARK MATTER

JAMES M. OVERDUIN
Space Sciences Laboratory, University of California, Berkeley, CA 94720
and
Physics Department, University of Waterloo, Waterloo, ON N2L 3G1

and

PAUL S. WESSON
Space Sciences Laboratory, University of California, Berkeley, CA 94720
and
Physics Department, University of Waterloo, Waterloo, ON N2L 3G1

ABSTRACT

We investigate possible contributions to the extragalactic background light from decaying dark matter of various kinds, concentrating on the model of Sciama, in which neutrinos decay and produce ultraviolet photons.

1. Introduction

The extragalactic background light (EBL) is the cosmological radiation field produced by the visible galaxies and possibly also by the dark matter required to close the universe and solve the "missing mass" problem. In this paper, we examine several varieties of decaying dark matter and calculate their contributions to the EBL.

To accomplish this, we adopt the approach of Wesson, Valle, and Stabell[1] and Wesson[2], who have developed equations for the intensity of the EBL in the visible spectrum, motivated by a desire to clarify the resolution of Olber's Paradox. In this paper, we consider the wider spectrum from radio waves to gamma rays.

The method is as follows: for a given source of radiation, one needs to obtain (1) a number density $n(t)$ of sources, and (2) an energy distribution function $F(\lambda, t)$ erg sec^{-1} Å^{-1}, which may be defined in terms of the bolometric luminosity $L(t)$ erg sec^{-1} of the source by the equation:

$$L(t) = \int_0^\infty F(\lambda, t) d\lambda. \tag{1}$$

Assuming a standard FRW cosmology, it is then possible to express the spectral intensity $I(\lambda_0)$ erg sec^{-1} cm^{-2} Å^{-1} at wavelength λ_0 Å due to all the sources out to a distance corresponding to some suitably early look-back time t_i by the formula:

$$I(\lambda_0) = c \int_{t_i}^{t_0} n(t) F(\lambda, t) \left(\frac{R}{R_0}\right)^5 dt. \tag{2}$$

where $R = R(t)$ is the cosmological scale factor, and the subscript zero denotes the present era.

2. Massive Neutrinos

Dark matter may take the form of massive τ-neutrinos distributed in galactic halos. These would produce radiation when they decay to lighter μ-neutrinos. In Sciama's[3] scheme, the energy of the decay photons is set at $E_\gamma = 13.8 \pm 0.2$ eV, just above the Lyman line at 13.6 eV. (This value is chosen by Sciama in order to account for the ionization characteristics of various regions in our galaxy.)

Thus, assuming that the decay photon wavelengths are distributed normally around $\lambda_\gamma = E_\gamma/hc = 898 \pm 13$ Å, we can write for the energy distribution function of the neutrino halo:

$$F(\lambda) = \sqrt{\frac{2}{\pi}} \frac{L}{\sigma} \exp\left[-\frac{1}{2}\left(\frac{\lambda - \lambda_\gamma}{\sigma}\right)^2\right]. \tag{3}$$

where we can take $\sigma = \Delta\lambda_\gamma = 13$ Å. Luminosity L of the halo can be obtained by integrating the source function $\phi = E_\gamma n_\tau/\tau$ over the volume of the halo, and reducing by an efficiency factor to account for absorption of decay photons within the halo. (τ is the decay lifetime of the τ-neutrinos, and n_τ is their number density, which is assumed to drop off as r^{-2} and take the value 5×10^7 cm^{-3} in the sun's vicinity.) If the halos are flattened by a factor of $\frac{3}{8}$, and if they are to comprise all the missing mass needed to close the universe (clearly an upper limit), then they will vary between 18 and 40 kpc in radius, and have luminosities in the range $L = (1-5) \times 10^{42}$ erg sec^{-1}.

The number density of the halos is simply the galactic number density n_0, normalized to account for expansion, so that (assuming zero pressure in our FRW model):

$$n(t) = n_0 \left(\frac{R_0}{R}\right)^3. \tag{4}$$

Substituting Eqs. 3 and 4 into Eq. 2 and carrying out the integration yields the results shown in Fig. 1. There, predicted intensities are compared with recent measurements of the ultraviolet EBL by Martin and Bowyer[4]. The existence of large neutrino halos (comprising more than half the needed "missing mass") is unlikely, but smaller halos are not ruled out. It should be noted, however, that Hurwitz, Bowyer and Martin[5] have since pointed out that most, if not all of the EBL in this waveband is probably due to the reflection of starlight from dust. If this is the case, then neutrino halos of the type envisioned by Sciama would be ruled out as significant dark matter candidates. More detailed observations from the Hopkins Ultraviolet Telescope are expected to resolve this issue shortly.

3. Other Candidates

3.1 Primordial Black Holes

Hawking[6] has shown that black holes emit radiation like black bodies with temperatures inversely proportional to their masses. Radiation from stellar-sized black holes is negligible but smaller, primordial black holes (PBHs) that could have formed from density fluctuations in the very early universe would emit considerably more. Carr[7] provides a plausible mass spectrum for such PBHs at the time of their formation, and Page[8] has modelled their evolution and calculated their luminosities. From

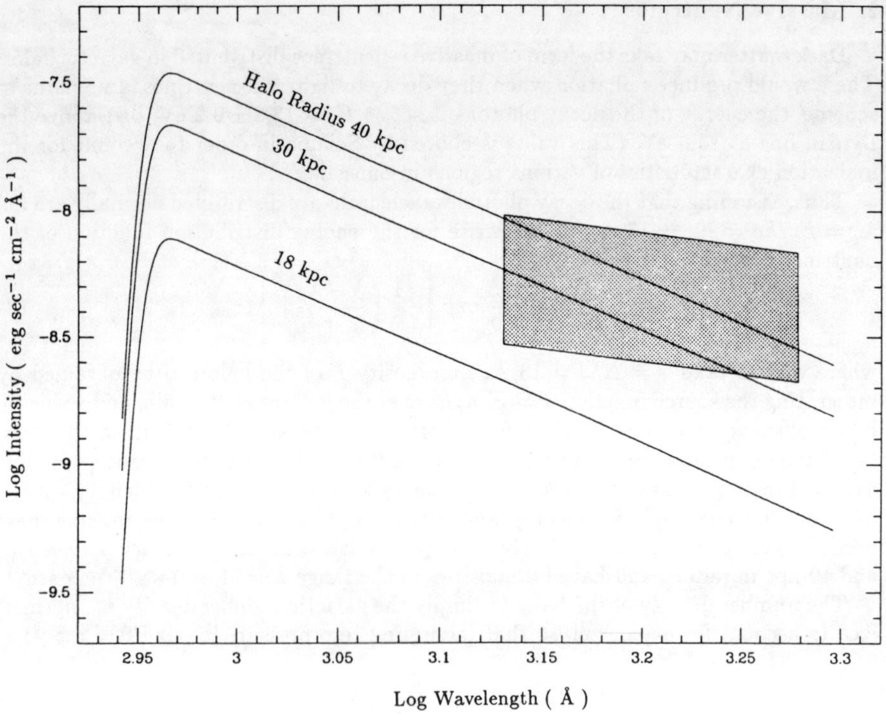

FIG. 1 - Comparison of expected EBL intensity (solid lines) from neutrino halos of various sizes with recent experimental upper limits (shaded area) obtained by Martin and Bowyer[4].

this information we are able to derive the forms of $F(\lambda, t)$ and $n(t)$, and compute the intensity by substituting into Eq. 2 and integrating. Tentative results support the findings of Page and Hawking[9], who obtain a peak in the gamma ray region around $\lambda_0 = 10^{-3}$ Å, with intensity of no more than about 10 erg sec^{-1} cm^{-2} Å^{-1}.

3.2 The Vacuum

In order to account for the observed low value of the energy density of the vacuum (which, according to contemporary quantum field theory ought to be filled with energetic virtual particles), some theorists have proposed that the vacuum naturally decays into photons and neutrinos. Freese et al.[10], in particular, have produced a detailed model, including predictions for the energy density of the radiation created.

Assuming that the photons are more or less thermalized, we can fit them to a Planck spectrum and hence obtain $n(t)$ and $F(\lambda, t)$. Substitution into Eq. 2 yields the expected intensity of such a process. Preliminary results are promising, and may lead to constraints on vacuum decay theories of the type proposed by Freese et al.

3.3 Axions

Axions are hypothetical particles intended to explain the P- and CP-conservation of strong interactions. They were first assigned very low ("invisible") rest-energies of less than 10^{-4} eV, but recent refinements in the physics of stellar interiors have relaxed this constraint to around 20 eV. At this value, axions will be unstable against decay into photons pairs. Our work here is still in progress, but it may be noted that Lodenquai and Dixit[11] predict that such decays will lead to an ultraviolet background with intensities close to the upper limits recently obtained by Martin and Bowyer[4].

4. Conclusion

We have investigated the hypothesis of Sciama[3] in which dark matter takes the form of decaying neutrinos, and are in the process of examining other theories in which primordial black holes, the vacuum, and axions decay to produce photons. The Sciama hypothesis is not decisively ruled out, but both it and the other theories can be constrained by measured limits on background radiation.

Acknowledgements

This work was carried out at the Center for Extreme Ultraviolet Astrophysics at the University of California, Berkeley. We are grateful to S. Bowyer for discussions, and to NSERC for financial assistance.

References

1. P. S. Wesson, K. Valle, and R. Stabell, *Ap. J.* **317** (1987) 601.
2. P. S. Wesson, *Ap. J.* **367** (1991) 399.
3. D. W. Sciama, *Ap. J.* **367** (1991) L39.
4. C. Martin and S. Bowyer, *Ap. J.* **338** (1989) 677.
5. M. Hurwitz, S. Bowyer, and C. Martin 1991, in preparation.
6. S. W. Hawking, *Commun. Math. Phys.* **43** (1975) 199.
7. B. J. Carr, *Ap. J.* **206** (1976) 8.
8. D. N. Page, *Phys. Rev.* **D13** (1976) 198.
9. D. N. Page and S. W. Hawking, *Ap. J.* **206** (1976) 1.
10. K. Freese, F. Adams, J. Frieman, and E. Mottola, *Nucl. Phys.* **B287** (1987) 797.
11. J. F. Lodenquai and V. V. Dixit, *Phys. Lett.* **B194** (1987) 350.

AFFINE CONFORMAL VECTORS IN SPACETIME

ALAN COLEY
Dept. of Maths., Stats. & Computing Science, Dalhousie University
Halifax, Nova Scotia, Canada B3H 3J5

BRIAN TUPPER
Dept. of Mathematics & Statistics, University of New Brunswick
Fredericton, New Brunswick, Canada E3B 5A3

ABSTRACT

All spacetimes admitting affine conformal vectors, i.e. vectors ξ^a satisfying $\mathcal{L} g_{ab} = 2\psi g_{ab} + K_{ab}$, where $\psi = \psi(x^a)$, $\psi_{;ab} \neq 0$, $K_{ab} = K_{ba}$, $K_{ab;c} = 0$, are found in principle.

1. Introduction

There has been much recent interest in the existence of symmetries in the spacetime manifold of general relativity and in the corresponding vector fields ξ^a. These symmetries are often discussed in the context of fluid (perfect, anisotropic, or viscous heat-conducting) spacetimes and the effect of these symmetries on the kinematic and dynamic properties of the fluids studied (see[1] and references therein).

The present investigation is concerned with affine conformal vectors (ACV), ξ^a, which are defined by

$$\mathcal{L} g_{ab} = 2\psi g_{ab} + K_{ab}, \tag{1.1}$$

where $\mathcal{L} g_{ab}$ denotes the Lie derivative of the metric tensor g_{ab} along ξ^a, $\psi = \psi(x^a)$ is the conformal scalar and K_{ab} is a covariantly constant symmetric tensor. When $\psi_{;ab} = 0$, ξ^a is a special affine conformal vector (SACV) and it is known[2] that few spacetimes admit SACV, none of which can represent perfect fluids or non-null Einstein-Maxwell fields, but they can represent viscous fluids or anisotropic fluids with particularly restrictive equations of state. Accordingly, we are interested here only in the case $\psi_{;ab} \neq 0$.

The problem of the existence of ACV's for positive definite manifolds has been completely solved[3] but no complete solution of the problem exists for spacetimes although ACV's have been the subject of a number of studies (see[4] and references therein). The complete solution for spacetimes is presented here without details. In particular we show that the Einstein static universe, which is known to admit an ACV[5], is the only perfect fluid spacetime and the only FRW spacetime to do so; this is contrary to an assertion made previously[4].

The existence of the covariantly constant tensor K_{ab} imposes strong restrictions on the possible spacetimes. Hall and da Costa[6] showed that if a simply connected spacetime admits a global nowhere zero, covariantly constant second-order symmetric tensor, then one of the following three possibilities must occur:

(a) There exists locally a timelike or spacelike nowhere zero covariantly constant vector field $\eta_a \equiv \eta_{,a}$ such that $K_{ab} = \eta_{,a}\eta_{,b}$ and the spacetime is locally decomposable into a 1 + 3 spacetime.

(b) There exists locally a null nowhere zero covariantly constant vector field $\eta_a \equiv \eta_{,a}$ such that $K_{ab} = \eta_{,a}\eta_{,b}$ and the spacetime is the generalized pp-wave spacetime[1,7] which, in general, is not decomposable.

(c) The spacetime is locally decomposable into a 2 + 2 spacetime and no covariantly constant vector exists unless the spacetime decomposes further.

In cases (a) and (b), the ACV can be shown to be of the form

$$\xi^a = \zeta^a + \tfrac{1}{2}\eta\eta_{,}{}^{a} \tag{1.2}$$

where ζ^a is a CKV. Thus the problem in these cases becomes one of finding spacetimes admitting both a CKV and a covariantly constant vector.

We consider the cases (a), (b), (c) in turn.

2. The Timelike Case

Consider the case in which these exists a timelike covariantly constant vector $\eta_{,a}$. Choosing co-ordinates such that $\eta = -t$, i.e. $\eta_{,a} = -\delta^0_a$, the metric takes the form

$$ds^2 = -dt^2 + g_{\alpha\beta}(x^\gamma)dx^\alpha dx^\beta, \tag{2.1}$$

where α, β take the values 1, 2, 3. It can be shown that any CKV admitted by such spacetimes must satisfy

$$\psi_{;\alpha\beta} = \psi_{,00}g_{\alpha\beta} \tag{2.2}$$

and an extension of a theorem given by Petrov[8] can be used to show that the spacetime metric must be of the form

$$ds^2 = -dt^2 + dx^2 + a^2(x)f^2(y, z)(dy^2 + dz^2) \tag{2.3}$$

where $f(y, z)$ is an arbitrary function and $a(x)$ satisfies

$$a_{xx} = \alpha a, \tag{2.4}$$

where α is a constant.

When $\alpha > 0$, the spacetimes (2.3) do not satisfy the dominant energy condition (DEC), and $\alpha = 0$ leads to $\psi_{;ab} = 0$, i.e. ζ^a is a SCKV and ξ^a is a SACV, so the only case to consider is $\alpha = -k^2$ which leads to

$$ds^2 = -dt^2 + dx^2 + \sin^2 kx\, f^2(y, z)(dy^2 + dz^2). \tag{2.5}$$

This represents a perfect fluid spacetime iff the (y, z) two-space is a space of constant positive curvature; this is the Einstein universe. Otherwise the metric (2.5) can represent an anisotropic fluid or a viscous heat-conducting fluid; it cannot represent an Einstein-Maxwell field.

The ACV is

$$\xi^a = (k^{-2} \cos kx \sin kt + \tfrac{1}{2}t,\ k^{-2} \sin kx \cos kt,\ 0,\ 0)$$
$$\psi = k^{-1} \cos kx \cos kt. \tag{2.6}$$

3. The Spacelike Case

In this case we have a spacelike covariantly constant vector $\eta_{,a}$ and, choosing co-ordinates such that $\eta_{,a} = \delta^3_a$, the metric is of the form

$$ds^2 = dz^2 + g_{\alpha\beta}(x^\gamma)dx^\alpha dx^\beta \tag{3.1}$$

where α, β take the values 0, 1, 2. Following arguments similar to those of the timelike case we find that there are two possible types of metric, viz.

$$ds^2 = dz^2 - dt^2 + b^2(t)g^2(x, y)(dx^2 + dy^2) \tag{3.2}$$

where $g(x, y)$ is an arbitrary function and $b(t)$ satisfies

$$b_{tt} = \beta b \tag{3.3}$$

where β is a constant, and

$$ds^2 = dz^2 + dy^2 + c^2(y)h^2(t, x)(-dt^2 + dx^2) \tag{3.4}$$

where $h(t, x)$ is an arbitrary function and $c(y)$ satisfies

$$c_{yy} = \gamma c \tag{3.5}$$

where γ is a constant. However, metrics of the form (3.4) cannot satisfy the DEC and so will be discarded.

If in (3.3) $\beta > 0$, the DEC is not satisfied, while if $\beta = 0$ we have an SCKV, so we consider only $\beta = -k^2$, i.e. the metric is

$$ds^2 = -dt^2 + \sin^2 kt\ g^2(x, y)(dx^2 + dy^2) + dz^2. \tag{3.6}$$

It is not possible for this metric to represent a perfect fluid or an Einstein-Maxwell field, but it can represent an anisotropic fluid or a viscous heat-conducting fluid. The ACV is

$$\xi^a = (k^{-2} \cos kz \sin kt,\ 0,\ 0,\ k^{-2} \sin kz \cos kt - \tfrac{1}{2}z)$$
$$\psi = k^{-1} \cos kz \cos kt.$$

4. The Null Case

When $\eta_{,a}$ is null, the metric is the generalized pp-wave metric of the form[7]

$$ds^2 = P^{-2}(dx^2 + dy^2) - 2du(dv - mdx + Hdu) \tag{4.1}$$

where H, m, P are each functions of u, x, y and $\eta_{,a} = (-1, 0, 0, 0)$. When the Ricci scalar, R, is zero the metric (4.1) can be put in the form

$$ds^2 = (dx^2 + dy^2) - 2du(dv + Hdu). \tag{4.2}$$

In the case of the metric (4.1), it can be shown that the conformal scalar ψ is a function of u only and any CKV is of the form

$$\zeta^a = [-A(u),\ \alpha v - D(u,x,y) + (2H + m^2P^2)A(u) + mP^2B(u,x,y),$$
$$P^2(mA + B),\ P^2C(u,x,y)], \tag{4.3}$$

$$\psi = \tfrac{1}{2}(\alpha - A_u), \tag{4.4}$$

where α is a constant and the functions A, B, C, D satisfy the system of differential equations

$$\begin{aligned}
D_u &= (H_u + mm_uP^2 + mH_xP^2)A + (m_u + H_x)P^2B + H_yP^2C - 2H\psi,\\
B_x &= -(P^{-3}P_u + mP^{-1}P_x + mx)A - P^{-1}P_xB - P^{-1}P_yC + P^{-2}\psi,\\
C_y &= -(P^{-3}P_u - mP^{-1}P_x)A + P^{-1}P_xB - P^{-1}P_yC + P^{-2}\psi,\\
D_x + B_u &= 2(H_x - mP^{-1}P_x)A - 2P^{-1}P_uB - m_yP^2C + 2m\psi,\\
D_y + C_u &= 2(H_y + \tfrac{1}{2}m_ymP^2)A + m_yP^2B - 2P^{-1}P_uC,\\
B_y + C_x &= -(m_y + 2mP^{-1}P_y)A - 2P^{-1}P_yB - 2P^{-1}P_xC.
\end{aligned} \tag{4.5}$$

Elimination of α, A, B, C, D from these equations leads to an expression connecting H, m, P and their derivatives, which delineates those members of the general set of spacetimes with metric (4.1) which admit proper CKV and thus ACV. The ACV is of the form

$$\xi^a = \zeta^a - \tfrac{1}{2}u\delta_1^{\ a} \tag{4.6}$$

where ζ^a is given by (4.3).

Note that the Einstein tensor G_{ab} and, consequently, the energy-momentum tensor T_{ab} for the metric (4.1) is of Segré type {(1, 1)(1 1)} if the following condition holds[1]

$$RR_{00} = 2P^2(R_{02}^{\ 2} + R_{03}^{\ 2}). \qquad (R \neq 0) \tag{4.7}$$

By taking the Lie derivative of this expression in the direction of ζ^a and using (4.4) we find that $A_{uuu} = 0$, i.e. $\psi_{;ab} = 0$. Hence those members of the class of spacetimes (4.1) with T_{ab} of Segré type {(1, 1)(1 1)} cannot admit a proper CKV, only a SCKV at best, and so do not admit ACV.

In the case when R = 0, i.e. metric (4.2), the problem of finding all such spacetimes admitting CKV has not yet been resolved. Recently Maartens and Maharaj[9] have discussed this problem and have given examples of a conformally flat pp-wave solution admitting seven proper CKV (null Einstein-Maxwell solution), thus providing an example of a spacetime admitting multiple ACV, and also an example of a non-conformally flat member of this class admitting only one CKV.

Since T_{ab} is of Segré type {2 (1 1)} for all metrics of the form

(4.2) and also for those of the form (4.1) that admit ACV, it follows that these metrics cannot represent perfect fluids, can represent anisotropic fluids only with $\mu = -p_{\parallel}$, $p_{\perp} = 0$, and can represent viscous fluids only with no heat conduction.

5. 2 + 2 Spacetimes

In this case there is no covariantly constant vector and we have to solve eqn. (1.1) directly. The metric can be written in the form

$$ds^2 = -2e^{2\mu(u,v)}dudv + 2e^{2\nu(\zeta,\bar{\zeta})}d\zeta d\bar{\zeta} \tag{5.1}$$

where u, v are real null co-ordinates and ζ, $\bar{\zeta}$ are conjugate complex co-ordinates. The only possible form for K_{ab} is

$$K_{01} = K_{10} = -\alpha e^{2\mu}, \quad K_{23} = K_{32} = \beta e^{2\nu} \tag{5.2}$$

where α, β are constants with $\alpha \neq \beta$ and, without loss of generality, we may put $\beta = 0$. Solving equations (1.1) we find that $\alpha = 0$ so that no ACV exists. Furthermore, we find that the spacetime (5.1) admits proper CKV iff each two-space has constant curvature σ_{01} and σ_{23} with $\sigma_{01} = -\sigma_{23}$, from which it follows that the spacetime is conformally flat. Of the two possibilities only that with $\sigma_{01} < 0$, $\sigma_{23} > 0$ satisfies the DEC and this is the Bertotti-Robinson solution.

6. Conclusion

We have shown that the only spacetimes admitting ACV are those given by (2.3) with (2.4), (3.2) with (3.3), (3.4) with (3.5) and those spacetimes with metrics (4.1) and (4.2) which admit a CKV. Not all of the spacetimes satisfy the DEC.

7. References

1. A. A. Coley and B. O. J. Tupper (1989), J. Math. Phys. 30, 2666.
2. A. A. Coley and B. O. J. Tupper (1990), J. Math. Phys. 31, 649.
3. Y. Tashiro (1960), Math. J. Okayama Univ. 10, 75.
4. K. L. Duggal (1989), J. Math. Phys. 30, 1316.
5. D. P. Mason and R. Maartens (1987), J. Math. Phys. 28, 2182.
6. G. S. Hall and J. da Costa (1988), J. Math. Phys. 29, 2465.
7. D. Kramer, H. Stephani, M. MacCallum, E. Herlt (1980), 'Exact Solutions of Einstein's Field Equations' (Cambridge U. P., Cambridge).
8. A. Z. Petrov (1969), 'Einstein Spaces' (Pergamon, Oxford).
9. R. Maartens and S . D. Maharaj (1991), Class. Quantum Grav. 8, 503.

Gravitational Entropy and Cosmology

Alan A. Coley

Department of Mathematics, Statistics and Computing Science
Dalhousie University, Halifax, Nova Scotia B3H 3J5

Abstract

By assuming that P, the ratio of the square of the Weyl curvature to the square of the Ricci tensor, is an appropriate measure of a gravitational entropy, we shall investigate the cosmological implications of the existence of gravitational entropy (i.e. the consequences of the 'thermodynamical' constraint $dP/dt \geq 0$) by studying the dynamics of a scalar field in a Bianchi I universe. The analysis suggests a possible connection between $dP/dt \geq 0$ and the decreasing of the cosmological 'constant' with time, and it is argued that the existence of such a cosmological arrow of time is not compatible with the inflationary scenario.

Penrose [1] has suggested the existence of a gravitational entropy, whereby the Universe is endowed with a distinct (cosmological) arrow of time. The idea is that the Universe is initially highly regular (that is, in some sense approximately FRW), and will become more inhomogeneous and more anisotropic due to the natural gravitational process of clumping of matter. In this scenario, also known as 'quiescent cosmology', the Universe is endowed with a thermodynamic boundary condition in that this entropy should be a minimum at an initial singularity. Using persuasive arguments Penrose [1] has taken this line of reasoning further, postulating that the gravitational entropy is related to some measure of the Weyl curvature tensor. Clearly cosmological models that are consistent with this scenario are of interest, and here we should like to investigate the cosmological implications of this postulate.

In order to do this we shall study the dynamics of a scalar field in a spatially homogeneous but anisotropic spacetime (we recall that the Weyl tensor is identically zero in an FRW model). In particular, we shall consider the simple (LRS) Bianchi I model, which is of particular interest since it contains the flat FRW model as a special case. We are interested in the dynamics of a scalar field since we particularly wish to investigate any possible relationship with the cosmological 'constant' and the inflationary scenario. There are many mechanisms which give rise to a varying cosmological constant (most often motivated by quantum field theoretical considerations - but $\Lambda(t)$ can also occur naturally in alternative theories of gravity), and its decay if often formulated in terms of a classical scalar field ϕ (where ϕ is a function of cosmic time); Peebles and Rhata [2] have described a cosmology based on a Λ-like term that decreases with time. In the inflationary scenario [3], again most naturally formulated in terms of a (weakly coupled) scalar field called the inflaton, whose associated mass-density acts like a cosmological constant that is decreasing in time, microphysical events may have led to a period of rapid expansion or inflation in which the mean expansion scale factor accelerated.

We start from the action (using standard notation [4] and appropriate geometric units)

$$I = \int (\mathcal{L}_g + \mathcal{L}_\phi) \sqrt{-g}\, d^4x; \; \mathcal{L}_g = \frac{R}{2}, \; \mathcal{L}_\phi = \frac{1}{2}\phi_{,\pi}\phi^{,\pi} + \Lambda(\phi), \tag{1}$$

the variation of which leads to the field equations

$$\Box\phi - \frac{\partial \Lambda}{\partial \phi} = 0; \; G_{\mu\nu} = \phi_{,\mu}\phi_{,\nu} - \left[\frac{1}{2}\phi_{,\pi}\phi^{,\pi} + \Lambda(\phi)\right] g_{\mu\nu} \tag{2}$$

Here $\Lambda(\phi)$ is the effective cosmological constant. Note that if $\phi \simeq \phi_0 = \text{constant}$, then $\mathcal{L}_\phi \simeq \Lambda(\phi_0) = \Lambda_0 \geq 0$, where Λ_0 is constant. In addition, $-\Lambda(\phi)$ is the potential for the scalar field (i.e., $V(\phi) = -\Lambda(\phi)$ contributes on "effective" cosmological "constant", which may vary with time).

We shall assume a Bianchi I (LRS) universe in which

$$ds^2 = -dt^2 + U^2(t)\, dx^2 + V^2(t)\, (dy^2 + dz^2); \; \phi = \phi(t) \tag{3}$$

The Einstein field equations yield

$$\frac{\dot{V}^2}{V^2} + \frac{2\dot{U}\dot{V}}{UV} = \frac{1}{2}\dot{\phi}^2 + \Lambda \equiv \rho_\phi, \quad -\frac{2\ddot{V}}{V} - \frac{\dot{V}^2}{V^2} = \frac{1}{2}\dot{\phi}^2 - \Lambda \equiv p_\phi, \tag{4}$$

the equation

$$\frac{\ddot{U}}{U} = \frac{\ddot{V}}{V} + \frac{\dot{V}^2}{V^2} - \frac{\dot{U}\dot{V}}{UV}, \tag{5}$$

and the equation for the scalar field

$$\ddot{\phi}+\left[\frac{\dot{U}}{U}+\frac{2\dot{V}}{V}\right]\dot{\phi}+\frac{\partial\Lambda}{\partial\phi}=0\,. \tag{6}$$

From (4) we have that

$$\Lambda(t)=\frac{\ddot{V}}{V}+\frac{\dot{V}^2}{V^2}+\frac{\dot{U}\dot{V}}{UV}\geq 0;\ \dot{\Lambda}=\frac{\partial\Lambda}{\partial\phi}\dot{\phi}=\frac{\dddot{V}}{V}+\left(\frac{\dot{U}}{U}+\frac{2\dot{V}}{V}\right)\frac{\ddot{V}}{V}-\left(\frac{\dot{V}}{V}+\frac{2\dot{U}}{U}\right)\frac{\dot{V}^2}{V^2}-\frac{\dot{U}^2\dot{V}}{U^2V}\,. \tag{7,8}$$

Calculation yields

$$\Psi\equiv\frac{1}{4}R_{ij}R^{ij}=\frac{3\ddot{V}^2}{V^2}+3\frac{\ddot{V}}{V}\frac{\dot{V}^2}{V^2}+\frac{\dot{V}^4}{V^4}+\frac{\dot{U}^2}{U^2}\frac{\dot{V}^2}{V^2}+\frac{\dot{U}}{U}\frac{\dot{V}^3}{V^3} \tag{9}$$

$$\Phi\equiv\frac{3}{16}C^{ijkm}C_{ijkm}=\frac{\dot{V}^2}{V^2}\left(\frac{\dot{V}}{V}-\frac{\dot{U}}{U}\right)^2 \tag{10}$$

Following Penrose [1] we shall assume the existence of a gravitational entropy, whereby the Universe is endowed with a cosmological arrow of time. From the work on isotropic singularities [5] (and noting that Φ (say) is not dimensionless and is zero in plane wave spacetimes, and that $\dot{\Phi}\leq 0$ here always) we take as the appropriate measure of gravitational entropy, P, the ratio of the square of the Weyl curvature to the square of the Ricci tensor, viz.,

$$P=C^{ijkl}C_{ijkl}/R_{ij}R^{ij};\ \text{where sign}(\dot{P})=\text{sign}\left[\Psi\dot{\Phi}-\Phi\dot{\Psi}\right], \tag{11a,b}$$

using (9) and 10). A direction calculation then yields

$$\Delta\equiv\Psi\dot{\Phi}-\Phi\dot{\Psi}=\mathcal{V}+\Theta; \tag{12}$$

$$\mathcal{V}=\frac{3\dot{V}^2}{V^2}\left(\frac{\dot{V}}{V}-\frac{\dot{U}}{U}\right)^2\left(\frac{2\ddot{V}}{V}+\frac{\dot{V}^2}{V^2}\right)\left[-\frac{\dddot{V}}{V}-\frac{\ddot{V}}{V}\left(\frac{\dot{U}}{U}+\frac{2\dot{V}}{V}\right)+\frac{\dot{V}^2}{V^2}\left(\frac{\dot{V}}{V}+\frac{2\dot{U}}{U}\right)+\frac{\dot{U}^2\dot{V}}{U^2V}\right]$$

$$= \frac{3\dot V^2}{V^2}\left(\frac{\dot V}{V} - \frac{\dot U}{U}\right)^2\left(2\frac{\ddot V}{V} + \frac{\dot V^2}{V^2}\right)\left[-\dot\Lambda\right] \tag{12a}$$

$$\Theta = -\frac{3}{2}\frac{\dot V}{V}\left(\frac{\ddot V}{V} + \frac{\dot V^2}{V^2} + \frac{\dot U\dot V}{UV}\right)\left[\frac{\dot U}{U} - \frac{\dot V}{V}\right]^2\left\{2\left(-\frac{\ddot V}{V} + \frac{\dot U\dot V}{UV}\right)\left(\frac{2\ddot V}{V} + \frac{\dot V^2}{V^2}\right)\right.$$
$$\left. + \frac{3\dot V^2}{V^2}\left(\frac{2\ddot V}{V} + \frac{\dot V^2}{V^2}\right) + \left(\frac{\dot V^2}{V^2} + \frac{2\dot U\dot V}{UV}\right)\frac{\dot V^2}{V^2}\right\} \tag{12b}$$

If we now assume that the entropy is non-decreasing, i.e. $\dot P \geq 0$, we have that

$$\Delta \geq 0. \tag{13}$$

Since we assume that the model is expanding with $\rho_\phi \geq 0$, $p_\phi \leq 0$ (which then implies that $\Lambda \geq 0$), and we have that $\rho_\phi + p_\phi = \frac{1}{2}\dot\phi^2 \geq 0$ (which then implies that $\dot\rho_\phi \leq 0$, so that we can interpret the expansion and entropy as being driven by ρ_ϕ), we have the following inequalities:

$$\frac{\dot V^2}{V^2} + 2\frac{\dot U\dot V}{UV} \geq 0, \quad 2\frac{\ddot V}{V} + \frac{\dot V^2}{V^2} \geq 0, \quad -\frac{\ddot V}{V} + \frac{\dot U\dot V}{UV} \geq 0 \tag{14a,b,c}$$

(where of course (14a) is implied by (14b,c). Inequalities (14) imply that the term in the curly brackets in (12b) is the sum of three positive definite terms, whence (for $\Lambda \geq 0$) it follows that

$$\Theta \leq 0. \tag{15}$$

Thus equation (13) implies that

$$\mathcal{V} \geq 0, \tag{16}$$

from which it follows that (using (12a))

$$\dot{\Lambda} \leq 0 \tag{17}$$

This analysis therefore suggests a possible connection between $\dot{P} \geq 0$ and $\dot{\Lambda} \leq 0$; we can speculate that the decreasing of the cosmological constant is connected to the cosmological arrow of time through the increase of entropy, and that the "observed" present low value of Λ might be understood in terms of a current high amount of entropy in the Universe (the number of background photons divided by the number of baryons ($\simeq$ entropy per particle in an expanding universe) $\sim 10^8$ [4]). That Λ is approximately constant now might be related to the fact that little entropy is presently being produced.

With $\dot{\Lambda} \leq 0$, Δ is the sum of a positive definite term ($\mathcal{V}$) and a negative definite term (Θ), so that $\dot{P}$ can, in principle, be of either sign. The sign of $\dot{P}$ will, in fact, depend on the explicit form for $\Lambda(\phi(t))$; indeed, the system of equations (4)-(6) is not closed unless $\Lambda(t)$ is specified (some specific forms for $\Lambda(t)$ have appeared in the literature; e.g. see Peebles and Ratra [2]). Clearly for some $\Lambda(\phi)$ the solution of the system will be such that $\dot{P} \geq 0$, while possibly for others $\dot{P} < 0$. We can therefore consider the entropy condition $\dot{P} \geq 0$ as a physical constraint on the potential $\Lambda(\phi)$ in that only such $\Lambda(\phi)$ will be regarded as physically acceptable.

If $\Lambda \cong \Lambda_0$ = constant, then the Bianchi I model will asymptotically approach the de Sitter model (in fact this is generally true for all models - the cosmic no hair theorem [6]), and the model is said to undergo an exponential inflationary period. Intuitively, since inflation drives the universe towards isotropy and homogeneity (i.e., towards regularity), and particularly since $\Phi \to 0$, this behaviour would seem to be incompatible with Penrose's notion of a cosmic arrow of time. Indeed, if we consider a general diagonal (non-LRS) Bianchi I model with energy-momentum tensor $T_{ab} = \Lambda_0 g_{ab}$, and attempt to find an approximate solution at late times by linearizing the resulting Einstein field equations about the de Sitter solution (in which P is identically zero), we find that $\dot{P} \cong -4\ell^2 \lambda e^{-6\lambda t}$ (where $\lambda = \sqrt{\Lambda_0/3}$ and ℓ^2 is a positive constant); that is, at late times P is indeed a decreasing function of time.

However, we should be more careful here and the full dynamics of the scalar field should be taken into consideration. By considering the second term (in the curly brackets) in Θ in (12b) we can write

$$\Delta = 3\frac{\dot{V}^2}{V^2}\left(\frac{\dot{V}}{V}-\frac{\dot{U}}{U}\right)^2\left(\frac{2\ddot{V}}{V}+\frac{\dot{V}^2}{V^2}\right)\Lambda\left[-\frac{\dot{\Lambda}}{\Lambda}-\frac{3}{2}\frac{\dot{V}}{V}\right]-Q^2, \tag{18}$$

where $-Q^2$ is a negative definite term. Now, if $\dot{\Lambda} \cong 0$ $(\Lambda \cong \Lambda_0)$, in an expanding universe the first term in the square brackets in (18) is negligible with respect to the second term and consequently $\dot{P} < 0$. [Indeed, for Λ to be approximately constant in an expanding universe it is meant that $\dot{\Lambda}/\Lambda$ is negligible with respect to the expansion rate.] That is, for $\dot{P} \geq 0$ it is not possible for Λ to be approximately constant and the conditions under which inflation can take place cannot occur. Consequently the analysis indicates that inflation is not compatible with the idea that the Universe evolves in such a way that the gravitational entropy, as represented by P, is an ever increasing quantity.

The work here is very speculative, and can be regarded as the first step in an investigation of the consequences of the existence of a cosmic arrow of time $(\dot{P} \geq 0)$ through the existence of a gravitational entropy. We have chosen to examine a simple (Bianchi I) model in which the consequences could be easily understood. The next step would be to consider more sophisticated models; that is, models in which the geometry is more complex (than Bianchi I), models in which the stress-energy tensor is more physical (incorporating ordinary matter, and, among other things, non-minimal coupling), or even models in other theories of gravity (e.g. scalar-tensor theories). Also, a more thorough investigation would include an analysis of other cosmological tests, such as, for example, the predicted age of the Universe.

1. Penrose, R., 1979, in General Relativity: an Einstein Centenary Survey, eds. S.W. Hawking and W. Israel (Cambridge University Press).
2. Peebles, P.J.E., and Ratra, B., 1988, Ap. J. Lett. 325, L17.
3. Guth, A., 1981, Phys. Rev. D 23, 347; Albrecht, A. and Steinhardt, P., 1982, Phys. Rev. Letts. 48, 1220; Linde, A., 1982, Phys. Letts. B 108, 369.
4. Weinberg, S., 1989, Rev. Mod. Phys. 61, 1.
5. Goode, S.W., and Wainwright, J., 1985, Class. Quantum Grav. 2, 99; see also Bonnor, W.B., 1986, Class. Quantum Grav. 3, 495, Goode, S.W., 1991, Class. Quantum Grav. 8, L1 and Tod, K.P., 1990, Class. Quantum Grav. 7, L13.
6. Wald, R.M., 1983, Phys. Rev. D 28, 2118.

DERIVATION OF THE SPECTRAL ENERGY DENSITY IN S^3

W. M. STUCKEY
Department of Physics, Elizabethtown College
Elizabethtown, Pennsylvania 17022, USA

and

GUST BAMBAKIDIS
Department of Physics, Wright State University
Dayton, Ohio 45435, USA

ABSTRACT

The spectral energy density as a function of temperature is derived for a photon gas in S^3. The photons are assumed to obey Bose-Einstein statistics. The frequency spectrum of the photon gas is obtained from the eigenvalues of the Laplacian on S^3. The result is precisely that of a like photon gas in R^3. Therefore, as widely accepted, one cannot use the blackbody distribution of the cosmic microwave background to identify our universe as spatially open or closed.

1. Introduction

The spectral energy density as a function of temperature for a photon gas, $du_f(T)$, in R^3 is derived in many introductory statistical mechanics texts. These derivations assume that the photons are indistinguishable and that there is no limit to the number of photons that may occupy any given energy state, i.e. the photon gas obeys Bose-Einstein statistics. Further, the photons are in thermal equilibrium with the walls of their enclosure, the shape of which does not effect the spectral distribution of the photons[1].

The cosmic microwave background (CMB) has been found to conform rather precisely to this blackbody distribution[2]. Does this mean that the universe is spatially flat? It is widely accepted that the answer is no, the blackbody distribution of the CMB is not affected by the spatial curvature of the present-day universe[3]. However, many texts do not explain the result, arguing only that the distribution is container invariant. Since the spatial hypersurfaces of our spacetime manifold in Big Bang Cosmology are not merely different containers of R^3, but different spaces altogether, it seems plausible that container invariance may not necessitate that the CMB adhere to the R^3 form of $du_f(T)$.

Therefore, this paper derives $du_f(T)$ for a photon gas in S^3, paralleling the derivation used in the R^3 case. This derivation is valid for an S^3 with radius much larger than the photons' wavelengths[4]. Thus the result is applicable to the CMB of our large present-day universe.

2. Derivation

We begin, as in the R^3 case, with the assumption that the photon gas obeys Bose-Einstein statistics. Thus,

$$dN_f = dG_f/(\exp(hf/kT) - 1) \tag{1}$$

where dN_f is the number of photons with frequencies between f and f + df, dG_f is the number of energy states available for photons with frequencies between f and f + df, $1/(\exp(hf/kT) - 1)$ is the fraction of dG_f which are occupied when the photon gas temperature is T, h is Planck's constant, and k is Boltzmann's constant. Once dN_f is computed, we can compute $du_f(T)$ simply from

$$du_f = (hf)\cdot(dN_f/V) \tag{2}$$

where V is the volume of the S^3.

In order to compute dG_f, we need the spectrum of ∇^2 on S^3. Schrödinger[5] and Müller[6] derived this result in their early work on expanding universes. They found the eigen frequencies of ∇^2 on S^3 to be

$$f = c(p(p + 2))^{1/2}/(2\pi r) \tag{3}$$

where r is the radius of the S^3, c is the speed of light in vacuum, and p is obtained from the eigenvalues of ∇^2. That is,

$$\nabla^2 Y_{pmn} = -p(p + 2)Y_{pmn} \tag{4}$$

where the Y_{pmn} are the spherical harmonics on S^3, p = 0,1,2,3..., $|n| + |m| \leq p$, and n + m must be odd/even if p is odd/even. [For frequencies of order 1 Hz in a typical closed model[7], equation 3 gives p of order 10^{18}, thus the large p approximation is employed for the remaining calculations.]

The requirement that $|n| + |m|$ be $\leq p$ yields a volume element in state space of $2p^2dp$. The second condition on m and n requires a density of 1/2 state per unit volume. Therefore, we have (for large p)

$$dG_p = p^2dp \tag{5}$$

Eq. 3 gives (for large p)

$$p = 2\pi rf/c \tag{6}$$

so

$$dp = (2\pi r)\cdot(df/c) \quad (7).$$

Eqs. 5, 6, and 7 then give

$$dG_f = 8\pi^3 r^3 f^2 df/c^3 \quad (8).$$

Taking into account the spin degeneracy of a photon we have

$$dG_f = 16\pi^3 r^3 f^2 df/c^3 \quad (9).$$

The volume of an S^3 of radius r is

$$V = 2\pi^2 r^3 \quad (10).$$

Therefore, equations 1, 2, 9, and 10 give

$$du_f(T) = 8\pi h f^3 df/(c^3(\exp(hf/kT) - 1)) \quad (11)$$

exactly the result for $du_f(T)$ in R^3.

3. Summary

The spectral energy density as a function of temperature, $du_f(T)$, in S^3 was derived. As expected, the resulting $du_f(T)$ in an S^3 with radius much larger than the photons' wavelengths is precisely the same as $du_f(T)$ in R^3. Therefore, the blackbody distribution of the cosmic microwave background cannot be used to determine whether our universe is spatially open or closed.

4. References

1. Richard Becker, *Electromagnetic Fields and Interactions* (Blaisdell Publishing Company, 1964) p. 277.
2. I. Peterson, *Science News* **137** (1990) 36.
3. B. Dolan, these proceedings.
4. Robert Wald, *General Relativity* (University of Chicago Press, 1984) p. 71.
5. Erwin Schrodinger, *Physica* **VI** (October 1939) 899.
6. Von Paul Otto Muller, *Physik. Zeitschr.* **40** (March 1939) 366.
7. Charles W. Misner, Kip S. Thorne, and John Archibald Wheeler, *Gravitation* (W. H. Freeman and Company, 1971) p. 738.

ELECTROMAGNETISM IN NONSYMMETRIC THEORIES OF GRAVITY: GRAVITATIONALLY INDUCED POLARIZATION SHIFTS

J. H. Palmer† and R. B. Mann
Department of Physics, University of Waterloo
Waterloo, Ontario, N2L 3G1, CANADA

and

M. D. Gabriel and M. P. Haugan
Department of Physics, Purdue University
West Lafayette, IN 47907, USA

ABSTRACT

Theories of gravity that include a totally antisymmetric component of the metric coupled to electromagnetism predict that light will propagate through a gravitational field with a polarization dependent velocity. The observed polarization depends upon the initial polarization and the path traveled by the light. For extended sources of polarized light within magnetically active regions on the Sun, this leads to an observed depolarization. Observations of the polarization of solar spectral lines near the limb of the Sun are used to place limits on the nonsymmetric charge of the Sun in Moffat's Nonsymmetric Gravitation Theory (NGT).

1. Introduction

Nonsymmetric theories of gravity introduce a totally antisymmetric tensor identified as the antisymmetric part of a general metric.[1,3] The nonsymmetric metric then determines the connection, including macroscopic torsion, through the application of metric compatibility. (Metric compatibility arising as a field equation from the variation of an action with respect to the connection, or through the requirement that the inner product of two vectors is invariant under parallel transport). These theories are strongly related to metric-tensor theories through interpreting the antisymmetric part of the metric as an additional tensor field. However, the way in which nonsymmetric theories introduce the antisymmetric tensor leads naturally to its interpretation as the antisymmetric part of a general metric. The interpretation of the antisymmetric field as a gravitational field is bolstered by coupling the full metric to matter. In this way, the antisymmetric part of the metric is coupled to mass/energy as expected in a theory of

† **Presenter. This paper is based upon work by Gabriel, Haugan, Mann and Palmer to be published in Phys. Rev. Lett. Modifications made here to reflect the presentation are the responsibility of J.H. Palmer.**

gravitation. All nonsymmetric theories which couple the full metric to electromagnetism predict polarization dependent bending of light and radar echo time delay.[2] Furthermore, the observed polarization of light depends upon the path of propagation, as will be shown here.

The existence of path dependent polarization is general for all nonsymmetric theories which couple the full metric to electromagnetism. In order to make a physical prediction, however, it is necessary to work within one theory. The obvious nonsymmetric theory to choose is Moffat's Nonsymmetric Gravitation Theory (NGT).[3] This theory has been studied through its evolution over the past ten years. It is the unique nonsymmetric theory with hermitian symmetry which survives a multitude of theoretical and physical criteria. Other nonsymmetric theories can be created by relaxing the hermitian symmetric condition, but these theories have yet to be fully investigated.[1]

2. General Formalism

We begin with the general action coupling electromagnetism with gravitation,

$$I_{em} = \frac{-1}{16\pi}\int d^4x\,\sqrt{-g}\mathcal{F}g^{\mu\alpha}g^{\nu\beta}\{ZF_{\mu\nu}F_{\alpha\beta} + (1-Z)F_{\alpha\nu}F_{\mu\beta} + YF_{\mu\alpha}F_{\nu\beta}\} \quad (1)$$

where $F_{\mu\nu} \equiv A_{\nu,\mu} - A_{\mu,\nu}$ and the matrix $g^{\mu\nu}$ is the inverse of the nonsymmetric gravitational field $g_{\mu\nu}$ defined by $g^{\mu\alpha}g_{\nu\alpha} = g^{\alpha\mu}g_{\alpha\nu} = \delta^{\mu}_{\nu}$. The new contractions allowed in nonsymmetric theories are parametrized by the constants Y and Z. In addition, a scalar coupling function, $\mathcal{F} = \mathcal{F}(\sqrt{-g}/\sqrt{-\gamma})$, is introduced where $g \equiv \det g_{\mu\nu}$ and $\gamma \equiv \det g_{(\mu\nu)}$.

The Sun's essentially static, spherically symmetric gravitational field admits, in theories like NGT, a representation in spherical coordinates $g_{00} = -T(r)$, $g_{(0i)} = 0$, $g_{(ij)} = H(r)\delta_{ij}$, $g_{[0i]} = L(r)n_i$, and $g_{[ij]} = 0$, where $n_i \equiv x_i/r$ and $r \equiv |\vec{x}|$. Note that we have chosen to diagonalize the symmetric part of the covariant metric rather than the symmetric part of the contravariant metric in order to retain a sensible limiting velocity for the motion of particles. Generally in nonsymmetric theories $g^{(\mu\nu)}$ and $g_{(\mu\nu)}$ cannot be diagonalized simultaneously when $g^{\mu\alpha}g_{\mu\beta} = \delta^{\alpha}_{\beta}$. Using this metric in Eq. (1) and identifying $E_i = F_{0i}$ and $F_{ij} = \epsilon_{ijk}B_k$ gives

$$I_{em} = (8\pi)^{-1}\int d^4x\,\left\{\mathbf{E}\cdot\mathbf{E} + X\alpha\epsilon(\hat{\mathbf{n}}\cdot\mathbf{E})^2 - \frac{1}{\mu}(\mathbf{B}\cdot\mathbf{B} - \Omega(\hat{\mathbf{n}}\cdot\mathbf{E})^2)\right\} \quad (2)$$

where $\epsilon = \mathcal{F}(H/T)^{1/2}(1-\Omega)^{-1/2}$, $\mu = \mathcal{F}^{-1}(H/T)^{1/2}(1-\Omega)^{1/2}$, $\alpha = 2\Omega(1-\Omega)^{-1}$, $\Omega = L^2/TH$, and $X = 1 - Z - Y$.

We now choose $\mathcal{F} = \sqrt{-g}/\sqrt{-\gamma}$ to remove the electrostatic energy dependence of the rate of free-fall analyzed by Will.[4] We also choose $X = 0$ to keep electrostatics unmodified. We expect that experiments using atomic transitions to test the isotropy of local Lorentz frames will tightly constrain the $X\alpha$ term of Eq. (2). With these choices, Eq. (2) becomes

$$I_{em} = (8\pi)^{-1} \int d^4x \left\{ \epsilon E^2 - \frac{1}{\mu}(B^2 - \Omega(\hat{n} \cdot \vec{B})^2) \right\}, \tag{3}$$

where $\epsilon = \mu = (H/T)^{1/2}$, and $\Omega = L^2/TH$.

Varying Eq. (3) with respect to A_μ yields the gravitationally modified Maxwell equations. Substituting locally plane electromagnetic waves, $\vec{E} = \vec{A}_E e^{i\Phi}$ and $\vec{B} = \vec{A}_B e^{i\Phi}$ where $\partial_\mu \Phi = (\partial\Phi/\partial t, \vec{\nabla}\Phi) = (-\omega, \vec{k})$, into the Maxwell equations and applying the geometric optics approximation gives the Eikonal equation,[2]

$$k^2 \vec{A}_B - \epsilon\mu\omega^2 \vec{A}_B - \Omega(\hat{n} \cdot \vec{A}_B)(k^2 \hat{n} - \vec{k}(\hat{n} \cdot \vec{k})) = 0. \tag{4}$$

The Eikonal equation describes the propagation of light in the spherical coordinate system being used. When $\Omega \neq 0$, the propagation is dependent upon the polarization through the $\hat{n} \cdot \vec{A}_B$ term and upon the direction of propagation through the $\hat{n} \cdot \vec{k}$ term. Linearly polarized light with $\hat{n} \cdot \vec{A}_B = 0$ has a coordinate phase velocity of $(1/\epsilon\mu)^{1/2}$. When the light is linearly polarized with $\vec{A}_B$ in the plane spanned by $\vec{k}$ and $\hat{n}$, the coordinate phase velocity is $((1 - \Omega \sin^2 \psi)/\epsilon\mu)^{1/2}$ where ψ is the angle between the propagation vector and the unit radial, $\hat{k} \cdot \hat{n} = \cos\psi$. By perturbing the Eikonal equation about the path taken by light in the vacuum, it is possible to derive the total phase shift between the two polarizations. To first order in Ω and $(1 - \epsilon\mu)$ the total phase difference integrated over the unperturbed path is

$$\Delta\phi = \frac{\omega}{2} \int \Omega \sin^2 \psi \, dt \tag{5}$$

where t parametrizes the path.

Eq. (5) is general for all nonsymmetric theories. Now substituting the NGT solution for $\Omega = L^2/(TH)$ into Eq. (5) will allow the integration to be performed. To post-Newtonian order, NGT gives $\Omega = l_\odot^4/r^4$ where $l_\odot$ is the nonsymmetric charge of the Sun. Choosing the path taken by light traveling from the surface of the Sun of radius R to the Earth at infinity we perform the integral and obtain

$$\Delta\phi(\mu) = \frac{\pi l_\odot^4}{\lambda R^3} \left\{ \frac{3\pi}{16(1-\mu^2)^{3/2}} - \frac{\mu}{4} - \frac{3\mu}{8(1-\mu^2)} - \frac{3}{8(1-\mu^2)^{3/2}} \arcsin(\mu) \right\} \tag{6}$$

where μ (not to be confused with the magnetic permeability of Eq. (3)) is the standard solar physics notation for the cosine of the heliocentric angle and indicates the location of the origin of the polarized light on the surface of the Sun.

When the surface of the Sun is observed with a small aperture (5 arcsec diameter), the observed area is filled with magnetic and nonmagnetic areas. The magnetic field exists in small fluxtubes sprinkled throughout the aperture. Wherever light is emitted or absorbed within the magnetic field, Zeeman splitting can be observed. The partially circularly polarized σ components of the Zeeman split line, as recorded in measurements of Stokes V, provides the source of polarized light for our study. The value of Stoke's V observed depends strongly upon complicated processes occurring at the point of emission or absorption as well as upon the structure of the magnetic fields. Imagine making a number of identical copies of one such magnetic structure and sprinkling those copies across the solar disk from the centre of the disk to the limb. Eq. (6) shows that the observed polarization would vary from structure to structure as μ changes. One structure might be observed to be linearly polarized, the next circularly polarized, even though the physics of emission is identical. If the variation in the type of polarization occurs quickly enough with change in μ to allow a number of polarizations to exist within our small aperture, then there will be a wash-out of polarizations and a resultant depolarization. This can be modeled mathematically by assuming an initial polarization, introducing the gravitationally induced phase shift, Eq. (6), and calculating the observed Stoke's parameters at any point within the aperture. Since we cannot resolve structures smaller than the aperture, we then make use of the additive property of Stokes parameters for incoherent addition of light and integrate the Stoke's parameters over the magnetic area within the aperture. This will give the observed Stokes parameters.

3. Results and Discussion

We have written code which performs trapezoidal integration over the magnetic areas within the aperture. Assumptions made on the initial polarization of the light were chosen to be as conservative as possible. Since nonsymmetric theories will cause a decrease in circularly polarized light, we assume in all cases that the initial light is 100% circularly polarized. This assumption is extremely conservative since it results in the largest possible upper bound on the theoretically observable values of Stokes V. Furthermore, in the most conservative case we assume that all of the circularly polarized light is coming from one fluxtube with 1 arcsec diameter and that the light from this fluxtube is so intense as to totally dominate the light output within the aperture. This has the effect of giving a small range of μ values over which the polarization is allowed to vary and ensuring that the initial light is strongly polarized. Thus, significant gravitationally induced depolarization will only be achieved with relatively large values of $l_{\odot}$. In the second case we make a less conservative assumption on the total size of the magnetically active area by using the estimates of Solanki *et al.*[5] They have studied the polarization measurements made by Stenflo *et al*[6] and have found that, for the observation in question, 12.9% of the area observed within the aperture emits polarized light. Once again we assume that the polarized light is so intense as to totally dominate the light output.

We have run the code to find the theoretical upper limit of observed Stokes V as a function of the cosine of the heliocentric angle μ. In Figure I we present the results of this calculation by plotting $P_e = \sqrt{U^2 + V^2}/I$ against μ. Case 1 uses the extremely conservative assumption of a single small fluxtube and is calculated for light of wavelength 5250Å. Case 2 uses the modeling of Solanki *et al*[5] to assign a larger area to the magnetically active region and is calculated for light of wavelength 5247Å. The two Stokes V data points are at the two different wavelengths and come from the polarization survey done by Stenflo *et al.*[6] In case 1, the limit on the solar NGT charge is found to be $l_\odot \leq 1200$km, and in case 2, $l_\odot \leq 600$km. These are factors of 2.7 and 5.4, respectively, below the presently accepted limit of 3230km. Since the static, spherically symmetric metric in NGT depends upon $l_\odot^4$, the limits on the metric are tighter by factors of 50 and 850 respectively. With an increase in the number of observations of Stokes V near the limb of the Sun, these limits may be further substantiated and improved.

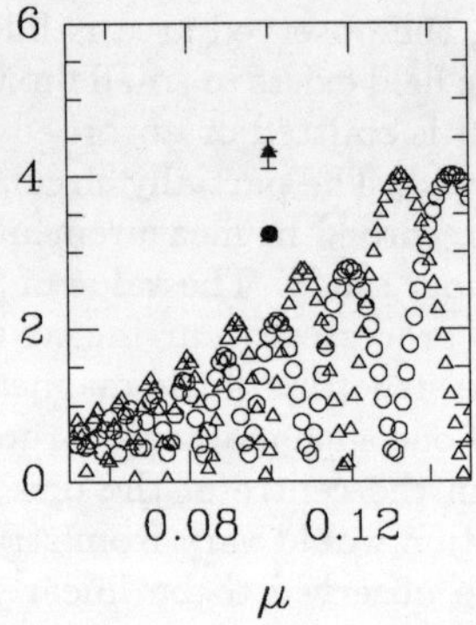

Fig. I. Triangles represent Case 1. Circles represent Case 2. Solid points represent observational data.

4. References

1. R.B. Mann, Class. Quantum Grav. **1**, 561 (1984); **6**, 41 (1989).
2. M.D. Gabriel, M.P. Haugan, R.B. Mann and J.H. Palmer, Phys. Rev D. **43**, 308 (1991).
3. J.W. Moffat in *Banff Summer Institute on Gravitation*, eds. P.S. Wesson and R.B. Mann (1990).
4. C.M. Will, Phys. Rev. Lett. **62**, 369 (1989); R.B. Mann, J.H. Palmer and J.W. Moffat, Phys. Rev. Lett. **62**, 2765 (1989).
5. J.O. Stenflo, S.K. Solanki and J.W. Harvey, Astron. Astrophys. **171**, 305 (1987).
6. S.K. Solanki, C. Keller, and J.O. Stenflo, Astron. Astrophys. **188**, 183 (1987).

Gravity-Spin Coupling and Neutrino Helicity Oscillations

G. Papini and Y.Q. Cai
Department of Physics, University of Regina
Regina, Saskatchewan, S4S 0A2, Canada.

Abstract

The covariant Dirac equation is solved exactly to first order in the metric deviation $\gamma_{\mu\nu}$. The solution yields general gravity-total angular momentum coupling terms and provides independent support for the existence of the rotation-spin effects for neutrons predicted by Mashhoon. The spin part of the general effect for fermions also applies to gravitational fields of arbitrary strength. Under specific conditions, the coupling is capable of flipping the helicity of the particles. This has, for neutrinos, interesting astrophysical implications.

In view of the fundamental importance of Einstein's theory, it is desirable to persue its testing at all levels. A new generation of particle interferometers made possible by recent technological advances may enhance the probability of success in this field[1]. With this in mind, we have recently devised ways of solving some of the basic equations of quantum mechanics and provided a more satisfactory basis for quantum interferometry in the presence of weak gravitational fields[2].

Fairly distinctive issues of interest at the quantum level involve the interaction of gravity with spin. Little is known on the subject. Here, however, the peculiar behaviour of particles that acquire entirely different properties in different helicity states can be turned to our advantage. This is the case with neutrinos (antineutrinos) that interact, according to the well established standard model, only as left-handed (right-handed) particles. It is shown below that a gravity induced neutrino spin-flip can affect the dynamical behaviour of rapidly rotating neutron stars and collapsing stars.

The general characteristic of the solutions of the covariant Klein-Gordon and Landau-Ginzburg equations previously given by us[2] is that they can be written in the form $\phi = e^{-i\chi}\Phi$ where Φ is a general solution of the field-free equations,

$$\chi = -\frac{1}{4}\wp \int_p^x dz^\lambda (\gamma_{\alpha\lambda,\beta}(z) - \gamma_{\beta\lambda,\alpha}(z)) L^{\alpha\beta}(z) + \frac{1}{2}\wp \int_p^x dz^\lambda \gamma_{\alpha\lambda} P^\alpha. \tag{1}$$

$\wp$ represents path ordering, $L^{\alpha\beta}(z)$, P^α are generators of the Poincaré group and the path integral is taken along the classical trajectory of the particle. For a closed path, the gravitational phase is related to Berry's phase[3] and can be rewritten as[4] $\chi = -\frac{1}{4}\iint_{\Sigma_p} R_{\mu\nu\alpha\beta} L^{\alpha\beta} d\tau^{\mu\nu}$, where Σ_p is the surface bound by the closed path p and $R_{\mu\nu\alpha\beta}$ represents the linearized Riemann curvature tensor. This result is needed in what follows.

Spin-1/2 particles. Consider the covariant Dirac equation

$$(\gamma^\mu(x) D_\mu - \frac{mc}{\hbar})\psi = 0, \tag{2}$$

where the matrices $\gamma^\mu(x)$ are obtained from the usual Dirac matrices γ^a by means of a vierbein field $h^\mu{}_a(x)$. In Eq.(2) $D_\mu \equiv \nabla_\mu + i\Gamma_\mu$, where ∇_μ is the usual covariant derivative. The spinoral connection Γ_μ is given by $\Gamma_\mu = \frac{i}{4}\gamma^\nu \gamma_{\nu;\mu} = -\frac{1}{4}h^\nu{}_a(x)h_{\nu b;\mu}(x)\sigma^{ab}$ with $\sigma^{ab} = -(i/2)[\gamma^a, \gamma^b]$. A new spinor ψ' defined by $\psi' \equiv \exp(i\wp \int_p^x \Gamma_\mu(z)dz^\mu)\psi$ satisfies the equation $(\gamma^\mu(x)\nabla_\mu - \frac{mc}{\hbar})\psi' = 0$, or $(g^{\mu\nu}\nabla_\mu\nabla_\nu - \frac{m^2c^2}{\hbar^2})\psi' = 0$. This can be solved in the weak field approximation for each component of ψ' and the gravitational field appears only in the phase factor of Eq.(1). One therefore has $\psi' = \exp(-i\chi)\Psi$, where Ψ satisfies the Dirac equation in Minkowski space, and finally[5]

$$\psi = \exp(-i\wp \int_p^x \Gamma_\mu(z)dz^\mu)\exp(-i\chi)\Psi \equiv \exp(-iS)\Psi. \tag{3}$$

Similar results can be obtained for spin-1 particles[5].

We now turn to a particular application of Eq.(3). Choose cylindrical coordinates (t, r, θ, z) for an inertial frame F_0. An observer at rest in a frame F' rotating with constant angular velocity Ω relative to F_0 will follow the world line $r = \text{const.}, \theta = \text{const.} + \Omega t, z = \text{const.}$. Consider an orthogonal tetrad consisting of the observer's four-velocity $\lambda^\mu_{(0)} = dx^\mu/ds$ and the triad $\lambda^\mu_{(i)}$ $(i = 1,2,3)$ normal to the world line. By choosing $\text{diag}\,\eta_{(\alpha)(\beta)} = (-1, 1, 1, 1)$ and using the local tetrad[6,7]

$$\lambda^\mu_{(0)} = (\gamma, 0, \frac{\gamma\Omega}{c}, 0) \quad \lambda^\mu_{(1)} = (0, 1, 0, 0) \quad \lambda^\mu_{(2)} = (\frac{\gamma\Omega r}{c}, 0, \frac{\gamma}{r}, 0) \quad \lambda^\mu_{(3)} = (0, 0, 0, 1), \tag{4}$$

where $\gamma \equiv (1 - r^2\Omega^2/c^2)^{-1/2}$, one can construct[7] a vierbein field $h^\mu{}_a(x)$ along the world line of the observer

$$h^\mu{}_{(0)} = (\gamma, 0, \frac{\gamma\Omega}{c}, 0) \quad h^\mu{}_{(1)} = (-\frac{\gamma\Omega r}{c}\sin\gamma\Omega t, \cos\gamma\Omega t, \frac{\gamma}{r}\sin\gamma\Omega t, 0)$$

$$h^\mu{}_{(2)} = (\frac{\gamma\Omega r}{c}\cos\gamma\Omega t, \sin\gamma\Omega t, \frac{\gamma}{r}\cos\gamma\Omega t, 0) \quad h^\mu{}_{(3)} = (0, 0, 0, 1). \tag{5}$$

It is then easy to calculate the spinor connection Γ_μ. By using the Dirac representation[8] for the γ-matrices, one obtains $\Gamma_0 = \frac{\gamma\Omega}{2c}\sigma_z$ and from Eq.(3) $\exp(-i\int \Gamma_0 dz^0)\Psi = \exp(-\frac{1}{2}\int \gamma\Omega\sigma_z dt)\Psi$, where Ψ has the usual plane wave form. Besides the contribution due to the coupling of the orbital angular momentum to rotation, which gives the Sagnac effect[2], one obtains the spin-rotation coupling $E' = E + \frac{\hbar}{2}\Omega\sigma_z$. *The present result is exact and follows as a special case from the general coupling of Eq.(3).* It also agrees with that obtained by Mashhoon[9] and by Hehl and Ni[10]. E' is the energy observed by the co-rotating observer. It should be multiplied by a factor γ when referred to a non-rotating observer. The same coupling also exists for spin-1 particles[5,11].

The solution (3) is exact to first order in $\gamma_{\mu\nu}$. It also is Lorentz invariant and invariant under gravitational gauge transformations[2] and provides, to the order considered, the general relativistic generalization of the notion of wave vector. Such a generalization appears intrinsically nonlocal, as indicated by the presence of the angular momentum operator in the expressions for χ and ξ. Unlike the standard approach, the result applies to both inertial and gravitational fields and does not depend on the eikonal approximation. In addition, the gravity-particle coupling becomes effectively vectorial, as previously found for the spin-0 case[2,3]. When $\gamma_{\mu\nu}$ represents rotation, the results are indistinguishable from those of Mashhoon and Hehl and Ni. Even though the solution (3) is only valid to first order in $\gamma_{\mu\nu}$, the contribution of the spin-curvature coupling to S in (3) *remains the same for gravitational fields of arbitrary strength.* This follows from the definition of ψ', which does not require any linearization.

One also finds that spin angular momentum couples to gravitation in just the same way as orbital angular momentum[5]. Of course, the relative magnitude of the two couplings for an interferometer of size ℓ is proportional to $[S^{\alpha\beta}]/[L^{\alpha\beta}] \sim \lambda/\ell$. The shift produced by the spin effect can therefore be neglected in some circumstances. This explains why the calculations performed for spin-0 particles[2] reproduce so well the existing experimental results obtained by using neutron interferometers.

The gravity-spin coupling found is capable, in principle, of flipping the helicities of particles. This has important consequences for neutrinos. In fact, left-handed neutrinos become unobservable when transformed by precession into right-handed ones. The effect is the analogue of the $\nu_L \to \nu_R$ conversion which has been conjectured to take place in a magnetic field if neutrinos possess a magnetic moment μ_ν. Focusing now on rotation, the addition of a term $\boldsymbol{\sigma} \cdot \boldsymbol{\Omega}$ to the Hamiltonian does in fact induce *neutrino oscillations* independently of those of magnetic origin. Neglecting the neutrino mass for simplicity, the new oscillations are described by the equation

$$i\frac{d}{dz}\begin{pmatrix} \nu_R \\ \nu_L \end{pmatrix} \approx -\frac{\Omega_\perp}{2c}\begin{pmatrix} \nu_L \\ \nu_R \end{pmatrix}.$$

Therefore if a beam of ν_L consists of N particles at $z = 0$, the corresponding number N_R of ν_R at $z = L$ is given by $N_R = N\sin^2\left(\frac{\Omega_\perp L}{2c}\right)$, where $\Omega_\perp$ is the angular velocity component transverse to the direction of propagation of the neutrinos and L is the characteristic length of the rotating source. For galaxies, white dwarfs and fast neutron stars one finds respectively $N_R \sim 10^{-6}N$, $10^{-4}N$, and $10^{-2}N$. Appreciable conversion can therefore occur within or in the neighborhood of the most rapidly rotating neutron stars at interior temperatures $T \leq 10^9$K when ν_e emission contributes significantly to the cooling of the star[12]. Here rotational conversion can even be larger than in the corresponding magnetic process. This can be seen by comparing the dimensionless parameter $P \equiv \frac{\Omega_\perp L}{2c}$, on which the transition probability depends, with the corresponding parameter $P' \equiv \mu_\nu B_\perp L/\hbar c$. One obtains $P/P' > 1$ for $B_\perp \sim 10^{12}G$, $\Omega_\perp \sim 10^3 s^{-1}$ and $\mu_\nu < 2\times 10^{-16}\mu_B$. This value of μ_ν is already a few orders of magnitude larger than the one allowed by the standard minimal $SU(2)_L \times U(1)$ electroweak theory, $\mu_\nu = (3m_e G_F/4\sqrt{2}\pi^2)\mu_B m_\nu$[13]! As with the magnetic process, the $\nu_L \to \nu_R$ helicity flip is unlikely to occur deep inside a neutron star because of its large density[14,15,16]. In the co-rotating magnetosphere, however

$$\frac{P}{P'} \sim \frac{\Omega_\perp m_e c}{e\mu_\nu B_\perp} \sim \frac{\Omega_\perp m_e c}{e\mu_\nu B_P}\left(\frac{r}{R}\right)^3 \sim \frac{2\times 10^{-16}}{\mu_\nu}\left(\frac{r}{R}\right)^3 \sim \frac{10^{-13}}{\mu_\nu} > 1$$

for $r \sim 8\times 10^6$cm, $R \sim 10^6$cm, $\mu_\nu < 10^{-13}\mu_B$ and a polar magnetic field $B_P \sim 10^{12}G$. In fact for a millisecond pulsar one finds $P \sim 0.6$, where $L \sim 8.5\times 10^6$cm includes part of the crust and the whole magnetosphere. For straight conversion in the magnetosphere one finds $P \sim 1/2$ independent of Ω for any rotating neutron star. The corresponding depletion factors are significant and indicate that the gravity induced helicity flip *can affect the dynamics of the star.*

Similar conclusions can be reached for supernovae. In the hot core of a supernova such as *SN* 1987*A* left-handed electron neutrinos are converted into right-handed ones by scattering on electrons and protons[17]. Unless reflipped, ν_e fluxes at Earth would be inconsistent with observations. To remove this difficulty, resonant $\nu_R \to \nu_L$ magnetic conversion has been conjectured to take place at $\sim$ 100km from the center of the star. This however requires a value of $\mu_\nu \sim 10^{-11}\mu_B$[17] which is well above current theoretical predictions. We believe a better solution is offered by the spin-flip process discussed. In fact a 10% to 50% rotational conversion could take place over distances $\sim$ 500km from the center of the star, if rotation occurred at the rates predicted by some models[18]. In this case, rotation-spin coupling could restore a measure of compatibility with *SN*1987*A* data even if $\mu_\nu \sim 3\times 10^{-19}\mu_B(m_\nu/1eV)$ as predicted by the electroweak theory[13].

This work was supported in part by the Natural Sciences and Engineering Research Council of Canada.

References

[1] S.A. Werner and H. Kaiser, Neutron Interferometry - Macroscopic Manifestation of Quantum Mechanics, in *Quantum Mechanics in Curved Space-Time*, Eds. J. Audretsch and V. DeSabbata, Plenum Press, New York, 1990, pp. 1-22.

[2] Y.Q. Cai and G. Papini, Class. Quantum Grav. **6**, 407(1989).

[3] Y.Q. Cai and G. Papini, Class. Quantum Grav. **7**, 269(1990).

[4] G. Papini, Nuovo Cimento **B52**, 136(1967).

[5] Y.Q. Cai and G. Papini, Phys. Rev. Lett. **66**, 1259(1991).

[6] J.F. Corum, J. Math. Phys. **18**, 770(1977).

[7] W.M. Irvine, Physica **30**, 1160(1964).

[8] C. Itzykson and J. Zuber, *Quantum Field Theory*, McGraw-Hill, New York, 1980.

[9] B. Mashhoon, Phys. Rev. Lett. **61**, 2639(1988).

[10] F.W. Hehl and W.-T. Ni, Phys. Rev. **D42**, 2045(1990).

[11] B. Mashhoon, Phys. Lett. **A139**, 103(1989).

[12] S.L. Shapiro and S.A. Teukolsky, *Black Holes, White Dwarfs and Neutron Stars*, John Wiley & Sons, New York, 1983.

[13] K. Fujikawa and R. Shrock, Phys. Rev. Lett. **55**, 963(1980).

[14] R. Barbieri and R.N. Mohapatra, Phys. Rev. Lett. **61**, 27(1988).

[15] R. Barbieri and G. Fiorentini, Nucl. Phys. **B304**, 909(1988).

[16] M.B. Voloshin, M.I. Vysotskii and L.B. Okun, Sov. Phys. JETP **64**, 446(1986).

[17] M.B. Voloshin, Phys. Lett. **B209**, 360(1988).

[18] E.M.D. Symbalisty, Astrophys. J. **285**, 729(1984).

KINKS IN 1+1, 2+1 AND 3+1 DIMENSIONS

K.A. DUNN
Department of Mathematics, Statistics and Computing Science, Dalhousie University, Halifax, Nova Scotia B3H 3J5, Canada

Tina A. HARRIOTT
Department of Mathematics, Mount Saint Vincent University, Halifax, Nova Scotia B3M 2J6, Canada

J.G. WILLIAMS
Department of Mathematics and Computer Science, Brandon University, Brandon, Manitoba R7A 6A9, Canada

ABSTRACT

The nature of kinks is discussed for spacetimes of a variety of dimensions. A number of examples of kink solutions of the gravitational field equations are presented, including some that are locally the same as, although globally distinct from, previously well known solutions.

Gravitational kinks,[1] that is to say homotopically nontrivial cross-sections of the metric tensor bundle, can occur in $(n+1)$-dimensional spacetimes for any $n \geq 1$. Not only the metric, but also the global topology of the spacetime manifold $\mathcal{M}^{n+1}$ and, in particular, the foliation of $\mathcal{M}^{n+1}$ into n-dimensional hypersurfaces M^n must be taken into account. Writing $\mathcal{M}^{n+1} = \mathbf{R}^1 \times M^n$, the hypersurface M^n will be assumed to be connected, orientable and compact – or at least to be *compactifiable* in the sense that the metric is required to take the same value everywhere on the (infinite) boundary of M^n so that the boundary could be identified to a point, if so desired. The relevant set of homotopy classes is[2] $[\, M^n \,,\, P^n \,]$, where P^n denotes real projective n-space. Manifolds M^n are classified according to *type*,[3,4] which depends on whether there exist mappings $\varphi \,:\, M^n \rightarrow P^n$ which have degree[5] equal to 1 (Type 1 manifolds) or whether no such mappings are possible (Type 2 manifolds).

In 1+1 dimensions, type 1 is the only possibility, namely: $M^1 = S^1$. A 1-kink metric is associated with a degree 1 mapping $\varphi : S^1 \rightarrow P^1$ (an example of which would be the usual homeomorphism between S^1 and P^1). The pertinent set of homotopy classes are in 1-1 correspondence with the group of integers: $[S^1, P^1] \approx$

$\pi_1(P^1) \approx \mathbf{Z}$. In 2+1 dimensions, because of the orientability assumption for M^2, type 2 is the only possibility. The set of homotopy classes of maps $\varphi : M^2 \to P^2$ has been discussed at length by Williams and Zvengrowski[6] and is not even a group for general M^2. In 3+1 dimensions, manifolds of type 1 (*e.g.* P^3) and type 2 (*e.g.* S^3 or $S^1 \times S^2$) are both possible. The set $[M^3, P^3]$ is a group and has been shown[3,7] to equal the direct sum of $\mathbf{Z}$ with a (possibly zero) number of copies of $\mathbf{Z}_2$. Note that the same classification scheme is applicable to vacuum tunnelling in quantum gravity[8] and to the study of line defects in glasses.[9]

Any Lorentz metric can be written[10] in terms of an arbitrary positive definite metric $h_{\mu\nu}$ and a line field:

$$g_{\mu\nu} = h_{\mu\nu} - 2\, h_{\mu\alpha} X^\alpha h_{\nu\beta} X^\beta \, (h_{\lambda\rho} X^\lambda X^\rho)^{-1},$$

and this expression specializes to a form that has been used extensively[11,12] as a starting point for the construction of kink metrics:

$$g_{\mu\nu} = \delta_{\mu\nu} - 2\phi_\mu \phi_\nu \qquad \mu, \nu = 0, 1, ..., n,$$

with $\|\delta_{\mu\nu}\| = \mathrm{diag}\,(1, ..., 1)$, and $\sum \phi_\mu \phi_\mu = 1$ so that the set $\{\phi_\mu\}$ defines the mapping φ. In particular, $\phi_0 := \cos\alpha$ determines the angle α of tilt of the light cones.

For 1+1 dimensions, the Einstein tensor is trivial, but Mann and his coworkers (see Ref. 13 and the references therein) have developed a (1+1)-dimensional theory of gravity whose field equation is $R - \Lambda = 8\pi G\, g_{\mu\nu} T^{\mu\nu}$ coupled with the requirement $T^{\mu\nu}{}_{;\nu} = 0$. These equations are satisfied by

$$ds^2 = -\left(\frac{t^2 - x^2}{t^2 + x^2}\right) dt^2 - \left(\frac{4tx}{t^2 + x^2}\right) dt\, dx + \left(\frac{t^2 - x^2}{t^2 + x^2}\right) dx^2$$

which describes a perfect fluid with $\rho = (t^2 + x^2)^{-1}$ and, since α changes by π over the range $-\infty < x < \infty$, there is one kink. The metric is locally, but not globally, transformable into one of the standard forms of 1+1 FLRW solutions listed by Sikkema and Mann.[13]

A number of (2+1)-dimensional kink solutions have been discussed previously,[6] including the following perfect fluid solution[14] which has no analog in 1+1 or 3+1 dimensions since it is constructed through a "combing" of the velocity field on circles about the origin of $M^2 = \mathbf{R}^2$:

$$ds^2 = -\left(\frac{a^2 - r^2}{a^2 + r^2}\right) dt^2 - \left(\frac{4a\,r^2}{a^2 + r^2}\right) dt\, d\theta + dr^2 + r^2 \left(\frac{a^2 - r^2}{a^2 + r^2}\right) d\theta^2.$$

The constant a is related to the width of the kink, and the solution has equation of state $\rho = 3p = 3a^2\pi^{-1}G^{-1}(a^2 + r^2)^{-2}$. (To be precise, the above metric has only half a kink and a whole kink can only be obtained by extending the coordinates beyond their usual range).

In 3+1 dimensions, the field equations $G_{\mu\nu} + \Lambda g_{\mu\nu} = 0$ are solved by the de Sitter kink metric,[12]

$$ds^2 = -\left(1 - \frac{2r^2}{A^2}\right) dt^2 - 4\left(\frac{r}{A}\right)\left(1 - \frac{r^2}{A^2}\right)^{1/2} dt\, dr + \left(1 - \frac{2r^2}{A^2}\right) dr^2 + r^2\, d\Omega^2.$$

The above form of the metric is only defined on the $0 \leq r \leq A$ coordinate patch and has only half a kink. An N-kink metric can be obtained by introducing an angular coordinate ξ defined by $r = A \sin N\xi$ and allowing ξ to vary from 0 to π. The spacetime manifold is $\mathbf{R}^1 \times S^3$ although, by making use of the stereographic projection $\zeta = A \tan(\xi/2)$, one can rewrite the metric so as to be asymptotically flat on $\mathbf{R}^1 \times \mathbf{R}^3$. For example, for $N = 1$,

$$ds^2 = -\Gamma\, dt^2 - 4A\Sigma\zeta\, dt\, d\zeta + A^2\Gamma\Sigma\, d\zeta^2 + \frac{4A^4\zeta^2}{(A^2+\zeta^2)^2}\, d\Omega^2,$$

where

$$\Gamma = \frac{(A^2-\zeta^2)^2 - 4A^2\zeta^2}{(A^2+\zeta^2)^2}, \qquad \Sigma = \frac{4A^2(A^2-\zeta^2)^2}{(A^2+\zeta^2)^4},$$

with $0 \leq \zeta < \infty$. Analogous solutions exist in both 1+1 dimensions[15] and 2+1 dimensions.[6]

A spacetime's kink nature is a consequence not only of the choice of metric but also of the way in which the spacetime manifold $\mathcal{M}^{n+1}$ is sliced into n-dimensional hypersurfaces. This is most easily appreciated in 1+1 dimensions where the change of coordinates

$$\tau = (t^2 + x^2)^{1/2}, \qquad \lambda = \arctan(x/t)$$

transforms our (1+1)-dimensional 1-kink metric into

$$ds^2 = -d\tau^2 + \tau^2\, d\lambda^2.$$

This is still a 1-kink metric for our original choice of hypersurfaces of constant t. It becomes a 0-kink metric if the original $\mathbf{R}^1 \times S^1$ manifold is changed to one that admits a global family of hypersurfaces of constant τ.

Acknowledgements

This work was supported in part by the Natural Sciences and Engineering Research Council of Canada.

References

1. D. Finkelstein and C.W. Misner, *Ann. Phys. (NY)* **6** (1959) 230.

2. N. Steenrod, *The Topology of Fibre Bundles* (Princeton U.P., Princeton, NJ, 1951) §40; C.J. Isham, *Modern Differential Geometry for Physicists* (World Scientific, Singapore, 1989) pp. 148–149.

3. A.R. Shastri, J.G. Williams and P. Zvengrowski, *Int. J. Theor. Phys.* **19** (1980) 1.

4. A.R. Shastri and P. Zvengrowski, "Type of 3-manifolds and addition of relativistic kinks," University of Calgary Preprint, 1990.

5. Strictly speaking, the concept of (Brouwer) degree is not always applicable and has to be replaced in certain dimensions with the idea of "twisted degree." Further details are given in Ref. 6 below.

6. J.G. Williams and P. Zvengrowski, "Kink metrics in (2+1)-dimensional spacetime," Preprint, 1991.

7. K. Bugajska, *J. Math. Phys.* **30** (1989) 619.

8. C.J. Isham, "Vacuum tunneling in static space-times," in *Old and New Questions in Physics, Cosmology, Philosophy, and Theoretical Biology - Essays in Honor of Wolfgang Yourgrau*, ed. A. van der Merwe (Plenum Press, NY, 1983).

9. N. Rivier and D.M. Duffy, *J. de Physique* **43** (1982) 293.

10. R. Geroch and G.T. Horowitz, "Global structure of spacetimes," in *General Relativity - An Einstein Centenary Survey* ed. S.W. Hawking and W. Israel (Cambridge U.P., Cambridge, 1979) p. 219.

11. J.G. Williams and R.K.P. Zia, *J. Phys. A* **6** (1973) 1; D. Finkelstein and G. McCollum, *J. Math. Phys.* **16** (1975) 2250; G. Clément, *Gen. Relat. Gravit.* **16** (1984) 131, 477; K.A. Dunn, *Gen. Relat. Gravit.* **22** (1990) 507.

12. T.A. Harriott and J.G. Williams, *J. Math. Phys.* **29** (1988) 179; K.A. Dunn and J.G. Williams, *J. Math. Phys.* **30** (1989) 87.

13. A.E. Sikkema and R.B. Mann, *Class. Quantum Grav.* **8** (1991) 219.

14. J.G. Williams, *Gen. Relat. Gravit.* **23** (1991) 181.

15. K.A. Dunn, T.A. Harriott and J.G. Williams, "Toy model for gravitational kinks," Preprint 1991.

2+1 GRAVITY KINKS FOR MULTIPLY CONNECTED SPACETIME MANIFOLDS

J.G. WILLIAMS
Department of Mathematics and Computer Science,
Brandon University, Brandon, Manitoba R7A 6A9 Canada

P. ZVENGROWSKI
Department of Mathematics and Statistics,
University of Calgary, Calgary, Alberta T2N 1N4 Canada

ABSTRACT

Lorentz metrics and their homotopy classification are discussed for multiply connected spacetime manifolds in 2+1 dimensions. Examples are presented of metrics having Finkelstein-Misner kinks, including a metric satisfying the Einstein equations with cosmological constant.

Homotopically nontrivial light cone configurations called *kinks*[1] exist for (2+1)-dimensional manifolds $\mathcal{M}^3 = \mathbf{R}^1 \times M^2$ and are enumerated by evaluating the set of homotopy classes $[M^2, P^2]$ where P^2 is real projective two-space and where, for simplicity, the two-dimensional manifold M^2 will be assumed to be compact, connected and orientable. The least complicated possibility for M^2 is the two-sphere S^2. The set $[S^2, P^2]$ is a group, namely the second homotopy group $\pi_2(P^2) \approx \mathbf{Z}$, and hence the various metric field configurations can be labelled by a single homotopy label $N \in \mathbf{Z}$. The case for which M^2 is a torus, $M^2 = S^1 \times S^1$, has been studied in the context of defects in ordered media by Garel[2] who notes that $[S^1 \times S^1, P^2]$ does not have a group structure. The possibilities for M^2 are summarized by $M^2 = F_g$, g $= 0, 1, 2, \ldots$, where F_g denotes the connected sum of g tori: $S^1 \times S^1 \# \ldots \# S^1 \times S^1$. The genus g is the "number of holes," so that the g $= 0$ case corresponds to the two-sphere, $F_0 = S^2$. The case g $= 1$ is the ordinary torus, $F_1 = S^1 \times S^1$. Although the homotopy classes $[F_g, P^2]$ are a group for g $= 0$, they are only a *set* for g ≥ 1. The set is evaluated by considering the short exact sequence:

$$H^2(F_g; \mathbf{Z}) \approx [F_g, S^2] \overset{\kappa_\#}{\rightarrowtail} [F_g, P^2] \overset{j_\#}{\twoheadrightarrow} [F_g, P^\infty] \approx H^1(F_g; \mathbf{Z}_2),$$

where $\kappa : S^2 \to P^2$ is the standard double covering, $P^\infty = P^1 \cup P^2 \cup \ldots$, and $j : P^2 \to P^\infty$ is the inclusion. The procedure for evaluating the cohomology groups $H^2(F_g; \mathbf{Z})$ and $H^1(F_g; \mathbf{Z}_2)$ is given by Massey.[3] For a compact, connected, orientable n-manifold M^n, $H^n(M^n; G) \approx G$, so that $H^2(F_g; \mathbf{Z}) \approx \mathbf{Z}$. For an arcwise connected manifold X, the one-dimensional cohomology group $H^1(X; G)$ is naturally isomorphic to the group of all homomorphisms of $\pi_1(X)$ into the coefficient group G. Denoting the abelianization of a group by $\mathcal{A}b$, since $\mathcal{A}b\{\pi_1(F_g)\} \approx \mathbf{Z}^{2g}$, one deduces that $H^1(F_g; \mathbf{Z}_2) \approx \mathbf{Z}_2{}^{2g}$. The exact sequence becomes

$$\mathbf{Z} \overset{\kappa_\#}{\rightarrowtail} [F_g, P^2] \overset{j_\#}{\twoheadrightarrow} \underbrace{\mathbf{Z}_2 \oplus \mathbf{Z}_2 \oplus \dots \oplus \mathbf{Z}_2}_{2g \text{ copies}}.$$

It follows that $[F_g, P^2]$ is a disjoint union of the 2^{2g} sets $j_\#^{-1}(x)$, $x \in H^1(F_g; \mathbf{Z}_2)$, and, furthermore,[4] that

$$[F_g, P^2] \approx \mathbf{Z} \,\dot{\cup}\, \mathbf{Z}_2 \,\dot{\cup}\, \mathbf{Z}_2 \,\dot{\cup}\, \dots \,\dot{\cup}\, \mathbf{Z}_2,$$

where $\mathbf{Z} = j_\#^{-1}(0)$. The following box diagram illustrates the $M^2 = F_1 \equiv S^1 \times S^1$ case.

$\mathbf{Z}_2$	$\mathbf{Z}_2$
$\mathbf{Z}$	$\mathbf{Z}_2$

To classify a given metric, one must first decide to which box it belongs. Metrics belonging to the $(0,0)$-box (i.e. the $\mathbf{Z}$ box) correspond to maps $M^2 \to P^2$ that can be lifted to $M^2 \to S^2 \to P^2$, and their further "within box" classification is decided by computing the degree of $M^2 \to S^2$. If a metric is found to belong to a box other than the $(0,0)$-box, then the within box classification becomes somewhat technical and leads to a decision between two possibilities that can be made by computing the Olum invariant,[5] or by taking advantage of a result of Adams[6] that concerns the "compressibility" of these maps into $P^1 \subset P^2$. Further details are given by Williams and Zvengrowski.[4] Note that increasing the genus leads to a considerable increase in complexity. The box diagram for $M^2 = F_2 \equiv S^1 \times S^1 \# S^1 \times S^1$ is:

$\mathbf{Z}_2$	$\mathbf{Z}_2$	$\mathbf{Z}_2$	$\mathbf{Z}_2$
$\mathbf{Z}_2$	$\mathbf{Z}_2$	$\mathbf{Z}_2$	$\mathbf{Z}_2$
$\mathbf{Z}_2$	$\mathbf{Z}_2$	$\mathbf{Z}_2$	$\mathbf{Z}_2$
$\mathbf{Z}$	$\mathbf{Z}_2$	$\mathbf{Z}_2$	$\mathbf{Z}_2$

Introducing a set of functions $\{\phi_\mu \,|\, \sum \phi_\mu \phi_\mu = 1\}$ and the Kronecker delta $\|\delta_{\mu\nu}\| = \mathrm{diag}\,(1,1,1)$, the expression

$$g_{\mu\nu} = \delta_{\mu\nu} - 2\phi_\mu \phi_\nu$$

is a starting point for the construction of kink metrics.[7,8] Consider a spacetime manifold $\mathcal{M}^3 = \mathbf{R}^1 \times F_1 \equiv \mathbf{R}^1 \times S^1 \times S^1$ with coordinates (t, θ, φ) where the angles $(\theta, \varphi) \in F_1$ have a range $0 \leq \theta, \varphi \leq 2\pi$. The metric $g_{\mu\nu}$ is determined by the $F_1 \to P^2$ mapping which is specified by giving (ϕ_0, ϕ_1, ϕ_2) as a function of (θ, φ). Since the mapping is into P^2, antipodal points $\pm(\phi_0, \phi_1, \phi_2)$ are identified. For example, the map

$$(\phi_0,\, \phi_1,\, \phi_2) = (\cos(\theta/2),\, \sin(\theta/2),\, 0)$$

clearly maps points $0 \leq \theta \leq 2\pi$ of the torus onto one of the homotopically nontrivial loops in P^2. [Recall that $\pi_1(P^2) \approx \mathbf{Z}_2$]. The map does not lift to $S^2 \to P^2$ and is thus a member of a $\mathbf{Z}_2$ box. The lack of a group structure is evident from the fact that the map

$$(\phi_0,\, \phi_1,\, \phi_2) = (\cos\theta,\, \sin\theta,\, 0),$$

which maps twice around a nontrivial loop in P^2, belongs to a completely different box, namely the $\mathbf{Z}$ box. The map is homotopic to the constant map $(\phi_0, \phi_1, \phi_2) = (1, 0, 0)$ and is a member of the zero homotopy class of $\mathbf{Z}$.

An example of a member of the generating class of $\mathbf{Z}$ (i.e. a 1-kink map) would be

$$(\phi_0,\, \phi_1,\, \phi_2) = (\cos\theta,\, \sin\theta\cos\varphi,\, \sin\theta\sin\varphi).$$

This can equally well be interpreted as a 1-kink map for $\mathcal{M}^3 = \mathbf{R}^1 \times S^2$ if one allows (θ, φ) to range over the 2-sphere: $0 \leq \theta \leq \pi$, $0 \leq \varphi \leq 2\pi$. The corresponding metric $g_{\mu\nu} = \delta_{\mu\nu} - 2\phi_\mu\phi_\nu$ can be used to calculate the Ricci tensor $R_{\mu\nu}$ and the scalar curvature R, although it is convenient to change from the original (t, x, y) coordinates to (t, θ, φ) coordinates, with $(x^2 + y^2)^{1/2} \equiv r = A \sin\theta$ where A is a positive constant. Then $R_{\mu\nu} = 4\,A^{-2} g_{\mu\nu}$, $R = 12\,A^{-2}$ and the Einstein tensor is $G_{\mu\nu} = -2\,A^{-2} g_{\mu\nu}$, so that the metric satisfies the Einstein equations, $G_{\mu\nu} + \Lambda\, g_{\mu\nu} = 0$ with $\Lambda = 2\,A^{-2}$. This is clearly the (2+1)-dimensional analog of the de Sitter kink metric[8] and is *locally* but not *globally* transformable to the non-kink form of the 2+1 de Sitter metric given by Deser and Jackiw.[9]

The usual ill-definedness of the angular coordinates at $\theta = 0$ (or $\theta = \pi$) could be removed by projecting onto the coordinate patch $r \leq A$ and transforming to the $(\partial/\partial t, \partial/\partial x, \partial/\partial y)$ basis, but such a coordinate patch would encompass only *half* of the kink. A more appropriate change of coordinates is given by the stereographic projection $\zeta = A\tan(\theta/2)$. The metric, which is then a full 1-kink metric, can be written

$$ds^2 = -\Gamma\, dt^2 - 4A\Sigma\zeta\, dt\, d\zeta + A^2\Gamma\Sigma\, d\zeta^2 + \frac{4A^4\zeta^2}{(A^2+\zeta^2)^2}\, d\varphi^2,$$

where

$$\Gamma = \frac{(A^2-\zeta^2)^2 - 4A^2\zeta^2}{(A^2+\zeta^2)^2}, \qquad \Sigma = \frac{4A^2(A^2-\zeta^2)^2}{(A^2+\zeta^2)^4},$$

and $0 \leq \zeta < \infty$. The spacetime manifold, $\mathcal{M}^3 = \mathbf{R}^1 \times \mathbf{R}^2$, is no longer compact but since the metric takes the same value everywhere on the infinite boundary of $\mathbf{R}^2$, it follows that $\mathcal{M}^3$ is equivalent to $\mathbf{R}^1 \times S^2$ from the standpoint of homotopy theory. As ζ increases from 0 to ∞, g_{tt} changes from -1 to 1 to -1, indicating that the angle of tilt of the light cones changes by π.

Acknowledgements

This work was supported in part by the Natural Sciences and Engineering Research Council of Canada.

References

1. D. Finkelstein and C.W. Misner, *Ann. Phys. (NY)* **6** (1959) 230.
2. A.T. Garel, *J. de Physique* **39** (1978) 225.
3. W.S. Massey, *Algebraic Topology: An Introduction* (Springer-Verlag, NY, 1967), p. 239.
4. J.G. Williams and P. Zvengrowski, "Kink metrics in (2+1)-dimensional spacetime," Preprint, 1991.
5. P. Olum, *Trans. Amer. Math. Soc.* **103** (1962) 30.
6. J.F. Adams, *Bull. London Math. Soc.* **14** (1982) 533.
7. J.G. Williams and R.K.P. Zia, *J. Phys. A* **6** (1973) 1;
   D. Finkelstein and G. McCollum, *J. Math. Phys.* **16** (1975) 2250;
   G. Clément, *Gen. Relat. Gravit.* **16** (1984) 131, 477.
8. T.A. Harriott and J.G. Williams, *J. Math. Phys.* **29**, (1988) 179;
   K.A. Dunn and J.G. Williams, *J. Math. Phys.* **30** (1989) 87.
9. S. Deser and R. Jackiw, *Ann. Phys. (NY)* **153** (1984) 405.

Participant List

Abhay Ashtekar	- - - Syracuse University
A. Beesham	- - - University of Zululand
Beverly K.Berger	- - - Oakland University
Patrick Brady	- - - University of Alberta
Robert Cassidy	- - - Elizabethtown College
Jose Castejon-Amenedo	- - - Dalhousie University
Alan Coley	- - - Dalhousie University
F. I. Cooperstock	- - - University of Victoria
Guillermo Covarrubias	- - - U. Autonoma Metropolitana-Iztapalapa
Bahman Darian	- - - University of Alberta
A. Das	- - - Simon Fraser University
Brian Dolan	- - - St. Patrick's College
Krishan Duggal	- - - University of Windsor
Ken Dunn	- - - Dalhousie University
R. Epp	- - - University of Manitoba
G. Fee	- - - University of Waterloo
John L. Friedman	- - - University of Wisconsin-Milwaukee
S. Fulling	- - - Texas A&M University
Claude Gauthier	- - - Universite de Moncton
Jack Gegenberg	- - - University of New Brunswick
Edward Glass	- - - University of Windsor

Tina Harriott	- - - Mount Saint Vincent University
J. B. Hartle	- - - I.T.P., Univ. of Calif.
Mark Haugan	- - - Purdue University
Geoff Hayward	- - - University of British Columbia
Paul Hickson	- - - Univ. Brit.Columbia
N. Kaiser	- - - CITA, Univ. of Toronto
Nemanja Kaloper	- - - University of Minnesota
Laine Keahey	- - - University of Winnipeg
Patrick Kelly	- - - University of Toronto
D. H. King	- - - Matkowski & Associates,Vancouver
Deborah Konkowski	- - - U.S. Naval Academy
Karel V. Kuchar	- - - University of Utah
Gabor Kunstatter	- - - University of Winnipeg
H. P. Kunzle	- - - University of Alberta
Jorma Louko	- - - University of Alberta
Garry Ludwig	- - - University of Alberta
Alex Lyons	- - - University of Alberta
M. A. H. MacCallum	- - - Q.M.C., Univ. of London
Alick MacPherson	- - - University of Alberta
R. B. Mann	- - - University of Waterloo
R. J. McKellar	- - - University of New Brunswick
R. G. McLenaghan	- - - University of Waterloo

C. Micheal --- Dalhousie University

Frank H. Molzahn --- University of Manitoba

Masahiro Morikawa --- University of British Columbia

M.S. Morris --- University of Waterloo

Sharon Morsink --- University of Waterloo

R. Myers --- McGill Univ.

Mark Naber --- University of Windsor

T. A. Osborn --- WITP, Univ. of Manitoba

J. Overduin --- University of Waterloo

Jody Palmer --- University of Waterloo

S. Richardson --- University of Victoria

D.S. Salopek --- Fermilab

Arnold Sikkema --- University of Guelph/U.B.C.

J. Stadel --- Brandon University

Mark Stuckey --- Elizabethtown College

R. Sussman --- U. Autonoma Metropolitana-Iztapalapa

Lev Tarasov --- University of Waterloo

N. Tariq --- University of New Brunswick

R.J. Torrence --- University of Calgary

B.O.J. Tupper --- University of New Brunswick

Jason Twamley --- University of Alberta

W. Unruh --- University of British Columbia

Dwight Vincent - - - University of Winnipeg

Jing Wang - - - University of Manitoba

Rachel Webster - - - CITA, Univ. of Toronto

J. G. Williams - - - Brandon University

Graeme Williams - - - University of Waterloo

W. Robert Wood - - - University of Regina

Eric Woolgar - - - Syracuse University